Applied Physical Geography

GEOSYSTEMS IN THE LABORATORY

Ninth Edition

Charles E. Thomsen
AMERICAN RIVER COLLEGE

Robert W. Christopherson
AMERICAN RIVER COLLEGE - EMERITUS

Pearson

Boston   Columbus   Indianapolis   New York   San Francisco   Upper Saddle River
Amsterdam   Cape Town   Dubai   London   Madrid   Milan   Munich   Paris   Montréal   Toronto
Delhi   Mexico City   São Paulo   Sydney   Hong Kong   Seoul   Singapore   Taipei   Tokyo

Senior Geography Editor: Christian Botting
Project Manager: Kristen Sanchez
Program Manager: Anton Yakovlev
Editorial Assistant: Amy De Genaro
Executive Marketing Manager: Neena Bali
Operations Specialist: Christy Hall
Supplement Cover Designer: Seventeeth Street Studios
Cover Photo Credit: *Waimea Canyon, Kauai, Hawai'i,* Bobbé Christopherson

Copyright © 2015, 2012, 2009, 2006, 2003, 2000, and 1997 Pearson Education, Inc. All rights reserved. Manufactured in the United States of America. This publication is protected by Copyright, and permission should be obtained from the publisher prior to any prohibited reproduction, storage in a retrieval system, or transmission in any form or by any means: electronic, mechanical, photocopying, recording, or likewise. To obtain permission(s) to use material from this work, please submit a written request to Pearson Education, Inc., Permissions Department, 1 Lake Street, Department 1G, Upper Saddle River, NJ 07458.

Many of the designations used by manufacturers and sellers to distinguish their products are claimed as trademarks.
Where those designations appear in this book, and the publisher was aware of a trademark claim, the designations have been printed in initial caps or all caps.

Library of Congress Cataloging-in-Publication Data
Thomsen, Charles E.; Christopherson, Robert W.
  Applies Physical Geography: Geosystems in the Laboratory/Charles E. Thomsen Robert W. Christopherson. -- Ninth edition.
       pages cm
ISBN-13: 978-0-321-98728-0
ISBN-10:   0-321-98728-4

1 2 3 4 5 6 7 8 9 10–V003–17 16 15

www.pearsonhighered.com

Contents

Correlation Grid iv
Preface v
Geography I.D. x
Prologue Lab xi

Laboratory Exercises:

1. Latitude and Longitude 01
2. The Geographic Grid and Time 07
3. Directions and Compass Readings 15
4. Map Projections, Map Reading, and Interpretation 23
5. Earth-Sun Relationships and Daylength 37
6. Insolation and Seasons 43
7. Temperature Concepts 53
8. Temperature Patterns 61
9. Temperature Maps 67
10. Earth's Atmosphere: Temperature and Pressure Profiles 75
11. Earth's Atmosphere: Pressure and Wind Patterns 83
12. Atmospheric Humidity 97
13. Stability and Atmospheric Processes 107
14. Weather Maps 117
15. Midlatitude Cyclones and Hurricanes 125
16. Water Balance and Water Resources 135
17. Global Climate Systems 143
18. Climate Change 161
19. Plate Tectonics: Global Patterns 173
20. Plate Tectonics: Faulting and Volcanism 183
21. The Rock Cycle and Rock Identification 197
22. Recurrence Intervals for Natural Events 207
23. Contours and Topographic Maps 221
24. Topographic Analysis: Fluvial Geomorphology 237
25. Topographic Analysis: Glacial Geomorphology 253
26. Topographic Analysis: Coastal and Arid Geomorphology 265
27. Topographic Analysis: Karst Landscapes 275
28. Soils 281
29. Biomes: Analyzing Global Terrestrial Ecosystems 295
30. An Introduction to Geographic Information Systems 309

Glossary 317
Graph Paper 329
Twelve Topographic Maps .... Inserts
Rock Chart .... Inserts

Correlation Grid

	Geosystems 9e	*Elemental Geosystems 8e*
1. Latitude and Longitude	1	1
2. The Geographic Grid and Time	1	1
3. Directions and Compass Readings	1	1
4. Map Projections, Map Reading, and Interpretation	1	1
5. Earth-Sun Relationships and Daylength	2	2
6. Insolation and Seasons	2	2
7. Temperature Concepts	5	3
8. Temperature Patterns	5	3
9. Temperature Maps	5	3
10. Earth's Atmosphere: Temperature and Pressure Profiles	3	2
11. Earth's Atmosphere: Pressure and Wind Patterns	6	4
12. Atmospheric Humidity	7	5
13. Stability and Atmospheric Processes	8	5
14. Weather Maps	8	5
15. Midlatitude Cyclones and Hurricanes	8	5
16. Water Balance and Water Resources	9	6
17. Global Climate Systems	10	7
18. Climate Change	11	8
19. Plate Tectonics: Global Patterns	12,13	9
20. Plate Tectonics: Faulting and Volcanism	13	10
21. The Rock Cycle and Rock Identification	12	9
22. Recurrence Intervals for Natural Events	15	12
23. Contours and Topographic Maps	1	1
24. Topographic Analysis: Fluvial Geomorphology	15	12
25. Topographic Analysis: Glacial Geomorphology	17	14
26. Topographic Analysis: Coastal and Arid Geomorphology	16	13
27. Topographic Analysis: Karst Landscapes	14	11
28. Soils	18	15
29. Biomes: Analyzing Global Terrestrial Ecosystems	19, 20	16, 17
30. An Introduction to Geographic Information Systems	1	1

Preface

Welcome to *Applied Physical Geography–Geosystems in the Laboratory*! During the ninth edition of this lab manual, important scientific research and global trends will unfold. Physical geography, as the essential Earth systems spatial science, integrates the latest information and discoveries from around the globe. This is an exciting time to be taking a lab class in which many of the principles of physical geography are applied in specific exercises. These exercises give you an opportunity to work with practices and tools of geographic analysis.

For example, imagine that toward the end of this century, the climate of Illinois in the summer will be more like that of present-day east Texas; that's the climate change forecast! You get to work with this forecast and its implications in Lab Exercise 17.

Our goal in this laboratory manual is to help you use geographic principles, methods, and tools to better relate your physical geography text to understanding Earth's systems. A specific source for figures and materials used in this manual is *Geosystems—An Introduction to Physical Geography*, by Robert W. Christopherson and Ginger H. Birkland, Pearson. Another source by these authors is *Elemental Geosystems*, although this lab manual does not follow the specific chapter sequence of either text.

NEW TO THE 9TH EDITION

- NEW: All lab exercises have been revised and reorganized into 30 shorter exercises for greater instructor flexibility.
- Over 50 NEW QR (Quick Response) codes link the print lab manual to Pre-Lab Videos, full-color figures, Interactive Animations, and Google Earth™ locations of the topographic maps at the back of the lab manual.
- NEW Pre-Lab Videos at the beginning of each lab exercise help students identify the key concepts and goals of the lab. Students can use their mobile devices to scan the QR links in the manual to view these videos.
- NEW Post-Lab quizzes in MasteringGeography give teachers an opportunity to assess student learning gains.
- NEW *Rock Cycle and Rock Identification* lab introduces students to the characteristics of igneous, sedimentary, and metamorphic rocks. This new lab has students identify common rocks, and includes a full-color rock identification chart.
- NEW *Climate Change* lab has students identify and analyze past, present, and future climate change.
- NEW *Biogeography Exercise*
- UPDATED *Soils* lab now includes an online component using the USDA Web Soil Survey.
- UPDATED *Geographic Information Systems* lab now incorporates the cloud-based ArcGIS Online system in a new volcanic hazards evaluation exercise.
- UPDATED lab manual website at www.mygeoscienceplace.com features Google Earth kmz files and optional Google Earth exercises.

THE SCIENCE OF GEOGRAPHY

Geography (from *geo*, "earth," and *graphein*, "to write") is the science that studies the interdependence among geographic areas, natural systems, society, and cultural activities over space. As a discipline that synthesizes knowledge from many fields, geography integrates spatial elements to form a coherent picture of Earth. The term *spatial* refers to the nature and character of physical space: to measurements, relations, locations, and the distributions of things. Geography is governed by a method—a spatial approach—rather than by a specific body of knowledge.

Physical geography, therefore, centers on the spatial analysis of all the physical elements and processes that comprise the environment: energy, air, water, weather, climate, landforms, soils, animals, plants, and Earth itself. Physical geography, as a spatial human-Earth science, is in a unique position among sciences to synthesize and integrate the great physical and cultural diversity facing us. Aspects of physical geography requiring spatial analysis are in the news daily, for these are dramatic times relative to human-Earth relations and spatial change.

As a science, geography uses the scientific method. Chapter 1 of both *Geosystems* and *Elemental Geosystems* briefly discusses this approach to problem solving. As you work through the laboratory manual, keep in mind this method of analyzing a problem, gathering data, organizing thought, and achieving the discovery of important principles.

Each lab exercise is organized in the following sequence. Each exercise and its steps will take varying lengths of time, depending on the emphasis taken in lab.

Organization of each lab exercise chapter:

- **Introduction**
  A brief overview of the subject and key terms and concepts involved in the exercise.
- Quick Response (QR) codes integrated into introductions lead students to online Pre Lab Lecture Videos, that ex lore the major concepts and goals of each lab exercise.
- **Key Terms and Concepts**
  Principal terms and concepts used in the exercise that are defined in the glossary and that you may want to check in a physical geography textbook.
- **Key Learning Concepts**
  The skills and knowledge you should have on completion of the exercise.
- **Materials/Sources Needed**
  Items that you will need to complete the exercise.
- **Lab Exercise and Activities**
  These headings denote parts of the exercise and include specific numbered questions and activities to complete.

This laboratory manual is designed for maximum utility—an asset to your learning activities. The standard symbols used on topographic maps appear inside the front cover. Inside the back cover are two world maps: the Köppen climate classification system and global biomes as Earth's terrestrial ecosystems. A fold-out flap, which you can deploy while you are working with the manual, features common metric-English conversions to assist you in converting between the two systems. The outside portion of the fold-out flap presents the four classes of map projections. The back cover features a genetic climate classification map, which identifies causal elements that produce the pattern of world climates.

Note: The exercises in this manual are set up in a manner sensitive to the fact that many of the U.S. students still use English units, whereas Canadian students use metric measures. Conversions are given throughout the manual wherever appropriate. Your instructor will guide you as to which system of units to use in the exercises. The graphs and other items are designed to use either system although metric is preferred.

The initial laboratory exercise, called the *Prologue Lab*, is unique and does not appear in other lab manuals. The Prologue Lab contains some activities that last throughout the term and will not be completed until the last class, acting as threads through the lab and helping you appreciate the relevancy of geography. We have also included a section with Google Earth activities. If you have not used this wonderful tool, we hope you find it useful and fascinating. If you have used Google Earth before, we hope you enjoy this new use for it. Some of these tasks work best as group activities, as guided by your instructor.

In order to use the QR codes, you will need a QR reader app on your mobile device. Apple's App Store and Google Play both have several free options. The Google Earth kmz files require Google Earth to be installed, either the desktop or mobile version. Both the desktop and mobile versions can be downloaded from http://www.google.com/earth/. The Google Earth website also has information about installing the programs, as well as tutorials on the programs.

PREFACE TO THE INSTRUCTOR

Applied Physical Geography: Geosystems in the Laboratory is for the student taking an introductory physical geography course, either concurrently or previously. Included are background materials, brief text explanations, and figures from the *Geosystems* and *Elemental Geosystems* texts, although this laboratory manual can be used with other current textbooks on physical geography.

A separate and complete *Answer Key* for this laboratory manual is available to adopters of this lab manual. Be sure to ask your Pearson sales representative (name on file at your bookstore) or download it from www.pearsonhighered.com/irc.

Contact your Pearson sales rep for examination copies of *Geosystems, Elemental Geosystems* and *Applied Physical Geography: Geosystems in the Laboratory.*

Course Equipment: We made a deliberate effort to write a manual requiring a minimum of equipment to keep your costs low. Students should provide their own atlas, pencils, colored pencils and pens, metric-English ruler, protractor, calculator, and dictionary. Along with the supplies and equipment you have on hand for labs to supplement this manual, the following items are needed: a sling psychrometer, a compass, weather instruments, and soil testing supplies, if available. You will also need stereolenses in order to properly view the stereopair images in the manual. Stereolenses can be purchased from a number of websites, including www.amazon.com. A sample stereopair photo is in the Prologue Lab for your students, with instruction to introduce this tool to them. There are eight sets of stereo photos in the lab manual and several stereo contour maps, demonstrating many concepts. Portions of 12 topographic sheets are included in the manual. Of course, you can supplement this brief presentation with your favorite topographic maps and stereoscopic and aerial photos featuring local examples. There is also a four-color rock identification key in the back of the lab manual, to help students identify and classify various rock types. This key will be helpful to students as they work through *Lab Exercise 21: The Rock Cycle and Rock Identification*. This edition features links to Google Earth KMZ files for several exercises so that students can actually fly through and experience 3-D landscapes as they work problems. The exercises can be completed in a campus computer lab if Google Earth is installed or outside of class if facilities are not available on campus.

Organization: The manual begins with a unique Prologue Lab. This initial lab includes several activities that span the entire term. You may choose only portions of the activities presented. These elements were designed for continuity and to help you initiate *group activities* early in the class, especially the "Hazard Identification: Natural and Anthropogenic" assignment.

The 30 exercises are subdivided into section segments. Depending on your emphasis, a lab meeting may include several sections as part of an entire exercise, leaving out other sections as you see fit. We have designed the sections so that they can stand alone in most cases.

Building on the success of previous editions, this ninth edition has been extensively updated and revised. There are now 30 exercises, including a new *Rock Cycle and Rock Identification* exercise and a new *Climate Change* exercise. Each lab exercise begins with a Quick Response (QR) code that students can scan on their mobile device to view a Pre-Lab Video. To view additional lab prep materials as well as post lab quizzes, students should visit www.masteringgeography.com. The topographic maps have all been revised for improved clarity and vibrance. Improved writing with tighter, more focused student questions and explanations are features of this edition. We continue the practice of placing some completed answers in each section to guide the student. We improved questions by requiring more interpretation of concepts under discussion. Please note that there are new materials, work, and examples in the lab exercises on global climates, recurrence intervals for natural events, coastal and arid geomorphology, and a geographic information systems exercise with data available for download, among many others. Note also the continuing improvement in production values with arts, maps, and overall reproduction.

The features that made the previous editions such a success are retained: the *Prologue* Lab as a term-length activity drawing the students together as a continuity thread; the stereopair photos and stereo contour lines; the selection of color topographic maps and related exercises; the section structure that allows you, the teacher, to edit and assign portions within a lab and the exclusion of other sections; and other features. We updated and improved the *Answer Key* with all the work completed, illustrations drawn, calculations made, and questions answered for quick reference when grading work. The answer key can be downloaded online at www.pearsonhighered.com/irc

ACKNOWLEDGMENTS

We extend our continuing gratitude to the editorial, production, and sales staff of Pearson. Thanks to Christian Botting, Senior Geography Editor, for willingness to risk new ideas and technologies, for his role in both *Geosystems* and *Elemental Geosystems*, and in geographic education in general. Our continuing compliments to the expertise of Project Manager Kristen Sanchez who oversees this *Applied Physical Geography* lab manual for her attention to detail and making the schedule work. We also extend our thanks to Program Manager, Anton Yakovlev for helping to oversee the production of the lab manual. And, our thanks go to the staff at Pearson for publishing what you hold in your hands.

Robert acknowledges the continuing invaluable production and photographic work of his wife Bobbé. Charlie offers thanks for the wonderful support from his wife Leslie and his children Emma and Finn in completing this work. Both of us acknowledge the role of students in our lives and the learning crucible that is the classroom.

In addition to all those geographers that contacted us by e-mail or at professional meetings, to Gail L. Hobbs and her work on several editions, and Pete Akers, Joy Mast, and Scott Mandia for their careful and thoughtful review of the manuscript. We thank the following colleagues for the time and effort they invested in content reviews and work on previous editions (a combined alphabetical list).

Maura Abrahamson, *Morton College*
Pete Akers, *University of Georgia*
Mark R. Anderson, *University of Nebraska, Lincoln*
Gregorey D. Bierly, *Indiana State University*
Mark Blumler, *SUNY, Binghamton*
Mark Bowen, *University of Wisconsin-Oshkosh*
Robin Buckallew, *Central Community College*
Peter U. Clark, *Oregon State University*
Woonsup Choi, *University of Wisconsin-Milwaukee*
Lisa DeChano-Cook, *Western Michigan University*
Carol Delong, *Victor Valley College*
John Dassinger, *Chandler-Gilbert Community College*
Robert H. Gorcik, *William Rainey Harper College*
Tracy Galarowicz, *Central Michigan University*
Sharon Johnson, *San Francisco State University*
Anil Kumar K Gangadharan, *Georgia State University*
Cecil S. Keen, *Mankato State University*
James Kernan, *SUNY Geneseo*
Ingrid Luffman, *East Tennessee State University*
Scott Mandia, *SUNY Suffolk*
Joy Mast, *Carthage College*
Raoul Miller
Abdullah F. Rahman, *Indiana University*
Philip P. Reeder, *University of Nebraska, Omaha*
Betty Rhodes, *University of Wisconsin-Milwaukee*
Bradley C. Rundquist, *University of North Dakota*
Donald C. Rundquist, *University of Nebraska*
Anne Saxe, *Saddleback College*
Susan Slowly, *Blinn College*
Tak Yung Susanna Tong, *University of Cincinnati*
Eugene Turne, *California State University, Northridge*
Forrest Wilkerson, *Minnesota State University*
Thomas Williams, *Western Illinois University*
Matthew R. Zorn, *Carthage College*
Henry J. Zintambilla, *Illinois State University*

Despite this expertise and support, we accept responsibility for whatever errors persist and for those areas needing further development toward future editions. We hope that *Applied Physical Geography* teaches principles effectively and enriches your experience with the science of physical geography. This is a remarkable moment in the human-Earth experience to be enrolled in physical geography.

CONCLUSION AND CONTACT INFORMATION

We live in an extraordinary era of **Earth systems science**. This science contributes to our emerging view of Earth as a complete entity—an interacting set of physical, chemical, and biological systems that produce a whole Earth. Physical geography is at the heart of Earth systems science as we answer the spatial questions concerning Earth's physical systems and their interaction with living things. Hopefully, this lab is a step along the path of understanding Earth.

Contact information: As before, please consider the ninth edition of *Applied Physical Geography* an evolving work in progress. Given this admission, we welcome you to send any feedback, suggestions, criticisms, and comments to help us improve future editions. **Thank you for any assistance!**

Send comments and inquiries to:

Charles E. Thomsen
American River College
4700 College Oak Drive
Sacramento, CA 95841
E-mail: thomsec@arc.losrios.edu

Robert W. Christopherson
P. O. Box 128
Lincoln, CA 95648
E-mail: bobobbe@aol.com

A Personal Geography I.D.

THE FIRST ASSIGNMENT

To begin: complete your personal Geography I.D. (see Preface, page vi). Use the maps in a physical geography text, an atlas, a college catalog, and additional library materials, if needed. Your laboratory instructor will help you find additional source materials for data pertaining to the campus.

On text or atlas maps find the information requested, noting the January and July monthly values (the small scale of such maps will permit only a general determination); air pressures are indicated by isobars, annual temperatures indicated by the isotherms, precipitation indicated by isohyets, climatic region, landform class, soil order, and ideal terrestrial biome, and the other information requested. Record the information from source maps in the spaces provided. The completed page gives you a relevant geographic profile of your immediate environment. As you progress through your physical geography class and physical geography lab, the full meaning of these descriptions will unfold. This page might be one you want to keep for future reference.

As you contemplate further schooling in colleges and universities in other parts of the world, or perhaps when you are preparing to move to another city, you may want to prepare a Geography I.D. for these new locations as part of your investigation into future options!

Geography I.D.

NAME: _____ LAB SECTION: _____

HOME TOWN: _____ LATITUDE: _____ LONGITUDE: _____

COLLEGE/UNIVERSITY: _____

CITY/TOWN: _____ COUNTY (PARISH): _____

    Standard Time Zone (for college location): _____

    Latitude: _____ Longitude: _____

    Elevation (include location of measurement on campus): _____

    Place (tangible and intangible aspects that make this place unique): _____

    Region (aspects of unity shared with the area; cultural, historical, economic, environmental):

_____

_____

    Population: Metropolitan Statistical Area (CMSA, PMSA, if applicable): _____

    Environmental Data: (information sources used: _____)

        January Avg. Temperature: _____ July Avg. Temperature: _____

        January Avg. Pressure (mb): _____ July Avg. Pressure: _____

        Average Annual Precipitation (cm/in.): _____

        Avg. Ann. Potential Evapotranspiration (if available; cm/in.): _____

    Climate Region (Köppen symbol and name description): _____

_____

    Main climatic influences (air mass source regions, air pressure, offshore ocean temperatures, etc.):

_____

    Topographic Region or Structural Region (including rock type, loess units, etc.): _____

_____

    Dominant Regional Soil Order _____

    Biome (terrestrial ecosystems description; ideal and present land use): _____

_____

_____

Prologue Lab

Physical Geography Logs, Hazard I.D. Group Activity, Weather Calendar, Google Earth™, and Stereophotos

Welcome to the Prologue Lab! All of us are geographers at one time or another. Our mobility is such that we cover great distances every day. Understanding both our lives and the planet's systems requires the tools of *spatial analysis* that are at the core of geography.

We personally interact with physical systems that are expressed in complex patterns across Earth. We need to develop continuity in our observations.

Throughout this term you may be using this Prologue Lab to record your observations of various environmental phenomena. Lab exercises that follow in later chapters explain various aspects of these observations in more detail. Your lab instructor will guide you as to which Prologue sections are required and the schedule for completion. These assignments are meant to span the entire term.

Key Terms and Concepts

environmental events log
hazard identification: natural and anthropogenic
hazard perception

weather and weather calendar
stereolenses
stereophotos

After completion of this lab you should be able to:

1. *Analyze* Internet, printed, and published media coverage of environmental events to *discern* the spatial implications of these happenings.
2. *Relate* these contemporary events to your physical geography textbook, lecture discussions, and lab, and to your daily experiences.
3. *Determine* natural and anthropogenic hazards on local and regional scales.
4. *Identify* and *utilize* local sources of environmental data, including broadcast and published media, instrument installations, and government sources, to *complete* a weather calendar for up to 30 days of the term, as assigned by your instructor.
5. *Use* Google Earth™ to open placemarks and navigate through landscapes.
6. *Utilize* stereolenses to view photo stereopairs and use the Bright Angel, Arizona, area for practice.

Name: _____ Laboratory Section: _____

Date: _____ Score/Grade: _____

Environmental Events Log

These are challenging times of significant global change relative to the environment, societies, and cultures, and it is hoped that this exercise will bring these important subjects to light! Studies estimate that annual weather-related damage losses (drought, floods, hail, tornadoes, derechos, tropical systems, storm surge, blizzard and ice storms, and wildfires) could exceed $1 trillion by 2040. The cost of weather-related destruction can be staggering—Hurricane Sandy in 2012 produced $65 billion in damages. In 2005, weather-related losses totaled $210 billion; Hurricane Katrina alone produced $146 billion (adjusted to current dollars) in damage that year. Weather-related destruction has risen more than 500% over the past three decades as population has increased in areas prone to violent weather and as climate change intensifies weather anomalies. 2010 was the warmest year ever recorded, followed by 2005 and 1998; 2013 and 2002 tied for the fourth-warmest year; 2006, 2009, and 2007 tied for the eighth warmest year; and 2004 and 2012 tied for the tenth warmest year. Only one year during the twentieth century, 1998, was warmer than 2013. Finally, 2013 was the thirty-seventh consecutive year of above-average global temperatures.

Use the spaces provided to record environmental events: weather events (heavy rain, drought, winds, freezes), earthquakes, volcanic eruptions, floods and coastal inundation, tsunami events (seismic waves), biodiversity issues and species extinctions, landslides, record icebergs, Antarctic ice-shelf disintegration, stratospheric ozone updates, air pollution occurrences, or other significant events involving the physical elements of the environment. Place the number of the item in its correct location on the accompanying map(s) at the end of this log, perhaps color-coding the number according to the type of event (blue for weather, brown for geomorphology, green for biogeography, etc.). A classroom bulletin board map might be used to compile a class world/North American map of environmental events.

Please access available media (Internet, broadcast, newspapers and magazines, and journals) as time permits. Use public television (PBS) and National Public Radio (NPR) shows such as *NOVA, Lehrer NewsHour, National Geographic, Nature,* and National Public Radio's *Morning Edition* and *All Things Considered,* or any other sources. *Geosystems,* ninth edition, lists more than 200 URLs for Internet sources. The Internet is particularly rich for learning about the latest occurrences.

You will see how much of physical geography is actually occurring out there in "the real world." Your list is not meant to be comprehensive; rather, it should reflect a growing interest in the physical systems that surround our lives. Restrict your list to events that occur throughout the term or semester. Spaces are provided for 18 items.

1. Date   **Source**   **Event**

_____; _____; _____

_____

_____

Tie-in to physical geography: _____

2. Date   **Source**   **Event**

_____; _____; _____

_____

_____

Tie-in to physical geography: _____

3. Date   **Source**   **Event**

_____ ; _____ ; _____

_____

_____

Tie-in to physical geography:_____

4. Date   **Source**   **Event**

_____ ; _____ ; _____

_____

_____

Tie-in to physical geography:_____

5. Date   **Source**   **Event**

_____ ; _____ ; _____

_____

_____

Tie-in to physical geography:_____

6. Date   **Source**   **Event**

_____ ; _____ ; _____

_____

_____

Tie-in to physical geography:_____

7. Date   **Source**   **Event**

_____ ; _____ ; _____

_____

_____

Tie-in to physical geography:_____

8. Date   **Source**   **Event**

_____ ; _____ ; _____

_____

_____

Tie-in to physical geography: _____

9. Date 　　　　　**Source** 　　　　　**Event**

_____; _____; _____

_____

_____

Tie-in to physical geography: _____

10. Date 　　　　　**Source** 　　　　　**Event**

_____; _____; _____

_____

_____

Tie-in to physical geography: _____

11. Date 　　　　　**Source** 　　　　　**Event**

_____; _____; _____

_____

_____

Tie-in to physical geography: _____

12. Date 　　　　　**Source** 　　　　　**Event**

_____; _____; _____

_____

_____

Tie-in to physical geography: _____

13. Date 　　　　　**Source** 　　　　　**Event**

_____; _____; _____

_____

_____

Tie-in to physical geography: _____

14. Date 　　　　　**Source** 　　　　　**Event**

_____; _____; _____

_____

_____

Tie-in to physical geography: _____

15. **Date**   **Source**   **Event**

_____; _____; _____

_____

_____

Tie-in to physical geography: _____

16. **Date**   **Source**   **Event**

_____; _____; _____

_____

_____

Tie-in to physical geography: _____

17. **Date**   **Source**   **Event**

_____; _____; _____

_____

_____

Tie-in to physical geography: _____

18. **Date**   **Source**   **Event**

_____; _____; _____

_____

_____

Tie-in to physical geography: _____

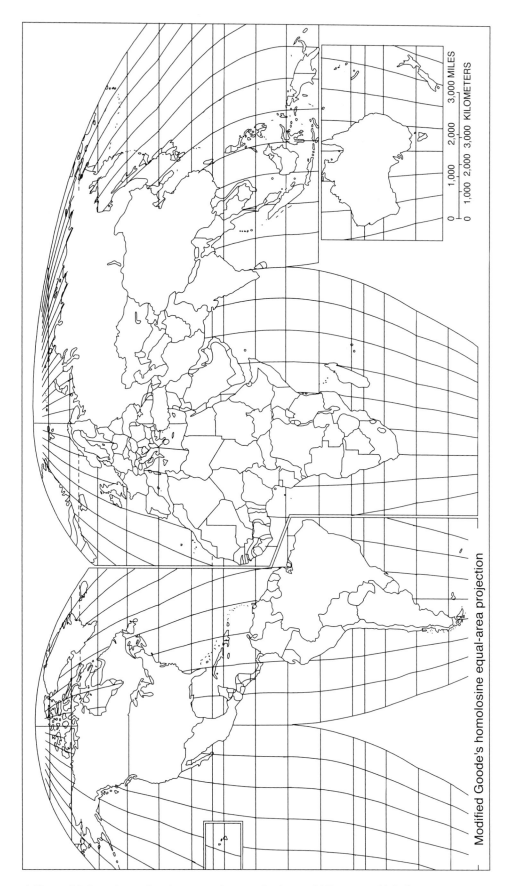

▲ Figure P.1 Locations of environmental events in the world (locate and label)

▲ Figure P.2 Locations of environmental events in the United States and Canada (locate and label)

Name:_____ Laboratory Section:_____
Date:_____ Score/Grade:_____

Hazard Identification

Natural Hazards and Anthropogenic (Human-Forced) Hazards
—A Group Activity—

A valid and applicable generalization is that humans are unable or unwilling or incapable of perceiving hazards in a familiar environment. Such an axiom of human behavior certainly helps explain why large populations continue to live and work in hazard-prone settings. Similar questions also can be raised about populations in areas vulnerable to natural disasters. Political and economic interests ride this wave of poor public perception and lack of physical geography's spatial perspective. Ideally, an informed public that acts on its knowledge can move to prevent development in hazardous areas. This poor perception applies to human-induced hazards as well, such as groundwater contamination, air pollution, and radiological hazards.

A consequence of poor hazard perception is that our entire society bears the financial cost, the subsequent bad planning, and irrational development—no matter where in the country it occurs. Of course, those directly affected are the victims who must shoulder the physical, emotional, and economic hardships of the disaster and for which society indirectly pays. This recurrent yet avoidable cycle—construction, devastation, reconstruction, devastation—continues. Recent events are reminders of this problem in our relationship with nature. A few of the many examples include the European heat wave of 2003 that killed more than 40,000 people, the Russian heat wave that killed 15,000, and the record-setting 2010 Northern Hemisphere heat waves that caused almost 18 million deaths; the devastating tornadoes of the past decade, with more than 1600 hitting the United States alone in 2004; the Midwest floods of 2011 and 2013, the Colorado floods of 2013, and the European floods of 2013 and 2014; the nor'easters that struck the east coast during January 1998 and January 2005; Hurricanes Katrina, Rita, and Wilma in 2005; Cyclone Sidr in 2007 and Hurricane Sandy in 2012; the October 2005 earthquake in Pakistan that killed 80,000, the China earthquake of 2008 that killed 87,000, the Haiti earthquake of 2010 that killed more than 316,000, and the more than 225,000 killed by the tsunamis caused by the magnitude 9.0 earthquake near the northwest coast of Sumatra, Indonesia in 2004; the 2010 eruption of Mount Merapi that displaced 350,000 people and the 2010 eruption of Eyjafjallajökull that disrupted air travel for months; and the devastating 2011 9.0 earthquake off of Japan, followed by the tsunami and the nuclear power plant meltdowns. The increasing pace of global climate change demands that we pay attention to our relationships with Earth systems.

A value in studying physical geography is a better knowledge of the environment and the workings of natural systems. This heightened awareness may lead to a more reasonable assessment of natural and anthropogenic hazards. Following group discussions with lab partners, describe and explain any natural or human-induced hazards that exist within these radii from campus. Note how they affect you.

10 km (6 mi) radius of your campus:

Natural–_____
_____
_____
_____

Anthropogenic–

100 km (60 mi) radius of your campus:

Natural–

Anthropogenic–

Your state or province:

Natural–

Anthropogenic–

Additional specific regional environmental issues:

Name:_____ Laboratory Section:_____
Date:_____   Score/Grade:_____

Weather Calendar

Weather is an important integrative subject within physical geography. Temperature, air pressure, relative humidity, wind speed and direction, day length, and Sun angle are important measurable elements that contribute to the weather. **Weather** is the short-term condition of the atmosphere, as compared to **climate**, which reflects long-term atmospheric conditions and extremes.

We go online or tune to a local station for the day's weather report from the National Weather Service (in the United States, **http://www.nws.noaa.gov**) or the Canadian Meteorological Center (in Canada, **http://weather.gc.ca/index_e.html**) to see the current satellite images and to hear tomorrow's forecast. Most cable systems carry the *Weather Channel*, which presents local weather information throughout the day (**http://www.weather.com**). Internationally, the World Meteorological Organization coordinates weather information (see **http://www.wmo.int/pages/index_en.html**).

Although this is early in the term and weather has not specifically been discussed in lecture, you will find this exercise an interesting, ongoing work-in-progress that will stimulate many questions and prepare you for those chapters when it is covered.

Meteorology is the scientific study of the atmosphere (*meteor* means "heavenly" or "of the atmosphere"). Embodied within this science is a study of the atmosphere's physical characteristics and motions; related chemical, physical, and geologic processes; the complex linkages of atmospheric systems; and weather forecasting. Computers permit the handling of enormous amounts of data on water vapor, clouds, precipitation, and radiation for accurate forecasting of near-term weather.

New developments in supercomputers and ground- and space-based observation systems are making this a time of dramatic developments. A few of the innovations include wind profilers that use radar to profile winds from the surface to high altitudes, computer-based modeling software for 3-D weather models and displays, improved international cooperation and weather data dissemination, and the standard Advanced Weather Interactive Processing System (AWIPS) at more than 150 stations. Data gathering is enhanced by the completed installation of 883 Automated Surface Observing System (ASOS) instrument arrays as a primary surface weather-observing network. By the end of 2013, 160 WSR-88D (Weather Surveillance Radar) Doppler radar systems as part of the NEXRAD (Next-Generation Radar) program were operational through the National Weather Service, in conjunction with the Federal Aviation Administration and the Department of Defense. In Canada, the National Radar Project will have 30 CWSR-98 radars in service.

NOAA operates the Earth System Research Laboratory, Boulder, Colorado, which is developing new forecasting tools (see **http://www.esrl.noaa.gov**). Research and monitoring of violent weather is centered at NOAA's National Severe Storms Laboratory in several cities (see **http://www.nssl.noaa.gov**). For tropical weather, see the National Hurricane Center, at **http://www.nhc.noaa.gov**.

Building a database of atmospheric conditions is key to *numerical weather prediction* and the development of weather forecasting models. *Synoptic analysis* involves the characterization of weather elements over a region within the same time frame, as shown on the weather map presented in Lab Exercise 14 of this manual. You will not be observing all these elements for your weather calendar, but a complete weather forecast is based on knowing the following nine items (the seven marked with asterisks are required for this exercise):

- Barometric pressure*
- Pressure tendency
- Surface air temperature*
- Dew-point temperature*
- Wind speed and direction*
- Type of clouds*
- Current weather*
- State of the sky*
- Precipitation since last observation

The purpose of this weather exercise is twofold. *First*, it will help you find sources of weather information. Our lives take place within an environment of variable weather, so it is important to have access to such information.

Second, you will gather data for up to 30 days (or less, as assigned by your instructor) during the term, not necessarily consecutive days, and record them on a weather calendar using official international symbols and designations. If your lab meets 15 or 16 times during the term and the department has weather instruments, then half your days of observation are easily completed on class days. The other days can be met by outside sources or visits to the department.

The National Weather Service and the Canadian Meteorological Center use a standardized system of recording and reporting weather data around a weather station symbol developed by the World Meteorological Organization (WMO). An example of this method is shown below; a complete explanation of symbols and data forms is in Lab Exercise 14.

WEATHER STATION SYMBOL

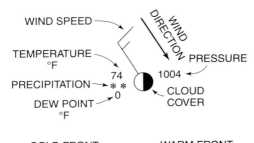

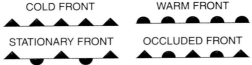

Attached is your "Weather Calendar," with spaces for up to 30 days of weather observations. These days need not be consecutive and may be spread out over the entire term. The completed calendar is due at the end of the term. Your instructor will give you more details. The circle in the middle of each square is the weather station symbol, and above and below are spaces to list the date and the place of observation. (If you are using a broadcast source, list the place the weather instruments are located that made the observation.)

Weather station symbol explanations are in Lab Exercise 14. We can refer to the weather station symbol as if it were a clock. Temperature is recorded in the 10:00 position, barometric pressure at 2:00, weather type at 9:00, and dew-point temperature (if available) at 8:00; wind speed and direction vary in location, depending on the wind itself. If you are observing clouds, high clouds are symbolized above the station symbol, and middle and low clouds below it.

The state of the sky (cloud cover) is the area inside the circle. In the example to the left, temperature is 74°F (23°C), pressure is 1004.0 mb, the sky is 50% (five-tenths) overcast, with slight continuous fall of snowflakes at time of observation. Winds are from the northwest (line drawn on the side of the station where the winds originate) at 15–20 mph (24–32 kmph), as noted by the "flags" on the line. For a complete listing of weather symbols, see Lab Exercise 14.

PRECIPITATION TYPE
- , DRIZZLE
- • RAIN
- * SNOW
- ▽ SHOWERS
- ⊺⎧ THUNDERSTORMS
- ≡ FOG
- ∞ DRY HAZE
- ⌒⌣ FREEZING RAIN
- △ HAIL
- △ SLEET

WIND SPEED (kph/mph)
- CALM — (O)
- 5–13 — 3–8
- 14–23 — 9–14
- 24–32 — 15–20
- 33–40 — 21–25
- 89–97 — 55–60
- 116–124 — 72–77

CLOUD COVER
- ○ NO CLOUDS
- ◐ ONE-TENTH OR LESS
- ◔ TWO-TENTHS TO THREE-TENTHS
- ◕ FOUR-TENTHS
- ◑ FIVE-TENTHS
- ◒ SIX-TENTHS
- ◕ SEVEN-TENTHS TO EIGHT-TENTHS
- ◍ NINE-TENTHS OR OVERCAST WITH OPENINGS
- ● COMPLETELY OVERCAST
- ⊗ SKY OBSCURED

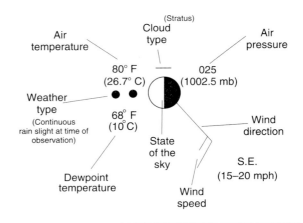

Name: _____ Laboratory Section: _____

Weather Calendar

Date/time_____	Date/time_____	Date/time_____	Date/time_____	Date/time_____
○	○	○	○	○
Place_____	Place_____	Place_____	Place_____	Place_____
Date/time_____	Date/time_____	Date/time_____	Date/time_____	Date/time_____
○	○	○	○	○
Place_____	Place_____	Place_____	Place_____	Place_____
Date/time_____	Date/time_____	Date/time_____	Date/time_____	Date/time_____
○	○	○	○	○
Place_____	Place_____	Place_____	Place_____	Place_____
Date/time_____	Date/time_____	Date/time_____	Date/time_____	Date/time_____
○	○	○	○	○
Place_____	Place_____	Place_____	Place_____	Place_____
Date/time_____	Date/time_____	Date/time_____	Date/time_____	Date/time_____
○	○	○	○	○
Place_____	Place_____	Place_____	Place_____	Place_____
Date/time_____	Date/time_____	Date/time_____	Date/time_____	Date/time_____
○	○	○	○	○
Place_____	Place_____	Place_____	Place_____	Place_____

Copyright © 2015 Pearson Education, Inc.

Google Earth™

Virtual Geography in *Applied Physical Geography*

Incorporated in this edition of the Lab Manual are exercises built around Google Earth™. This technology is the most exciting and useful geographic tool since GPS (Global Positioning System) technology became available to the public. This program allows you to view a photo-realistic 3-D model of Earth. You can zoom in and out, tilt and rotate the view, adjust the vertical exaggeration of the landscape, and add your own image overlays. Some of the exercises in the Lab Manual involve simply navigating around terrain features to gain a better understanding of how they look, while other exercises involve the application of critical thinking skills.

To use this wonderful free tool, you need to download Google Earth™ from **http://google.com/earth**. The Google Earth™ website has information on the hardware and network requirements for the program. It also has an excellent tour of the features of the program and a gallery of downloadable placemarks called KMZ or KML files.

Once you have installed the program, you should go through Google's tour of the program. The tour covers the basics of how to navigate around the virtual world, how to search for places, and how to open and save placemarks.

The exercises in this manual all involve downloading KMZ files from the Geosystems website and opening them in Google Earth™. This introductory exercise will help acquaint you with the process of opening a KMZ file and using it with the questions of a lab exercise. The first KMZ file is called intro.kmz. For the KMZ file and the exercise questions, go to **http://mygeoscienceplace.com**. This web page will have the KMZ files and the questions for each assignment. Depending on how your web browser is set up, the KMZ file will automatically open up in Google Earth™ or you may have to download it to your local machine and open it up manually.

Once you have opened the KMZ, it will appear in the sidebar in the Places box, in the Temporary Places folder. When you double-click on the Intro placemark, it will fly you to the first location and open up a pop-up window with information about that place. Some of the placemarks will only have one location; others may have several locations that you will visit, as well as overlay images that you can turn on and off. You can control the transparency of an image overlay with the transparency slider control. This will be useful when you want to interpret information from multiple overlays. After you have found your house (which is the first thing almost everyone does), go through each of the questions in the Intro assignment. The questions for each assignment are found on the same website as the KMZ files. Your instructor will tell you how to submit your assignments.

Aerial Photo Stereopairs

3-D Photography in This Manual

If you cover one eye and look about you, you see length and width but not depth. The view with one eye is *monocular vision*. The normal depth perception we experience is due to using two eyes and seeing with *binocular vision*, or *stereoscopic vision*. Each eye sees objects at a slightly different angle. Our brain blends these two signal inputs into a three-dimensional image that features the dimensions length, width, and depth.

The photo stereopairs in this manual are made by a camera platform that takes aerial photos from slightly different positions along a flight path. The U.S. Geological Survey's National Aerial Photography Program (NAPP) is the source for the photographs we use, all at a 1:40,000 scale.

Along north-south NAPP flight paths, adjoining frames of aerial photographs overlap by 60%. In these overlap areas, the camera views the same scene from different angles. When we place any two overlapping photos side-by-side and view them through stereolenses or a stereoscope, each eye assumes the position of the camera as it took each picture. These stereolenses straighten your line of sight (to infinity) so that each eye looks ahead to one of the photographs. If you relax your eyes, the overlapping area of the photographs will spring forth in three dimensions. This stereoscopic technique assists in landscape analysis and enhances photo interpretation.

To give you some practice with this technique, we present a demonstration photo stereopair for you. On the next page is a pair of NAPP photographs showing the Bright Angel area, in the Grand Canyon of Arizona. We chose this scene because the topographic relief is dramatic and the Grand Canyon is a familiar place. Because of the north-south flight path, these photos are oriented in such a way that north is to the right, south to the left, and west at the top of the page.

Hold the lenses to your eyes and relax. If you are doing it correctly, you will see two light-colored lines (the single margin between the photos). As your eyes adjust, the area that appears to be between the lines is where you will see stereovision. Experiment until you find the right distance from the photos, usually about 25 cm (10 in.) above the photo pair. Initially, it might take a few minutes to adjust your eyes to stereoscopic viewing; give yourself time.

The vertical depth that you see is several times more than is actually in Bright Angel Canyon. This is called vertical exaggeration and is one of the traits of stereophotography. Vertical exaggeration helps when viewing some scenes of gentle, rolling topography and low relief.

In the stereophoto, observe the intricately dissected Colorado Plateau cliffs, slopes, vista points, and a deep inner V-shaped canyon occupied by Bright Angel Creek. Study the landscape and compare your results with those of others in your lab section.

You can purchase stereolenses from a variety of websites, such as Amazon.com.

For more information about NAPP, go to the USGS and the EROS Data Center, at **http://eros.usgs.gov.**

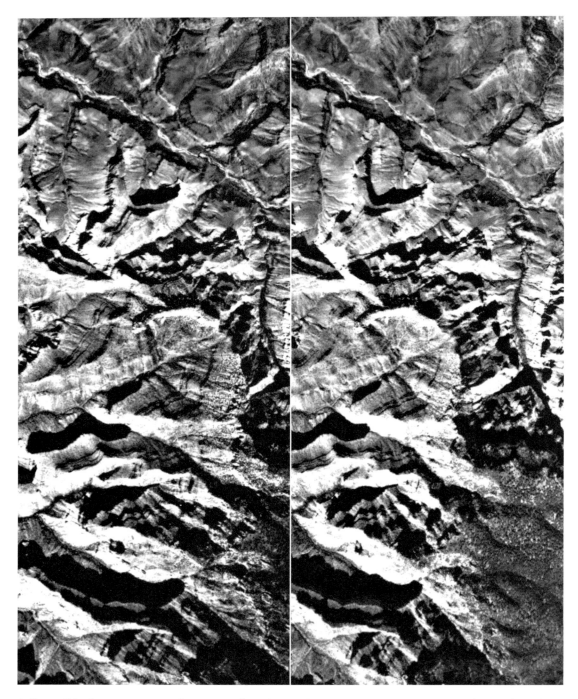

▲ Figure P.3 Photo stereopair of a portion of Bright Angel Canyon, Arizona. Note that north is to the right, south to the left, and west to the top. (NAPP photos by USGS, EROS Data Center.)

Name: _____ Laboratory Section: _____ Video
Date: _____ Score/Grade: _____ Exercise 1
Pre-Lab Video

LAB EXERCISE

Scan to view the Pre-Lab video

http://goo.gl/pQk8vC

1 Latitude and Longitude

To know specifically where something is located on Earth's surface, a coordinated grid system is needed, one that is agreed to by all peoples. The terms *latitude* and *longitude* were in use on maps as early as the first century A.D., with the concepts themselves dating back to Eratosthenes and others.

As you complete your *Geography I.D.* in the Preface, you will eventually determine the latitude and longitude for your campus. We examine great circles, latitude, and longitude in this exercise. Lab Exercise 1 features three sections.

Key Terms and Concepts

great circle
latitude
longitude
meridian

parallel
prime meridian
small circle

After completion of this lab you should be able to:

1. *Define* great circle and small circle and *describe* the relationship between a great circle and travel.
2. *Define* latitude and parallel, longitude and meridian, and *use* them in simple sketches to *demonstrate* how Earth's reference grid is established.

Materials/Sources Needed

globe
string
ruler
protractor
world atlas

Copyright © 2015 Pearson Education, Inc. Lab Exercise 1   1

Lab Exercise and Activities

SECTION 1

Great and Small Circles

Great circles and small circles are important concepts relating to latitude and longitude—concepts that we will refer to in later activities. A **great circle** is any circle of Earth's circumference whose center coincides with the center of Earth. An infinite number of great circles can be drawn on Earth. On most flat maps, airline and shipping routes appear to arch their way across oceans and landmasses following great circles. Despite their curved appearance on flat maps, great circle routes are the shortest distance between two points on Earth.

Using a globe and a piece of string, place one end on San Francisco and stretch the string tautly to London. This direct route is a portion of a great circle. **Small circles** are circles whose centers do not coincide with Earth's center. In Figure 1.1 properly draw and label a great circle and a small circle, other than those noted by the latitude and longitude grid.

▲ Figure 1.1 Label great circles and small circles

Again using the globe and piece of string, show the great circle routes linking the following cities, listing at least four or five prominent geographical features (e.g., mountains, seas, rivers, nations, cities, etc.) that you would cross as you traveled along those routes:

1. London, England, and Colombo, Sri Lanka (English Channel, Germany, Carpathian Mtns.)

2. Vancouver, British Columbia, and Sydney, Australia

3. Your hometown and Beijing, China

---

2 **Lab Exercise 1** Copyright © 2015 Pearson Education, Inc.

SECTION 2

Latitude and Parallels

Adapting from the Babylonians, Ptolemy divided the circle into 360 *degrees* (360°), with each degree subdivided into 60 *minutes* (60′), and each minute further subdivided into 60 *seconds* (60″). This method of dividing degrees into minutes and seconds is often referred to as DMS coordinates. Geographic Information Systems, GPS, and other computer-based systems often use decimal degrees or DD coordinates. In this method, each degree is divided into a more familiar base ten system. We'll now examine each of these grid coordinate elements.

Latitude is an angular or arc distance north or south of the *equator* (the line running east to west halfway between the poles), measured from the center of Earth (Figure 1.2a). On a map or globe, the lines designating these angles of latitude run east and west, parallel to the equator, whose latitude is 00.00°.

The North Pole is 90.00° north latitude, and the South Pole is 90.00° south latitude. "Lower latitudes" are those nearer the equator, whereas "higher latitudes" refer to those nearer the poles.

A line connecting all points at the same latitudinal angle is called a **parallel**. Only one parallel—the equatorial parallel—is a great circle; all others are small circles, diminishing in circumference north and south toward the poles, which are merely points. As shown in Figures 1.2a and 1.2b, *latitude is the name of the angle* (49° north latitude), *parallel names the line* (49th parallel), and both indicate arc distance north of the equator.

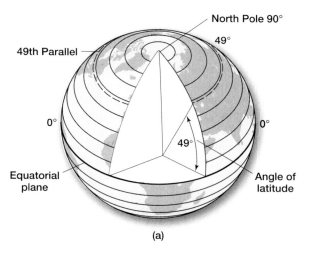
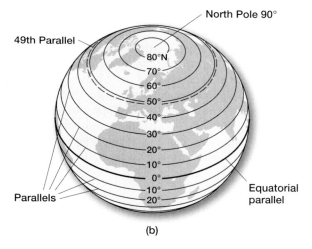

▲ Figure 1.2 a and b Latitude and parallels

1. Parallels have often been used to demarcate political boundaries. The 49th parallel north forms a portion of the border between which two countries?

2. Parallels made famous by wars in the last century include the parallel dividing the two Koreas and the parallel that divided Vietnam until 1975. (The border was approximately 80 km [50 mi] north of the city of Hue). What are these two parallels?

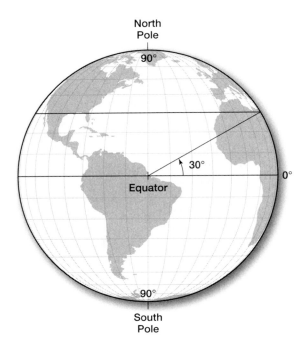

▲ Figure 1.3 Measuring latitudes on Earth

3. Figure 1.3 is a half-Earth section through the poles. An angle of 30° has been measured north of the equator, with the 30th parallel drawn on the globe. Following this example, use your protractor to measure an angle of 23.5° north of the equator and draw and label the corresponding parallel, which is known as the Tropic of Cancer. Measure another angle 66.5° north of the equator and draw and label the parallel known as the Arctic Circle. Repeat this in the Southern Hemisphere, drawing and labeling the Tropic of Capricorn (23.5° south) and the Antarctic Circle (66.5° south). We will be working with these parallels while studying Earth–Sun relationships in Lab Exercise 5. Finally, measure, draw, and label one last parallel to show your own latitudinal location.

4. Using a globe and your atlas or other world map, locate three cities that are located at approximately the 23rd parallel in the Northern Hemisphere; note their location in degrees (and minutes, if your map is detailed enough to estimate minutes). Use the globe first, then refer to the atlas maps to better determine specific latitudes. Be sure and list their country names as well.

   City and Country Name   Longitude (degrees and minutes, if possible)

   a) _____   _____

   b) _____   _____

   c) _____   _____

5. List the names and longitudes of three cities that are located at approximately your latitude.

   City and Country Name   Longitude

   a) _____   _____

   b) _____   _____

   c) _____   _____

SECTION 3

Longitude and Meridians

Longitude is an angular or **arc distance** east or west of a point on Earth's surface, measured from the center of Earth (Figure 1.4a). On a map or globe, the lines designating these angles of longitude run north and south at right angles (90°) to the equator and to all parallels. A line connecting all points at the same longitude is a **meridian** (Figure 1.4b). Every meridian is one-half of a great circle that passes through the poles. As shown in Figures 1.4a and 1.4b, longitude is the name of the angle, meridian names the line, and both indicate arc distance east or west of the prime meridian.

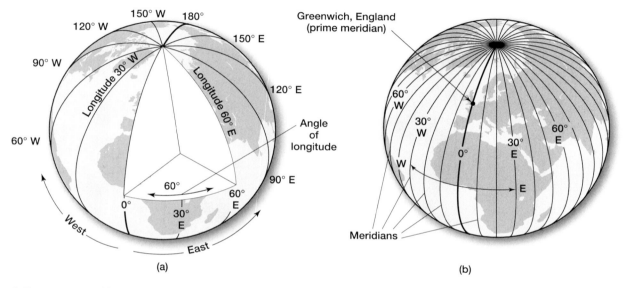

▲ Figure 1.4 a and b Longitude and meridians

1. On a political globe or world map follow the International Date Line across the Pacific Ocean. Why do you think the International Date Line is not straight but zigs and zags?

2. Examine an atlas or a political globe and in the spaces marked a through h list the provinces and states through which the 100th meridian in the Western Hemisphere passes—north to south. The first answer is provided for you in bracketed italics.

    a) _____[Nunavut, Canada]_____   e) _____

    b) _____   f) _____

    c) _____   g) _____

    d) _____   h) _____

3. Figure 1.5 is a view of Earth from directly above the North Pole; the equator is the full circumference around the edge. A line has been drawn from the North Pole to the equator and labeled 0°, representing the prime meridian. Earth's prime meridian through Greenwich, England, was not generally agreed to by most nations until 1884. To the right of 0° on the diagram is the *Eastern Hemisphere*—label this—and to the left of 0° is the *Western Hemisphere*—label this.

Extend another line from the North Pole to the other side of Earth, opposite the prime meridian, and label it 180°. You now have marked the line that is the International Date Line, which extends from North to South Poles on the opposite side of Earth from the prime meridian. Using your protractor, measure, draw, and label the meridians that are 100° east and 60° west of the Greenwich meridian. Finally, locate, draw, and label the meridian that marks your present longitude.

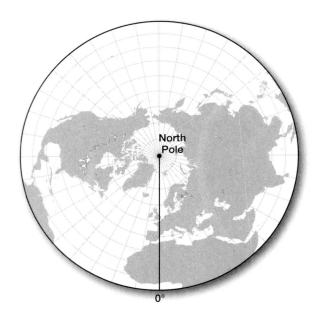

▲ Figure 1.5 Measuring longitudes on Earth

The 98th meridian is roughly the location of the 51 cm (20 in) isohyet (a line connecting points of equal precipitation), with wetter conditions to the east and drier conditions to the west. In North America, tallgrass prairies once rose to heights of 2 m (6.5 ft) and extended westward across the Great Plains to about the 98th meridian, with short-grass prairies farther west. The deep sod formed beneath these tall grasslands posed problems for the first settlers, as did the climate. The self-scouring steel plow, introduced in 1837 by John Deere, allowed the interlaced grass sod to be broken apart, freeing the soils for agriculture. Keep this information about the region around the 100th meridian, and close to the 98th, in mind as a "location reference" later on when studying precipitation and vegetation.

MasteringGeography™ | Looking for additional review and lab prep materials? Go to www.masteringgeography.com for pre lab videos and pre and post lab quizzes.

Name: _____ Laboratory Section: _____

Date: _____ Score/Grade: _____

Video
Exercise 2
Pre-Lab Video

LAB EXERCISE 2

Scan to view the Pre-Lab video

The Geographic Grid and Time

In the previous exercise we studied latitude and longitude. In this exercise we will examine latitude and longitude together as they make the geographic grid that we use to locate places. As you complete your *Geography I.D.* in the Preface, you will determine the time zone you live in. A knowledge of longitude and Earth's rotation forms the basis of standard time and the world system of Coordinated Universal Time (UTC). We examine both this essential geographic location grid and time in this exercise. Lab Exercise 2 features three sections.

Key Terms and Concepts

antipode
Coordinated Universal Time (UTC)
daylight saving time
geographic grid

Greenwich Mean Time (GMT)
International Date Line
local Sun time

KEY LEARNING concepts

After completion of this lab you should be able to:

1. *Use* latitude and longitude to *locate* places on Earth's surface.
2. *Relate* the time where you are with world standard time and *calculate* differences in standard and Sun time.
3. *Contrast* the prime meridian with the International Date Line.

Materials/Sources Needed

protractor
world atlas

Copyright © 2015 Pearson Education, Inc. Lab Exercise 2 7

SECTION 1

Earth's Geographic Grid

Latitude and longitude form Earth's **geographic grid**. Every location on our planet is described with this Cartesian coordinate system—where location is described on a plane by two intersecting lines. The *Geography I.D.* asks you to identify the general coordinates of your hometown and college campus. **Antipodes** are points on the globe that are diametrically opposite each other that would be at either end of a line drawn through the center of Earth—such as the North Pole and South Pole. Determine the following geographic grid coordinates and cities from an atlas, wall map, or large political world globe. While wall maps and globes will help you see points in relation to the rest of the world, you should use regional maps in an atlas to be more precise in your readings.

1. Locate and give the geographic coordinates for the following cities (to a tenth of a degree if your atlas maps are detailed enough) or identify the cities from the given coordinates. The answers to a) and d) are provided for you in bracketed italics.

City	Latitude and Longitude
a) Greenwich, London, England	*[51.5°N 0°]*
b) Rio de Janeiro, Brazil	
c) Your state's/province's capital city	
d) *[Tokyo, Japan]*	35.7°N 139.7°E
e)	8.8°S 13.2°E
f)	21.3°N, 157.8°W

On the map grid in Figure 2.1, plot the coordinates in items 1 (a) through (f) above, and label the city names.

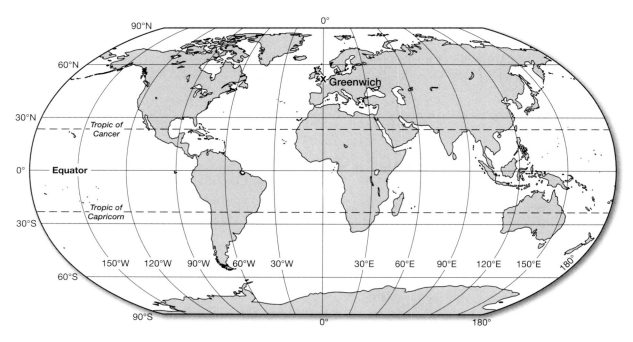

▲ Figure 2.1 Plotting coordinates

8 Lab Exercise 2

2. Using your knowledge of latitude and longitude, find and circle the errors in the following geographic grid coordinates. Rewrite the coordinates correctly in the space to the right. You do not have to locate these on a map. The first error is identified with a box.

   a) DMS format Lat. 57° 86′ 24″ S, Long. 149°02′63″ N _____ [*minutes cannot exceed 60′*] _____

   b) DD format Lat. 105.03° W, Long. 93.99° E _____

3. If you were halfway between the equator and the South Pole and one-quarter of the way around Earth to the west of the Prime Meridian, what would be your latitude and longitude?

4. You are at 10°N and 30°E; you move to a new location that is 25° south and 40° west of your present location. What is your new latitudinal/longitudinal position?

5. You are at 20°S and 165°E; you move to a new location that is 45° north and 50° east from your present location. What is your new latitudinal/longitudinal position?

The antipode is the point on the opposite side of Earth from another point. You may have heard that if you dig straight down, you'll eventually reach China. Setting aside the difficulties of digging through the molten iron outer core of Earth, you wouldn't end up in China if you started digging from the United States.

To find the antipode of a location, you'll need to find both the latitude and longitude of the antipode. To find the antipodean latitude, convert the location's latitude to the opposite hemisphere. If the latitude is 50°N, it's antipodean latitude is 50°S. With longitude, the antipode is always 180° away. To find the antipodean longitude, simply subtract the location's latitude from 180 and change it to the other hemisphere. To find the antipodean longitude of 120°W, subtract 120° from 180°, and change it to the Eastern Hemisphere to find 60°E.

6. What is the antipode of your current location?

7. If you wanted to dig through the center of the Earth and come up in Beijing, China, where should you start digging?

SECTION 2

Latitude and Longitude Values

Because meridians of longitude converge toward the poles, the actual distance on the ground (linear distance) spanned by a degree of longitude is greatest at the equator (where meridians separate to their widest distance apart) and diminishes to zero at the poles (where they converge). The linear distance covered by a degree of latitude, however, varies only slightly, due to the oblateness of Earth's shape. Table 2.1 compares the linear or physical distance of a degree of latitude and longitude at selected latitudinal locations. It also shows the similarity of ground distance for a degree of latitude and a degree of longitude at the equator.

Latitudinal Location	Latitude Degree Length km	Latitude Degree Length (mi)	Longitude Degree Length km	Longitude Degree Length (mi)
90° (poles)	111.70	(69.41)	0	(0)
60°	111.42	(69.23)	55.80	(34.67)
50°	111.23	(69.12)	71.70	(44.55)
40°	111.04	(69.00)	85.40	(53.07)
30°	110.86	(68.89)	96.49	(59.96)
0° (equator)	110.58	(68.71)	111.32	(69.17)

At North Pole: 1° of latitude = 111.70 km, 1° of longitude = 0 km
At 40°: 1° of latitude = 111.04 km, 1° of longitude = 85.40 km
At Equator: 1° of latitude = 110.58 km (1 arc of a meridian), 1° of longitude = 111.32 km (1 arc of a parallel)

▲Table 2.1 Physical Distances Represented by Degrees of Latitude and Longitude

1. From the table, you can see that latitude lines are evenly spaced, approximately 111 km (69 miles) apart at any latitude. Using these values as the linear distance separating each degree of latitude, the distance between any given pair of parallels can be calculated. (*Note: Locations must be due north-south of each other.*) For example, Denver is approximately 40° north of the equator (arc distance). The linear distance between Denver and the equator can be calculated as follows:

    40° N to 0° = 40° × 111 km/1° = 4440 km

    or

    40° N to 0° = 40° × 69 miles/1° = 2760 miles

    Using these same values for a degree of latitude and an atlas for city location, calculate the linear distance in km and miles between the following sets of points (along a meridian):

    a) Mumbai, India, and the equator _____

    b) Miami, Florida, and 10° south latitude _____

    c) Edinburgh, Scotland, and the 5th parallel north _____

    d) Your location and the equator _____

2. The table also shows that the linear distance separating each 1° of longitude decreases toward the poles. For example, at 30° latitude each degree of longitude is separated by slightly more than 96 km (nearly 60 miles), and at 60° latitude, the linear distance is reduced to approximately half that at the equator. For each of the following latitudes, determine the linear distance in km and in miles for 15° of longitudinal arc (along a parallel): The first answer is provided for you in bracketed italics.

10   Lab Exercise 2

	km	miles
a) 30° latitude:	[15° × 96.49 km = 1447 km]	[15° × 59.96 mi = 899 mi]
b) 40° latitude:	_____	_____
c) 50° latitude:	_____	_____
d) 60° latitude:	_____	_____

3. Again using Table 2.1, what is the linear distance in km and miles along the parallel at your latitude from your location to the prime meridian?

4. What is the approximate linear distance of the following angular distances, at your present latitude?

	km	miles
One degree	_____	_____
One minute of longitude	_____	_____
One second of longitude	_____	_____
A tenth of a degree	_____	_____
One hundredth of a degree	_____	_____

5. John Harrison's chronometer (a clock giving Coordinated Universal Time), lost 5 seconds during the 81-day voyage from England to Jamaica. Given that Earth rotates through 15° in 1 hour, how many degrees of longitude would the ship be off, with an error of 5 seconds? How many kilometers and miles would that be, assuming 111 km per degree of longitude?

Precision and accuracy are two related concepts regarding measurements. *Precision* refers to the number of significant digits in a measurement, while *accuracy* refers to how close a measurement is to the actual figure. Giving a distance measurement as 2 km is less precise than giving it as 2.0 km or 2.00 km. If you were going to meet a friend in the Lake Nakuru National Park in Kenya, and that person's location were 0°, 36°E, how many square kilometers would you have to look through? Your friend's location is precise to 1°, so you would have to look for the person in an area 1° by 1°. At this location, 1° of both longitude and latitude are approximately 111 km (69 mi), so you would have to look for your friend in 12,321 km² (4761 mi²).

6. How large an area would you have to look through if your friend's location were given 10 times as precisely, as 0.0°, 36.0°E?

7. Write your friend's location with sufficient precision so that you would only have to look in an area 111 m by 111 m.

SECTION 3

Time, Time Zones, and the International Date Line

Today we take for granted international standard time zones and an agreed-upon prime meridian. If you live in Oklahoma City and it is 3:00 P.M., you know that it is 4:00 P.M. in Baltimore and Toronto, 2:00 P.M. in Salt Lake City, and 1:00 P.M. in Seattle and Los Angeles. You also probably realize that it is 9:00 P.M. in London and midnight in Riyadh, Saudi Arabia. (The designation A.M. is for *ante meridiem*, for "before midday/noon," whereas P.M. is for *post meridiem*, meaning "after midday/noon.") Coordination of international trade, airline schedules, business activities, and daily living depends on a common time system.

In 1884 at the International Meridian Conference in Washington, DC, the prime meridian was set as the official standard for the world time zone system—**Greenwich Mean Time (GMT)** (see http://wwp.greenwich-meantime.com/). Coordinated Universal Time For decades, GMT was determined using the Royal Observatory's astronomical clocks and was the world's standard for accuracy. In 1972, the **Coordinated Universal Time (UTC)** time-signal system replaced GMT and became the legal reference for official time in all countries. UTC is based on average time calculations from atomic clocks collected worldwide. You might still see official UTC referred to as GMT or Zulu time.

Because Earth revolves 360° every 24 hours, or 15° per hour (360° ÷ 24 = 15°), a time zone of 1 hour is established for each 15° of longitude. Each time zone theoretically covers 7.5° on either side of a controlling (standard) meridian (0°, 15°, 30°, 45°, 60°, 75°, 90°, 105°, 120°, etc.) and represents 1 hour. Traveling eastward, one would set the clock 1 hour later, and going westward, 1 hour earlier for each time zone.

Local Sun time, or solar time, is based on the actual longitudinal arc distance (in degree, minutes, and seconds) between a location and the prime

Figure
Modern International Time Zones

http://goo.gl/YvJmY9

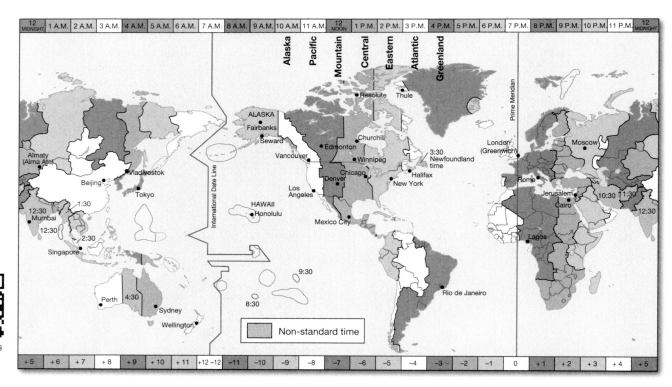

▲ Figure 2.2 **Modern international standard time zones.**
The numbers at the bottom indicate how many hours each zone is earlier (plus sign) or later (minus sign) than Greenwich Mean Time, today known as Coordinated Universal Time (UTC) or Zulu Time.

12   Lab Exercise 2

meridian. There could be as much as 30 minutes' difference in time (or more, depending on the actual time zone boundaries) between local Sun time and standard time.

Opposite the globe from the prime meridian is the **International Date Line**, the starting point for each new day. Since the Date Line is the standard (controlling) meridian for a time zone, no change in time occurs when crossing the line, only the date. When traveling westward, one will set the calendar 1 day ahead, and when crossing the line heading eastward, the calendar will be reset 1 day back.

1. From the map of global time zones in Figure 2.2, determine the present time in the following cities: (For your time, use the starting time of the lab.)

   Moscow _____   Los Angeles _____

   London _____   Honolulu _____

   Chicago _____   Mumbai _____

2. You may not always have a time zone map available, but by remembering the relationship of 1 hour for every 15° of longitude, you can easily calculate the difference in time between places. <u>Indicating and using the standard meridians to determine time zones</u>, solve the following problems. The first answer is provided for you in bracketed italics. Show your work:

   a) If it is 3 A.M. Wednesday in Vladivostok, Russia (132°E), what day and time is it in Moscow (37°E)? [*The controlling meridian for Moscow is 105° away from Vladivostok's controlling meridian of 135°E (135° − 30° = 105° difference). Since Earth rotates 15° per hour, Moscow is 7 hours earlier than Vladivostok (105° difference / 15° rotation per hour = 7 hours time difference), therefore if it is 3 A.M. Wednesday in Vladivostock it is 8 P.M. Tuesday in Moscow.*]

   b) If it is 7:30 P.M. Thursday in Winnipeg, Manitoba, Canada (97°W), what day and time is it in Harare, Zimbabwe (31°E)?

   _____

   c) If you depart from San Francisco International Airport at 10:00 P.M. on Tuesday, what day and time will you arrive in Auckland, New Zealand (175°E), assuming a flight time of 14 hours?

   _____

3. If there is a difference of 15° of longitude for each hour of time, how much difference in time is there for 1° of longitude? for 1′ of longitude?

4. What is the standard (controlling) meridian for your time zone (75°—Eastern, 90°—Central, 105°—Mountain, 120°—Pacific, 135°—Alaska, other)?

   How many degrees of longitude separate you from this standard controlling meridian?

   How does your distance from the standard meridian affect the difference between the time on your clocks and actual Sun (solar) time? (Calculate the difference between standard and Sun time using the answer you determined in #3 above).

Knowing your standard meridian is useful because it will tell you how many minutes off (fast or slow) your watch is in your time zone from the actual position of the Sun in the sky.

Just as you can use longitudinal distance to calculate time, you can also use time difference to calculate your longitude. This is the method that was often used to determine longitudinal location when sailing (if they didn't happen to have a GPS unit on board!). Again, the relationship between time and longitude (15° = 1 hour of time) is the key. If you know the difference in time between two places, and you know the longitude of one place, you can calculate the longitude of the other. For example: Your watch (kept on your hometown standard time) reads 10:30 A.M., and you see that a clock in the town you are visiting reads 7:30 P.M. Your home near St. Paul, Minnesota, uses 90°W as its standard meridian. Therefore:

- Time difference between 10:30 A.M. and 7:30 P.M. = 9 hours
- 9 hours × 15°/hour = 135° of longitude separating the two locations
- Since you are at a place later in time than home, you must be east of home
- 135° east of your hometown at 90° W would be 45° E longitude

5. Assume the time on your watch, showing local standard time, is 4:15 P.M. A chronometer reads 2:15 A.M. What is your longitude?

Daylight saving time is the practice of setting time ahead 1 hour in the spring and back 1 hour in the fall in the Northern Hemisphere. Clocks in the United States were left advanced an hour continuously from 1942 to 1945 during World War II, and again from January 1974 to October 1975, in order to conserve energy during an oil embargo. Daylight saving time increased in length in the United States and Canada in 1986. Under the Energy Policy Act of 2005, Daylight saving time begins 3 weeks earlier than previously, at 2 A.M. on the second Sunday in March. DST has been extended by 1 week and now lasts until 2 A.M. on the first Sunday of November.

6. Does your community adopt daylight saving time? What are the dates for adjusting clocks in the spring and fall?

7. What time does your physical geography lab start
   a) according to standard time? _____
   b) according to daylight saving time? _____
   c) in UTC? _____
   (24-hour clock time in Greenwich, England; e.g., 3:00 P.M. = 15:00 hours)

MasteringGeography™ | Looking for additional review and lab prep materials? Go to www.masteringgeography.com for pre lab videos and pre and post lab quizzes.

Name: _____ Laboratory Section: _____

Date: _____ Score/Grade: _____

Video
Exercise 3
Pre-Lab Video

Scan to view the Pre-Lab video

LAB EXERCISE 3

Directions and Compass Readings

Earth's coordinate grid system was covered in Lab Exercise 2, so you are familiar with using latitude and longitude to find a point's absolute location on Earth's surface. There are times when relative direction between two places is desired—for example if you want to navigate from one point to another. There are two systems used to indicate relative direction on maps: compass points and azimuths.

Understanding and being able to use these direction-finding methods is important, whether you are reading a street/road map or are interpreting specialized maps such as topographic maps, which are discussed and used in later exercises. These systems are all measurements of horizontal direction, as if you were in the air looking down at Earth from above.

Compasses are useful for finding directions, but corrections must be made for deviation from true or geographic north. Lab Exercise 3 has four sections.

Key Terms and Concepts

azimuths
cardinal points
compass points
geographic north
geographic south
grid north
isogons

isogonal map
magnetic compass
magnetic declination
magnetic north
topographic map
true north
true south

KEY LEARNING
concepts

After completion of this lab, you should be able to:

1. *Describe* the method of indicating direction using compass readings and *apply* that method to compass directions.
2. *Define* magnetic declination and *determine* differences in "north" readings on a map.
3. *Explain* any difference between a local topographic map and the isogonal map relative to magnetic declination.

Materials/Sources Needed

color pencils
compass
local topographic map

protractor
ruler

Lab Exercise and Activities

SECTION 1

Compass Points

The directional system familiar to most people is that of **compass points**. The symbol below is a *compass rose*. Cartographers often included elaborate compass roses as a decorative, as well as a functional, part of antique maps. The compass rose has four **cardinal points**—in a clockwise direction—north, east, south, and west, separated by four intermediate points: NE, SE, SW, NW. These are split into 16 and then, sometimes, again into 32 compass points. You may not be familiar with compass points such as NNE and ENE. These points are used when greater precision is needed. NNE (read north northeast), is north of NE, while ENE (read east northeast), is east of NE.

1. Label 16 compass points on Figure 3.1, using color pencils on the compass rose to distinguish the categories of division. (Include a legend for your color scheme.) In each division, the points are listed clockwise.

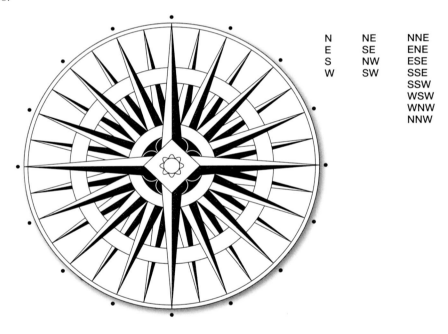

N	NE	NNE
E	SE	ENE
S	NW	ESE
W	SW	SSE
		SSW
		WSW
		WNW
		NNW

▲ Figure 3.1 Compass points

SECTION 2

Azimuths

While useful, compass points are not extremely accurate, and most often the direction between two points does not fall exactly on one of the 32 points. Degrees of an arc are used for more precise directions.

Azimuths are read from the north in a clockwise direction, from 0° to 360° (with 0° and 360° being the same point—north). For further precision, the degrees can be broken down into minutes (′).

1. Figure 3.2 has a few azimuths drawn and labeled as examples. Using your protractor, determine the azimuth readings for A and B, and label the value for each on the diagram.

2. Measure, draw, and label the following azimuths on the diagram: 230°, 78°, 145°.

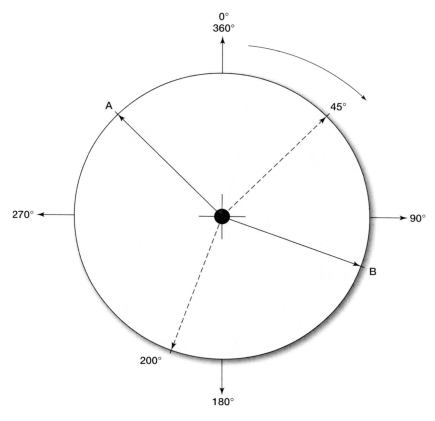

▲ Figure 3.2 Azimuths

3. Using Figure 3.1, Figure 3.2, and your protractor, answer the following questions.

   Which azimuth is found at the following compass points: NE __[45°]__ SW _____

   W _____ ESE _____ NNW _____

   Which compass point is found at the following azimuths: 135° __[SE]__ 247.5° _____

   180° _____ 315° _____ 225° _____

SECTION 3

Compass Declination

Earth rotates around an axis that is marked by the **geographic North** and **South Poles**, also referred to as **true north** and **true south**. Meridians of longitude converge at these geographic points.

A **magnetic compass**, an instrument used to determine direction, points to the **magnetic North Pole**, which does not coincide with true north and the geographic pole. The difference between the north arrow on a magnetic compass and true north is the **magnetic declination**. A compass needle that points west of true north indicates a *west declination*, and a compass needle that points east of true north indicates an *east declination*. When using a magnetic compass, you must make adjustments for the difference between magnetic north and true north.

Isogons, lines connecting points of equal magnetic declination, are shown on *isogonal maps*, such as in Figure 3.3. The magnetic poles change over time, slowly migrating, so that adjustments for these magnetic pole changes need to be considered if map navigation is to be up-to-date.

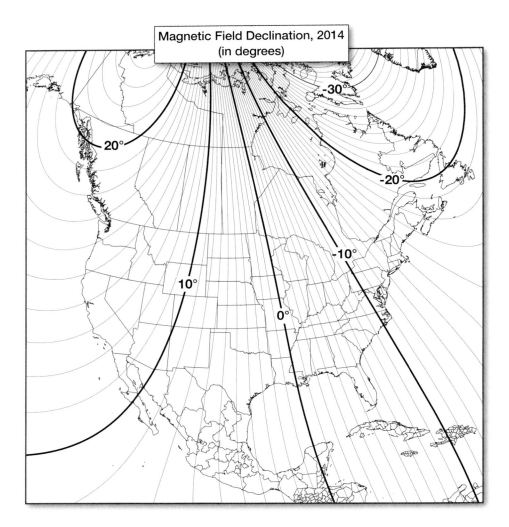

▲ Figure 3.3 Magnetic field declination, 2014 (in degrees)

1. Using the isogonal map in Figure 3.3 and your atlas, in what city would you have to stand during 2014 to get a compass reading with zero magnetic declination (where magnetic north aligns with true north, or an agonic line). Mark the city with a dot and a label on the map.

2. Using your atlas, turn to the pages featuring the polar regions and find the latitude/longitude coordinates for the:

    a) Magnetic North Pole _____

    b) Magnetic South Pole _____

3. Find the linear distance in km and miles between the latitude of the:

   a) Magnetic North Pole and true north _____

   b) Magnetic South Pole and true south _____

Topographic maps (or topo maps) are a set of maps covering the entire United States that show the horizontal position (latitude/longitude) of boundaries, land-use aspects, bodies of water, and economic and cultural features. Topographic maps also have a vertical component to show topography (configuration of the land surface), including slope and relief (the vertical difference in local landscape elevation). These fine details are shown through the use of elevation contour lines. A contour line connects all points at the same elevation. Elevations are shown above or below a vertical datum, or reference level, which usually is mean sea level. The contour interval is the vertical distance in elevation between two adjacent contour lines. Inside the front cover of this manual are the standard symbols commonly used on these topographic maps. These symbols and the colors used are standard on all USGS topographic maps: black for human constructions, blue for water features, brown for relief features and contours, pink for urbanized areas, and green for woodlands, orchards, brush, and the like. The margins of a topographic map contain a wealth of information about its concept and content. In the margins of topographic maps, you find the quadrangle name, names of adjoining quads, quad series and type, position in the latitude-longitude and other coordinate systems, title, legend, datum plane, symbols used for roads and trails, the dates and history of the survey of that particular quad, magnetic declination (alignment of magnetic north) and compass information, and more.

Magnetic declination is shown by a declination diagram or arrow in the margin of a topographic map, as shown in Figure 3.4. True north is indicated by the star, and the magnetic declination is shown by the arrow marked MN. You work further with magnetic declination in Lab Exercise 23 (topographic map reading). **Grid north (GN)** refers to the relation of the topo map to the Universal Transverse Mercator system (UTM). GN is another reference system based on rectangular map zones, so that as meridians taper toward the pole, the GN shows the relation of these converging lines and the rectangular grid.

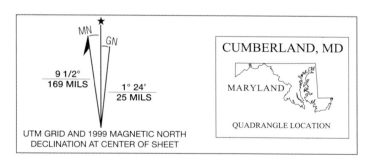

▲ Figure 3.4 Declination arrow on topographic maps

4. According to Figure 3.4, what is the magnetic declination between:

   a) True north and magnetic north (MN)? _____

   b) True north and grid north (GN)? _____

5. Look for the declination arrow in the margin of the topographic map provided by your instructor. What is the magnetic declination between true north and magnetic north on the topographic map provided by your instructor? What is the magnetic declination between true north and magnetic north shown for this location in Figure 3.3? Any discrepancies you find relate to the migrating magnetic pole and the dates of the topo map and the isogonal map (2014). (Keep this topo map handy for the next section.)

SECTION 4

Compass Bearing

The compass bearing between two points on a map may be determined by using a protractor or a magnetic compass. To accomplish this, lightly draw a line through the specified points and extend the line to the edge of the map. To use a protractor, place the straight edge of the protractor along the margin of the map, with the origin of the protractor at the point where the line intersects the map margin. The extension of the drawn line indicates the compass bearing (Figure 3.5a). To use the compass, simply align the margin of the map with the 0°/360° and 180° points on the compass. The bearing can be read from the compass dial (Figure 3.5b). (Pay no attention to the compass needle.)

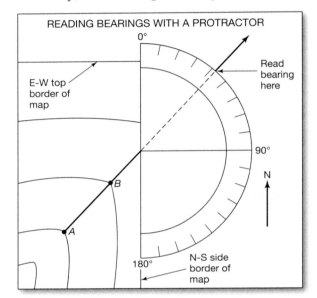
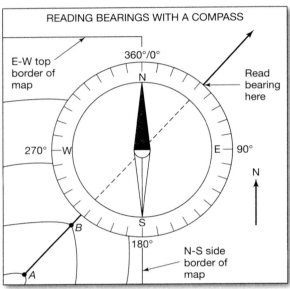

▲ Figure 3.5 Reading compass bearings on a map using a protractor (a); using a magnetic compass (b). (From Busch, Richard M. editor, *Laboratory Manual in Physical Geology*, 3rd ed., Macmillan Publishing Company © 1993.)

Using the local topographic map from the last section, complete the following.

1. Find the compass bearing between two points on the map indicated by your instructor. Give the reading as an azimuth.

On the circle provided below, sketch your determination of the azimuth.

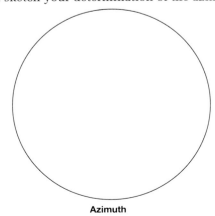

Azimuth

2. What is the highest elevation on the local topographic map? (Indicate a geographic feature that is at or adjacent to that elevation.)

3. What is the lowest elevation on the map? (Indicate a geographic feature that is at or adjacent to that elevation.)

4. What is the relief (highest elevation minus lowest elevation in a local landscape) of the area portrayed on this map? Perhaps compare two or three areas of the topographic map if relief varies across the region.

   (Show your work.)

MasteringGeography™ | Looking for additional review and lab prep materials? Go to www.masteringgeography.com for pre lab videos and pre and post lab quizzes.

Name: _____ Laboratory Section: _____

Date: _____ Score/Grade: _____

Video
Exercise 4
Pre-Lab Video

LAB EXERCISE 4

Scan to view the Pre-Lab video

http://goo.gl/BppTYe

Map Projections, Map Reading, and Interpretation

A **map** is a generalized view of an area, usually some portion of Earth's surface, as seen from above and greatly reduced in size. The part of geography that embodies mapmaking is called **cartography**. Maps are critical tools with which geographers depict spatial information and analyze spatial relationships. We all use maps at some time to visualize our location and our relationship to other places, or maybe to plan a trip, or to coordinate commercial and economic activities.

In this lab exercise, we begin with an overview of various map projections that are used for a variety of purposes. An important decision in making maps is the proper scale to use. Maps, like architectural blueprints, are produced at a greatly reduced size compared to reality. Then, we specifically work with a U.S. mapping program—the township and range system. Lab Exercise 4 has four sections.

Key Terms and Concepts

cartography
conic
cylindrical
equal area (equivalent)
gnomonic projection
great circle
map
map projection

Mercator projection
planar
rhumb lines
scale
standard line
township and range
true shape (conformal)

KEY LEARNING
concepts

After completion of this lab, you should be able to:

1. *Define* the map projection concept and *identify* the four classes of map projections.
2. *Recognize* distortions that are characteristic of selected map projections and *compare* distortions in order to *select* the appropriate map for a given use.
3. *Plot* a great circle route from a gnomonic projection to a Mercator projection.
4. *Compare* methods of expressing map scale and *calculate* map scales.
5. *Explain* the township and range system employed in the U.S.

Materials/Sources Needed

tracing (or wax) paper or notebook paper
world globe
scissors

color pencils
calculator
ruler

Copyright © 2015 Pearson Education, Inc.   Lab Exercise 4   23

Lab Exercise and Activities

SECTION 1

Map Projections

We worked with Earth's coordinate grid system in Lab Exercise 2. Cartographers prepare large-scale flat maps—two-dimensional representations (scale models) of our three-dimensional Earth. This transformation of spherical Earth and its latitude–longitude coordinate grid system to a flat surface in some orderly and systematic realignment is called a **map projection**.

A globe is the only true representation of distance, direction, area, shape, and proximity, and preparation of a flat version requires decisions as to the type and amount of distortion that will be acceptable. To understand this problem, consider these important physical properties of a globe:

- Parallels always are parallel to each other, always are evenly spaced along meridians, and always decrease in length toward the poles.
- Meridians converge at both poles and are evenly spaced along any individual parallel.
- The distance between meridians decreases toward poles, with the spacing between meridians at the 60th parallel equal to one-half the equatorial spacing.
- Parallels and meridians always cross each other at right angles.

The problem with preserving these properties is that all globe qualities cannot be reproduced on a flat surface. The larger the area of Earth depicted on a map, the greater the distortion will be. Simply taking a globe apart and laying it flat on a table illustrates the problem cartographers face in constructing a flat map (Figure 4.1).

In Figure 4.1, you can see the empty spaces that open up between the sections, or gores, of the globe when it is flattened. Flat maps always possess some degree of distortion—much less for large-scale maps representing a few kilometers, and much more for small-scale maps covering individual countries, continents, or the entire world.

The best projection is always determined by its intended use. The major decisions in selecting a map projection involve the properties of **equal area** (equivalence) and **true shape** (conformality). If a cartographer selects equal area as the desired trait, as for a map showing the distribution of world climates, then shape must be sacrificed by *stretching* and *shearing* (having parallels and meridians cross at other than right angles).

If, on the other hand, true shape is desired, as for a map used for navigational purposes, then equal area must be sacrificed, and the scale will actually change from one region of the map to another. Two additional properties related to map projections are true direction and true distance.

The fold-out flap of the back cover shows four classes of map projections and geometric-surface perspectives from which three classes of maps—**cylindrical**, **planar** (or *azimuthal*), and **conic**—are generated. Another class of projections that cannot be derived from this physical-perspective approach is the non-perspective *oval*-shaped. Still others are derived from purely mathematical calculations.

With all projections, the contact line (or point) between the globe and the projection surface—called a **standard line**—is *the only place where all globe properties are preserved.* A *standard parallel* or *standard meridian* is true to scale along its entire length, without any distortion. Areas away from this critical tangent line or tangent point become increasingly distorted. Consequently, this area of optimum spatial properties should be centered on the region of immediate interest so that greatest accuracy is preserved there.

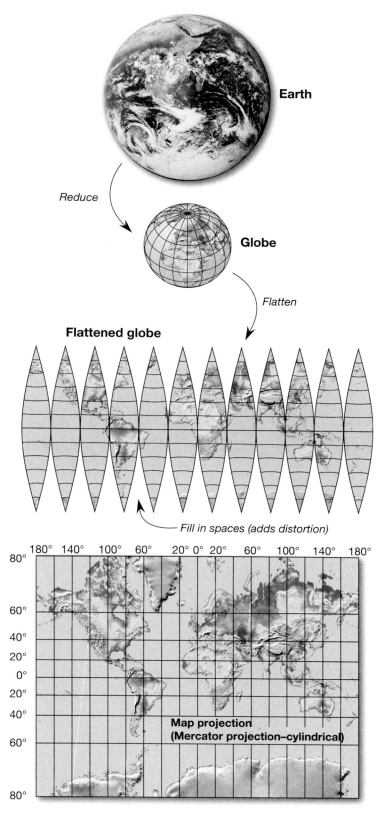

▲ Figure 4.1 Conversion of a globe to a flat map requires decisions about the desired properties to preserve, how much distortion is acceptable, and which projection best serves the purpose of the map.

SECTION 2

Cylindrical, Planar, and Conic Projections

This step may be done by working in groups. Your instructor will direct you.

1. Using a globe and tracing (or wax) paper—notebook paper works with an illuminated globe—trace outlines of North America, South America, and Greenland. You can do this quickly; a rough outline showing size and shape is sufficient.

   Using a globe and your tracings, compare the relative sizes of North America, South America, and Greenland. Which is larger and by approximately how many times? Describe your observations.

2. Examples of two cylindrical projections are presented in Figure 4.2, the **Mercator projection** (conformal, true-shape), and Figure 4.3, the *Lambert projection* (equivalent, equal-area). These two map projections—the Mercator and the Lambert—are simply being used to examine *relative* size and shape in each portrayal of Earth.
   Trace around the outlines of the same continents and island as indicated above—North America, South America, and Greenland. You might want to mark and color code the tracings according to their source (Mercator or Lambert) for future reference. Use your tracings to answer the following questions. Keep in mind that these two sets of map outlines compared to the globe you used are not at the same scale: the distance measured along their respective equators will not be equal.

   a) Using the outlines from the Mercator and Lambert map projections on the next pages, make the same comparisons among North America, South America, and Greenland. Which is relatively larger and by approximately how many times? Again, describe your observations.

   Mercator _____
   _____

   Lambert _____
   _____

   b) Now compare the relative shapes of Greenland, North America, and South America from the globe (shows true shape) with those from the Mercator and Lambert projections. Is there distortion in terms of shape? If so, briefly explain how they are distorted.

   Mercator _____
   _____

   Lambert _____
   _____

   Figure 4.4a and Figure 4.4b once again show the Mercator and Lambert cylindrical projections. Diagrams to the right of each projection help you to identify the distortion in each. The circles along the equator—which is the standard line, or line of tangency between the globe and the cylinder and is without distortion—are the "reference standard." The shape and size of the other "circles" is proportionate to the distortion at various latitudes on the maps.

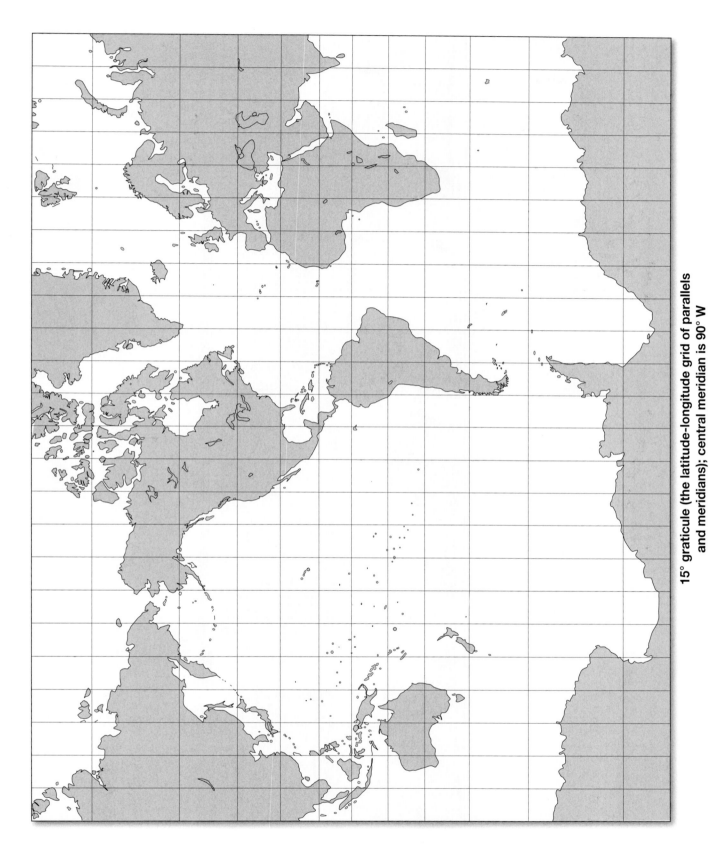

▲ Figure 4.2 Mercator projection—conformal, true-shape map

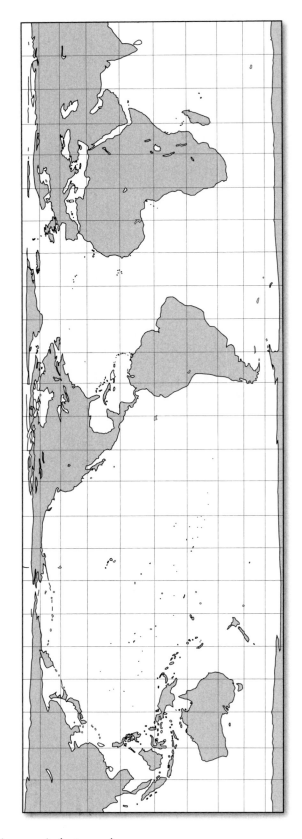

▲ Figure 4.3 Lambert projection—equivalent, equal-area map

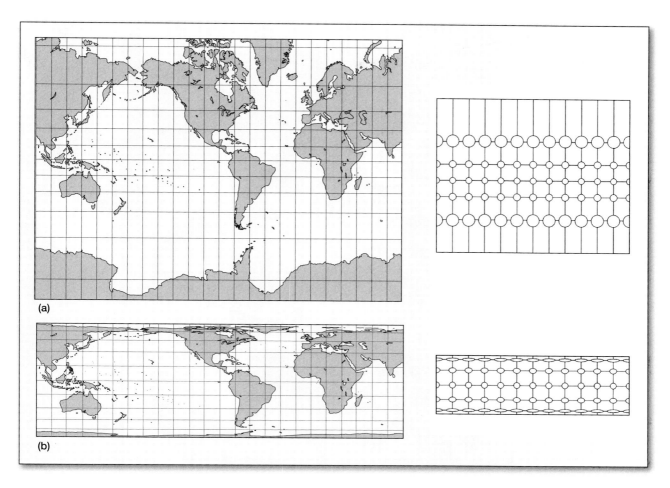

▲ Figure 4.4 (a) Mercator cylindrical true shape (conformal) projection, 30° graticule; standard parallel 0° central meridian 90°W. (b) Lambert cylindrical equal area (equivalent) projection, 30° graticule; standard parallel 0° central meridian 90°W.

3. What happens to the circles in the Mercator projection (and, therefore, to the landmasses) as latitude increases? What causes this distortion of the circles away from the equator? (*Hint:* Compare parallels and meridians with those on the globe.)

4. What happens to the circles in the Lambert projection (and, therefore, to the landmasses) as latitude increases? What causes this distortion of the circles away from the equator? (Once again, compare parallels and meridians.)

5. Cylindrical projections such as Mercator and Lambert would best be used in mapping what areas of the globe? Why?

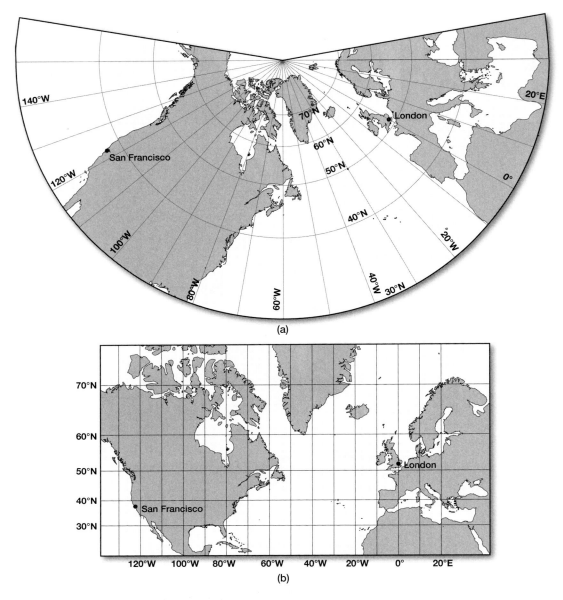

▲ Figure 4.5 (a) Gnomonic/planar projection; (b) Mercator/cylindrical projection.

6. Figure 4.5a shows Earth on a *planar* projection, in which the globe is projected onto a plane. This is a **gnomonic projection** and is generated by projecting a light source at the center of a globe onto a plane that is tangent to (touching) the globe's surface. The resulting increasingly severe distortion as distance increases from the standard point prevents showing a full hemisphere on one projection.

   a) Where is the projection surface tangent to (touching) the globe? (See end flap of this lab manual.)

   _____

   _____

   b) What kind of distortion, if any, occurs on a planar projection, and where?

   _____

   _____

30 Lab Exercise 4

c) Can you show the entire Earth on a single gnomonic projection? If not, why not?

_____

_____

d) Which areas of the globe would likely be mapped on a planar projection, and why?

_____

_____

The Mercator projection is useful in navigation and has become the standard for nautical charts since Gerardus Mercator, a Flemish cartographer, devised it for navigational purposes in 1569. The advantage of the Mercator projection is that lines of constant compass direction, called **rhumb lines**, are straight and thus facilitate the task of plotting compass directions between two points. However, these lines of constant direction, or bearing, are not the shortest distance between two points.

From the *planar* class of projections, a gnomonic projection possesses a valuable feature: All straight lines are **great circles**—the shortest distances between two points on Earth's surface. A gnomonic projection is used to determine the coordinates of a great-circle (shortest) route; these coordinates are then transferred to a Mercator (true-direction) projection for determination of precise compass headings—the route along which the pilot or captain steers the airplane or ship.

Complete:

7. Use the two maps in Figure 4.5 to plot a great circle route between San Francisco (west coast of the United States, where you can see small details of San Francisco Bay) and London (southern England at the 0° prime meridian). The straight line on the gnomonic projection will show the shortest route between the two cities. Transfer the coordinates of this route over to the Mercator map and connect with a line plot to show the route's track. Note the route arching over southern Greenland on the Mercator.

8. Assume you board a plane in San Francisco for a flight to London. Briefly describe your great-circle flight route between the two cities (plotted on the map in Figure 4.5b). What are some of the features over which you fly? Describe the landscape and water below.

SECTION 3

Map Scale

The ratio of the measured distance on a map to the actual distance on the ground is called **scale**; it relates a unit on the map to a similar unit on the ground. A 1:1 scale would mean that a centimeter on the map represents a centimeter on the ground, or an inch on the map represents an inch on the ground. A more appropriate scale for a local map is 1:24,000, in which 1 unit on the map represents 24,000 identical units on the ground.

Map scales may be presented in several ways. A *written* (or verbal) *scale* simply states the ratio using words—for example, "one centimeter to one kilometer" or "one inch to one mile." A *representative fraction* (RF, or fractional scale) can be expressed with either a : or a /, as in 1:125,000 or 1/125,000. No actual units of measurement are mentioned because any unit is applicable as long as both parts of the fraction are in the same unit: 1 cm to 125,000 cm, 1 in. to 125,000 in., or even 1 arm length to 125,000 arm lengths, and so on. A *graphic (bar) scale* is a line or a bar divided into segments illustrating the correct map-to-ground ratio (Figure 4.6).

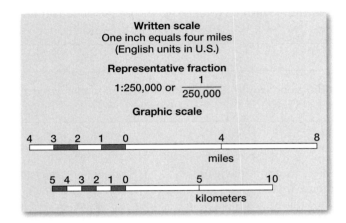

▲ Figure 4.6 Three expressions of map scale: written scale, representative fraction, and graphic (bar) scale

Scales are called "small," "medium," and "large," depending on the ratio described. Thus, a scale of 1:24,000 is a large scale, whereas a scale of 1:50,000,000 is a small scale. The greater the size of the denominator in a fractional scale (or number on the right in a ratio scale), the smaller the scale and the more abstract the map must be in relation to what is being mapped.

Examples of selected representative fractions and written scales are listed in Table 4.1 for small-, medium-, and large-scale maps.

System	Scale Size	Representative Fraction	Written Scale
English	Small	1:3,168,000	1 in. = 50 mi
		1:2,500,000	1 in. = 40 mi
		1:1,000,000	1 in. = 16 mi
		1:500,000	1 in. = 8 mi
		1:250,000	1 in. = 4 mi
	Medium	1:125,000	1 in. = 2 mi
		1:63,360 (or 1:62,500)	1 in. = 1 mi
		1:31,680	1 in. = 0.5 mi
		1:30,000	1 in. = 2500 ft
	Large	1:24,000	1 in. = 2000 ft
System		Representative Fraction	Written Scale
Metric		1:1,000,000	1 cm = 10.0 km
		1:50,000	1 cm = 0.50 km
		1:25,000	1 cm = 0.25 km
		1:20,000	1 cm = 0.20 km

▲ Table 4.1 Sample representative fractions and written scales for small-, medium-, and large-scale maps

Assume:

If a world globe is 61 cm (24 in.) in diameter and we know Earth has an equatorial diameter of 12,756 km (7926 mi), then the scale of the globe is the ratio of 61 cm to 12,756 km. Therefore, in order to determine the globe's representative fraction:

$$\mathbf{RF} = \frac{\text{diameter of globe}}{\text{diameter of Earth}} = \frac{61 \text{ cm}}{12{,}756 \text{ km}} = \frac{61 \text{ cm}}{1{,}275{,}600{,}000 \text{ cm}} = \frac{1 \text{ cm}}{20{,}911{,}000 \text{ cm}}$$

Thus, the representative fraction for the 61 cm globe is expressed in centimeters as 1:20,900,000 (rounded off). This representative fraction can now be expressed in *any* unit of measure, metric or English, as long as both numbers are in the same units.

Complete:

1. Several companies manufacture large world globes for display in museums, corporate lobbies, and science exhibit halls. If such a globe featured a *diameter* of 4 m (13 ft), what is the scale of this globe expressed as a representative fraction? (Show your work.) (*Hint:* Earth's equatorial diameter = 12,756 km [7926 miles].)

2. Given the large globe in question #1 and your calculation of its representative fraction, convert the RF to a *written scale* for this globe (1 cm on the globe =? km on Earth). (Show your work and remember that there are 100,000 cm in 1 km.)

3. Use the topographic maps in the section at the back of the lab manual and a ruler (inches or cm) to measure the map distance between the points indicated. Noting the RF scales posted on each map, calculate the approximate ground distance (miles or km). Keep in mind that there are 63,360 inches in a mile: 12 in/ft × 5280 ft/mile. Compare results with the graphic scale. (Show your work.)

    a) **Kilauea Crater, HI Topographic Map #1:** from the Park Headquarters to Sand Hill, BM 3700.
    _____

    b) **Mt. Rainier National Park Topographic Map #6:** from Columbia Crest to McClure Rock
    _____

    c) **Point Reyes, CA Topographic Map #9:** from McClures Ranch to Chimney Rock
    _____

4. Find three different maps or globes as follows. Briefly describe the content of each and record the scale as a representative fraction or other scale noted on each map or globe.

    a) Classroom globe _____

    b) Map from your atlas _____

    c) Topo map at back of this manual _____

5. Give an example of a specific scale for each of the following:

    a) Small scale: _____

    b) Medium scale: _____

    c) Large scale: _____

6. What is an appropriate scale to use in preparing a world map for a classroom (pull-down "wall" map)? Why?

7. If you wanted a map to help you plan a local trip within your state or province, what scale would be best?

8. **Challenge question:** If you were helping to plan the painting of a world map on a local school playground that is approximately 10 meters along the equator, what scale would be best?

   RF: _____

   Written scale: _____

9. Using the metric-to-English conversion chart on the inside of the fold-out back flap of this manual, what would be the approximate length of the equator, in feet?

SECTION 4

Township and Range Survey System

The Public Lands Survey System (1785) delineated locations in the United States. After modifications and adjustments, the final plan called for the establishment of the **township and range** grid system, based on an initial survey point. For those portions of the United States surveyed under this system, the *initial points* are presented in Figure 4.7. Places are designated as north or south of the base line and east or west of the principal meridian. For example, Township 3 N means 3 "tiers" north of the base line and Range 4 W means 4 "columns" west of the principal meridian. Note that due to the latitudinal extent of California (southern border: 32°40′N; northern border: 42°N), three initial points are used.

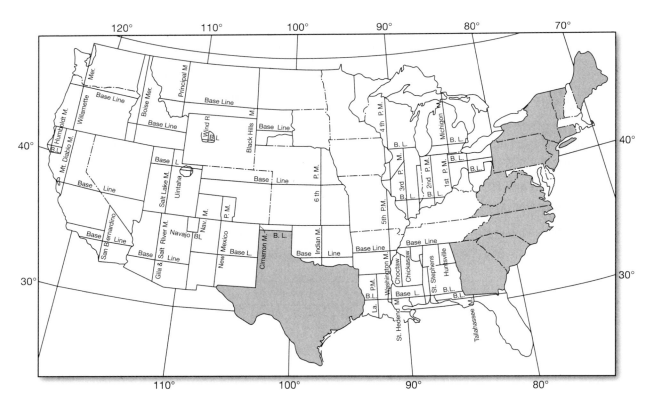

▲ Figure 4.7 Principal meridians and base lines used for the township and range public-land surveys.
Shaded states were not surveyed with the PLSS. (Adapted from the Bureau of Land Management, Manual of Surveying Instructions, Washington, DC, 1947, p. 168.)

The township and range grid system is based on 6-mile square *townships*, subdivided into 1-mile square *sections*, and quarter-section *homesteads*. Figure 4.8 portrays this system and shows the derivation of a parcel of land described as "S1/2, NW1/4, NE1/4, SW1/4, Sec. 26, T2S, R4E." Note that you start from the smallest parcel of land ("the south half, of the northwest quarter, of the northeast quarter ..."). Some land deeds still reflect this system.

Questions and analysis about the township and range system.

1. Complete the fill-ins presented in Figure 4.8 on the next page.

2. Using Figure 4.7 as your guide, which meridian and base line pair would be used to survey the area you are in?

3. If your campus was surveyed using the public-lands system, give a full description of the campus in standard township and range format (consult a topographic map for your campus area).

4. If the state or region where your campus is located was surveyed under a different system (for example, parts of Texas, the original 13 colonies), describe it relative to this system.

MasteringGeography™ | Looking for additional review and lab prep materials? Go to www.masteringgeography.com for pre lab videos and pre and post lab quizzes.

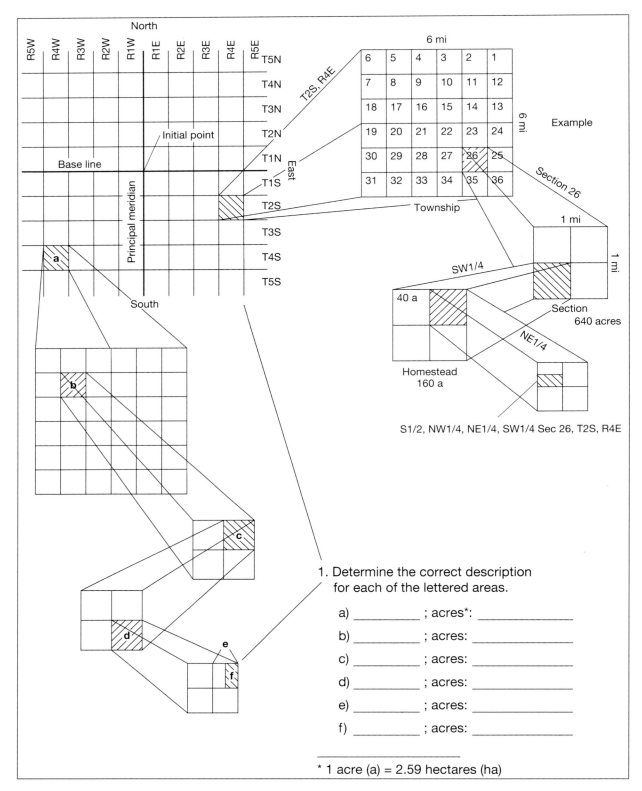

1. Determine the correct description for each of the lettered areas.

a) _____ ; acres*: _____
b) _____ ; acres: _____
c) _____ ; acres: _____
d) _____ ; acres: _____
e) _____ ; acres: _____
f) _____ ; acres: _____

* 1 acre (a) = 2.59 hectares (ha)

▲ Figure 4.8 Township and range rectangular survey grid system illustrating a parcel of land at S1/2, NW1/4, NE1/4, SW1/4, Sec 26, T2S, R4E.

Name: _____ Laboratory Section: _____

Date: _____ Score/Grade: _____

Video
Exercise 5
Pre-Lab Video

LAB EXERCISE 5

Earth–Sun Relationships and Daylength

Scan to view the Pre-Lab video

http://goo.gl/8yj0IS

For more than 4.6 billion years, solar energy has traveled across interplanetary space to Earth, where a small portion of the solar output is intercepted. Because of Earth's curvature, the arriving energy is unevenly distributed at the top of the atmosphere, creating energy imbalances over Earth's surface. Also, the annual pulse of seasonal change varies the distribution of energy during the year.

Earth's distinct seasons are produced by interactions of revolution (annual orbit about the Sun) and rotation (Earth's turning on its axis), Earth's axial tilt (at about 23.5° from a perpendicular to the plane of the ecliptic), axial parallelism (the parallel alignment of the axis throughout the year), and sphericity. As Earth rotates, the boundary that divides daylight and darkness is the **circle of illumination**.

Earth rotates about its axis, an imaginary line extending through the planet from the geographic North Pole to the South Pole. The Tropic of Cancer and Tropic of Capricorn mark the farthest north and south the subsolar point migrates during the year, about 23.5°N and 23.5°S latitude. The **solstices** occur when the Sun is directly over the tropics, and the **equinoxes** occur when the Sun is directly over the equator. The **march of the seasons** refers to the procession of the March equinox, June solstice, September equinox, and the December solstice.

Daylength, the interval between sunrise and sunset, varies during the year, depending on latitude. The equator always receives equal hours of day and night. At 50°N or S latitude, people experience almost 8 hours of annual daylength variation.

At the North and South Poles, the range of daylength is extreme, with a 6-month period of no insolation, beginning with weeks of twilight, then darkness, then weeks of predawn. Following sunrise, daylight lasts for a 6-month period of continuous 24-hour insolation—literally, the poles experience one long day and one long night each year! Lab Exercise 5 has two sections.

Key Terms and Concepts

circle of illumination
daylength
equinox

march of the seasons
solstice
subsolar point

After completion of this lab, you should be able to:

1. *Label* diagrams representing Earth–Sun relations in the annual march of the seasons.
2. *Calculate* daylength for selected locations for the solstices and equinoxes and *analyze* the relationship between latitude and seasonality.

Materials/Sources Needed

pencil
calculator

color pencils
ruler

Lab Exercise and Activities

SECTION 1

Earth–Sun Relations—Seasonality

Because Earth is spherical, its curved surface presents a continually varying angle to the incoming parallel rays of insolation. Differences in the angle of solar rays at each latitude result in an uneven distribution of insolation and heating. The place receiving maximum insolation is the point where insolation rays are perpendicular to the surface (radiating from directly overhead), called the **subsolar point**. All other places receive insolation at less than a 90° angle and thus experience more diffuse energy receipts.

Solar beam angles become more pronounced at higher latitudes. As a result, during a year's time, the top of the atmosphere above the equatorial region receives 2.5 times more insolation than that received above the poles.

Animation
Animation of the Seasons

http://goo.gl/LLDpgQ

1. On the accompanying diagrams in Figure 5.1, complete the following items. June has been started for you. Also, you may want to consult your physical geography text for reference.

   a) Extend the Sun's rays until they *intersect* (pass through) or are *tangent* to Earth's surface (touch at only one point). (See the Earth profile diagram for examples.)

   b) Where the rays intersect or are tangent to Earth's surface, draw a short line tangent to the surface, indicating the angle at which the Sun's rays are intercepted.

   c) Label the rays with the appropriate term—*vertical ray* (striking Earth at a 90° angle), *oblique ray* (striking Earth at less than a 90° angle), or *tangent ray* (parallel to Earth)—and mark and label the *subsolar point*.

   d) Draw the following on the Earth profiles and label:

   North and South Poles
   Equator
   Tropics of Cancer and Capricorn
   Circle of Illumination
   Arctic and Antarctic Circles

   e) Lightly shade the portion of Earth that is experiencing night, or the night half of the circle of illumination. Half of Earth is in sunlight and half is in darkness at any moment. The traveling boundary that divides daylight and darkness is called the circle of illumination—it is the day–night dividing circle.

2. Describe the changing position of the subsolar point and vertical rays throughout the four key seasonal positions of Earth.

3. On the **equinoxes**, the subsolar point is at the equator. Which latitudes does the circle of illumination run through on the equinoxes? On the June **solstice**, the subsolar point is at the Tropic of Cancer. Which latitudes does the circle of illumination pass through on the June solstice? On February 26 the subsolar point is at 9°S. Which latitudes does the circle of illumination pass through on this date?

38 Lab Exercise 5 Copyright © 2015 Pearson Education, Inc.

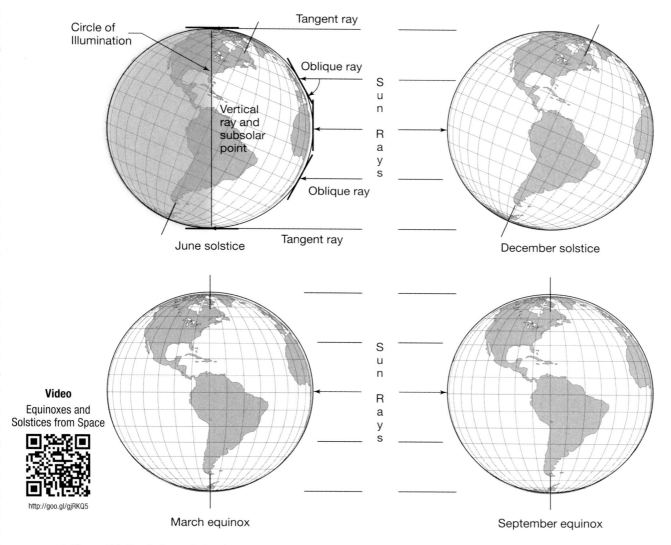

▲ Figure 5.1 Earth–Sun relationships

4. What is the relationship between the location of the subsolar point and the circle of illumination?

Lab Exercise 5

SECTION 2

Daylength

The length of daylight hours, the time between sunrise and sunset, varies both with latitude and the season of the year. Seasonal shifting of the subsolar point results in a realignment of the **circle of illumination** relative to the poles. See the diagrams in Section 1 of this exercise and note the proportions of each latitude line (equator, Tropics of Cancer and Capricorn, Arctic and Antarctic Circles) that are in daylight vs. darkness at the various seasons.

1. Complete Table 5.1 (below), filling in the daylength at selected latitudes.

	Winter Solstice (December Solstice) December 21–22			**Vernal Equinox (March Equinox) March 20–21**			**Summer Solstice (June Solstice) June 20–21**			**Autumnal Equinox (September Equinox) September 22–23**		
	A.M.	P.M.	Daylength	A.M.	P.M.	Daylength	A.M.	P.M.	Daylength	A.M.	P.M.	Daylength
0°	6:00	6:00	___	6:00	6:00	___	6:00	6:00	___	6:00	6:00	___
30°	6:58	5:02	___	6:00	6:00	___	5:02	6:58	___	6:00	6:00	___
40°	7:26	4:34	___	6:00	6:00	___	4:34	7:26	___	6:00	6:00	___
50°	8:05	3:55	___	6:00	6:00	___	3:55	8:05	___	6:00	6:00	___
60°	9:15	2:45	___	6:00	6:00	___	2:45	9:15	___	6:00	6:00	___
90°	No sunlight			Rising Sun			Continuous sunlight			Setting Sun		

▲Table 5.1 Daylength—the time between sunrise and sunset—at selected latitudes for the Northern Hemisphere

2. Explain how changes in the Sun's altitude (angle above the horizon) and daylength vary with the seasons at the equator, 30°, 60°, and 90°.

3. Estimate the approximate length of daylight for the following locations:
   a) Dawson, Yukon Territory, Canada (64° N), on December 21 _____
   b) Adelaide, South Australia (35° S), on June 21 _____
   c) Bangkok, Thailand (14° N), on March 20 _____
   d) Your location on June 21 _____
   e) Your location on December 21 _____

4. Complete the following graph, using the values for daylength calculated in question #1 for the following latitudes: 0°, 30°, 40°, 50°, 60°, and 90°. Use a different color for each latitude. The line for 30° has already been done for you.

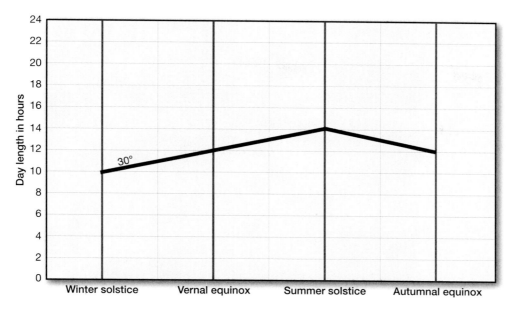

▲ Figure 5.2 Daylength and latitude

5. What is the relationship between seasonal changes in daylength and latitude?

| MasteringGeography™ | Looking for additional review and lab prep materials? Go to www.masteringgeography.com for pre lab videos and pre and post lab quizzes. |

Name: _____ Laboratory Section: _____ **Video**
Date: _____   Score/Grade: _____   Exercise 6
                                                                    Pre-Lab Video

LAB EXERCISE 6

Insolation and Seasons

Scan to view the Pre-Lab video

http://goo.gl/mcZcpV

Earth's systems are powered by a constant flow of **insolation**—incoming solar radiation. Earth intercepts only one two-billionth of the entire solar output, yet this is the singular significant source of energy for living systems. This exercise examines the nature of this energy and contrasts the solar output with energy re-radiated by Earth's surface back to space. The Sun's rays arrive at the top of the atmosphere in parallel beams, but the curvature of Earth presents surfaces at differing angles to this incoming radiation: Lower latitudes receive more direct (vertical) illumination, whereas higher latitudes receive more indirect (oblique) rays. This pattern of uneven heating produces an energy imbalance in the atmosphere and on Earth's surface below—namely, equatorial and tropical energy surpluses and polar energy deficits.

Earth's systems are further influenced by shifting seasonal rhythms produced by changing day-length and Sun altitude (height of the noon Sun above the horizon at a given location). These seasonal changes become more noticeable toward the poles. Seasons occur as a result of Earth–Sun relationships produced by rotation, revolution, tilt of the axis, axial parallelism, and Earth's sphericity. Lab Exercise 6 has two sections.

Key Terms and Concepts

altitude
analemma
declination
insolation

revolution
rotation
zenith

KEY LEARNING concepts

After completion of this lab, you should be able to:

1. *Identify* the pattern of daily energy receipts at the top of the atmosphere at different latitudes throughout the year and *graph* the observations.
2. *Determine* the angle of incidence of insolation and the resultant intensity of the solar beam for various latitudes.
3. *Utilize* the analemma to *determine* the subsolar point, latitude, and noon Sun angle (altitude) for several locations using the analemma.

Materials/Sources Needed
pencil
calculator

color pencils
ruler

Copyright © 2015 Pearson Education, Inc.   Lab Exercise 6   43

SECTION 1

Distribution of Insolation at the Top of the Atmosphere

As discussed in Lab Exercise 5, the spherical Earth presents a curved surface to the Sun's parallel rays, producing an uneven distribution of energy across the latitudes. Only the subsolar point receives insolation from directly overhead: All other locations receive the Sun's rays at increasingly lower angles of incidence as their distance increases from the subsolar point. A lower angle of insolation is less intense at the surface, that is, the energy arriving is more diffuse. The Sun's rays striking a surface at a 30° angle are diffused over twice the surface area as that receiving a perpendicular beam. Figure 6.1 shows that these oblique rays, striking Earth's surface at a 30° angle at point B, cover twice as much surface area (and are thus more diffuse and only one-half as intense) as the direct rays, which strike at a 90° angle at the subsolar point A.

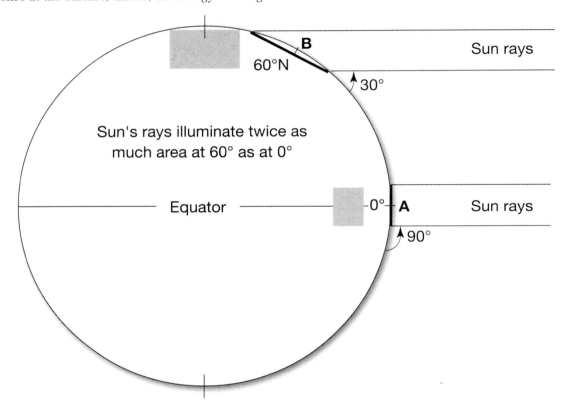

▲ Figure 6.1 Insolation angles and concentration of the Sun's energy

In addition, the greater length of travel through the atmosphere at higher latitudes reduces solar intensity due to increased reflection, absorption, and scattering within the atmosphere. This energy receipt also varies daily, seasonally, and annually as Sun angle and daylength vary—less toward the equator, more toward the poles.

A useful altitude at which to characterize insolation is the top of the atmosphere (480 km, 300 mi). The graph in Figure 6.2 plots the daily variation in insolation for selected latitudes. Latitudes are marked along the left side in 10° intervals. The months of the year are marked and labeled across the top and bottom. The dashed curved line shows the declination of the Sun (the latitude of the subsolar point) throughout the year.

To read the graph, select a latitude line and follow it from left to right through the months of the year. The daily insolation totals are read from the curved lines, given in watts per square meter per day (W/m^2 per day). For example, at the top of the atmosphere above 30°N latitude the insolation values (in W/m^2 per day) are as follows (readings are approximate—extrapolate when necessary): 240 on January 1, 250 on January 15, and 275 on February 1.

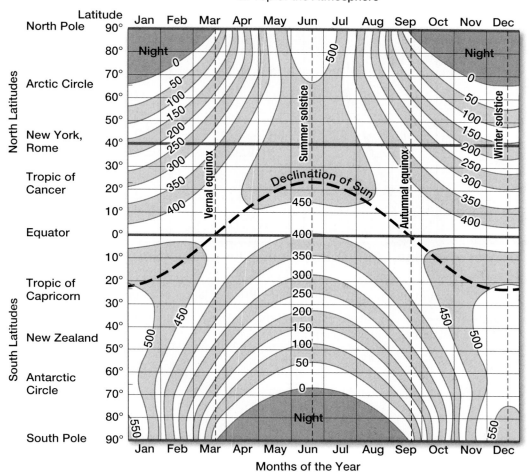

▲ Figure 6.2 Total daily insolation received at the top of the atmosphere charted in watts per square meter by latitude and month (1 watt/m² = 2.064 cal/cm²/day). Adapted from the Smithsonian Institution Press from *Smithsonian Miscellaneous Collections: Smithsonian Meteorological Tables*, vol. 114, 6th Edition. Robert List, ed. Smithsonian Institution, Washington, DC, 1984, p. 419, Table 134.

1. Using the data in Figure 6.2 and the graph provided below, plot data for specific latitudes for the first of each month. The graph you create will enable you to note the changes throughout the year in the amount of insolation received at a given latitude. Compare the differences in annual insolation patterns at various latitudes. Plot all seven latitudes on one graph, using color pencils to distinguish each. (Be sure to include a legend.)

   - North Pole (started on the graph with xs)
   - New York (started on the graph; city at 40°N with dots)
   - Tropic of Cancer
   - Equator
   - South Pole

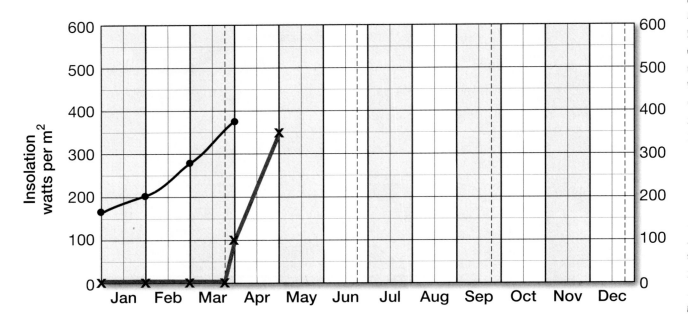

After completing the graph of the energy receipt at these different latitudes, complete the following:

2. Compare and contrast the plotted insolation values at the equator and either of the poles. What factors explain these different patterns of energy receipt?

3. What factor results in the June (summer) solstice energy receipt at the North Pole exceeding that received on the same day at the equator?

4. What is happening at the poles on the March (vernal) equinox?

5. What kind of generalization can you make about the relationship between latitude and annual variation in insolation?

6. **Challenge question:** Why do you think the South Pole receives over 550 W/m² at the December solstice, whereas the North Pole receives over 500 W/m² during the June solstice?

SECTION 2

Seasonal Variation in the Sun's Declination and Altitude

Seasonal variations are a response to changes in the Sun's **altitude**, the angle between the horizon and the noon Sun. The Sun's **declination**, the latitude of the subsolar point, migrates annually through 47° of latitude—between the *Tropic of Cancer* at 23.5°N and the *Tropic of Capricorn* at 23.5°S. The **analemma** is a convenient device to track the passage of the Sun's path and declination throughout the year (Figure 6.3). The horizontal lines are latitudes from 25°N to 25°S. The "figure 8" is a calendar, with each day of the year represented by a black or white segment.

Find August 20 on the analemma; note that it lies on the 12°N line—the subsolar point on that date. While you should remember the subsolar points for the solstices and equinoxes, an analemma will help you determine the declination for the other days of the year.

The analemma also is useful for ascertaining the positive and negative equations of time and periods of "fast-Sun times" (greatest in October and November) and "slow-Sun times" (greatest in February and March). A 24-hour (86,400-second) average day determines *mean solar time*. However, an *apparent solar day* is based on observed successive passages of the Sun over a given meridian. Any difference between observed solar time and mean solar time is called the *equation of time*. On successive days, if the Sun arrives overhead at a meridian *after* 12:00 noon local standard time (taking longer than the 24 hours of a mean solar day)—like an airliner arriving later than scheduled—the equation of time is *negative*, and the Sun is described as "slow." If, on the other hand, the Sun arrives overhead *before* 12:00 noon local time on successive days (taking less time than 24 hours)—like an airliner arriving ahead of schedule—the equation of time is *positive*, and the Sun then is described as "fast." The combination of the tilt and the eccentricity of Earth's orbit around the sun produces this difference between mean solar time and apparent solar time. Because Earth's orbit is elliptical rather than circular, Earth actually travels more quickly when we are at perihelion and more slowly when we are at aphelion. The solar day increases in length by up to 7.9 seconds for several months and then decreases in length for several months. This difference accumulates from day to day and results in the equation of time ranging from up to 16 minutes slow and 14 minutes fast at different times during the year. The equation of time is across the top of the analemma.

Find January 17 on the analemma (marked with a dot). It lies on the line labeled 9 minutes, on the "Sun slow" side. This means that the Sun will reach its "noon" **zenith** (highest point in the sky on that day) at 12:09 P.M., 9 minutes "late."

1. Use the analemma to find:

    a) The subsolar point on November 10 (marked) _____[17° south latitude]_____

    b) The subsolar point on May 11 _____

    c) The subsolar point today _____

    d) The date(s) when the declination is 9°S _____

    e) The date(s) when the declination is 21°N _____

2. At what clock time does the Sun actually reach zenith on:

    October 13? 11:47 A.M. ____[The Sun is 13 minutes fast, so it will reach its zenith at 11:47 A.M.]____

    March 8? _____

    May 20? _____

    Today? _____

As we have noted, the noon Sun angle changes at any given latitude throughout the year. Using the analemma to determine the subsolar point (Sun's declination), you can calculate the altitude of the noon Sun (Sun ∠) for any location by using the following formula:

$$\angle = 90° - (\text{arc distance between your latitude and subsolar point})$$

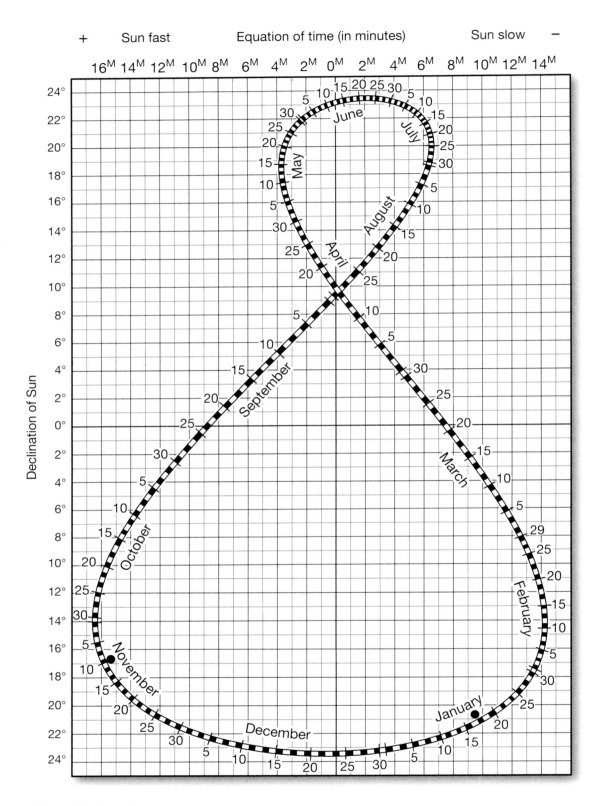

▲ Figure 6.3 The analemma

EXAMPLES:

You are at 30°N on December 21. You know that on the December solstice, the subsolar point of the noon Sun is 23.5°S. Using the above formula:

∠ = 90° − (arc distance between your latitude and subsolar point)
∠ = 90° − (30°N ↔ 23.5°S)(*the symbol* ↔ *is to indicate* ↔ *"arc distance between"*)
∠ = 90° − 53.5°
∠ = 36.5° above the southern horizon from your location

You determined that on December 21, if you are standing at 30° north latitude, you observe the Sun's noon altitude at 36.5° above the southern horizon (since you are north of the subsolar point on this day).

If you are at 40°N on April 20, you determine from the analemma that the subsolar point is 10°N; therefore:

∠ = 90° − (arc distance between your latitude and subsolar point)
∠ = 90° − (40°N ↔ 10°N)
∠ = 90° − 30°
∠ = 60° above the southern horizon*

*Note: You would observe the same Sun altitude on August 25, when the subsolar point is once again at 10°N. If you end up with a negative value for the Sun altitude, then the Sun is below the horizon and will not be visible on that day—as someone might experience north of the Arctic Circle in the Northern Hemisphere winter.

3. You are vacationing at Disney World near Orlando, Florida (28.5°N), on June 3. At what altitude will you observe the noon Sun? Include correct horizon. (Show work.) The problem has been started for you.

   ∠ = 90° − (arc distance between your latitude and subsolar point)

   ∠ = 90° − (28.5°N ↔ subsolar point)

4. On that same day, friends of yours are sightseeing at Iguaçu Falls in Brazil (25.7°S). At what altitude will they observe the noon Sun?

5. Calculate the altitude of the noon Sun if you are vacationing in Kuala Lumpur, Malaysia (3.2°N), on July 25.

6. Suppose you were going to install photovoltaic solar panels to generate electricity from the Sun at your house, and you wanted the Sun to shine on them at a 90° angle on the equinoxes. At what angle would they need to be placed, based on the latitude of your home? Should they face north or south?

Whether you view the noon Sun above your northern or your southern horizon depends upon your latitude. The subsolar point is limited to the latitudinal belt between the Tropics of Cancer and Capricorn. An observer who is poleward of the tropics will see the noon Sun only above the "equatorward" horizon (above the southern horizon if north of the Tropic of Cancer and above the northern horizon if south of the Tropic of Capricorn). An observer on or between the tropics will measure the noon Sun above the northern horizon part of the year, above the southern horizon part of the year, and directly overhead on 1 or 2 days.

Use the analemma in Figure 6.3 to answer the following questions:

7. a) If you are at 5°N latitude, on what dates will the noon Sun be directly overhead?
   _____

   b) Approximately how many days will you view the noon Sun above your northern horizon? _____

   ... above your southern horizon? _____

   c) If you are at 18°S latitude, on what dates will the noon Sun be directly overhead?
   _____

   d) Approximately how many days will you view the noon Sun above your northern horizon? _____

   ... above your southern horizon? _____

Using the same relationship between latitude and Sun altitude, you can also determine your latitudinal location by measuring the solar noon altitude, which can be done with a sextant or more simply with a protractor and a stick. For example:

Using a sextant, you measure the noon Sun angle at 59° above the southern horizon on March 11. From the analemma, you know that the subsolar point is 4°S; therefore:

$$\angle = 90° - (\text{arc distance between your latitude and subsolar point})$$
$$59° = 90° - (\text{your latitude} \leftrightarrow 4°S)$$
$$59° = 90° - 31°$$

If the arc distance between your latitude and the subsolar point = 31°, and the subsolar point is 4°S, then your latitude must be 27°N (31° − 4°). (You must be north latitude because the Sun is viewed above your southern horizon.)

This method of determining latitude, used along with chronometers and local time to determine longitude (Lab Exercise 1), allowed sailors or other navigators to obtain their absolute global location in the days before GPS (Global Positioning System) units were devised.

8. Calculate the latitude of a mariner who has measured the noon Sun altitude as 60° above the southern horizon on August 13.

9. What would the mariner's latitude be if the Sun had been measured above the northern horizon?

10. Using the graphs provided, plot the seasonal change in the Sun's noon altitude for:

   a) North Pole (90°N)

   _____

   b) La Ronge, Saskatchewan (55°N)—completed for you

   _____

   c) Atlanta, Georgia (34°N)

   _____

   d) Dunedin, New Zealand (46°S).

   _____

Complete these calculations only for each of the four seasonal anniversary dates at the solstices and equinoxes (March 20, June 20, September 22, and December 21). If needed, refer to your geography text to determine the Sun's declination on these dates for each of the four locations. Then use the procedure above to determine the Sun's altitude in the sky at noon that you would observe if you were standing at each location.

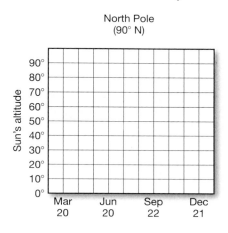
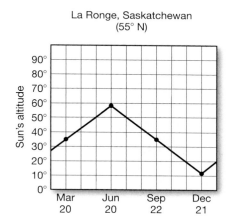
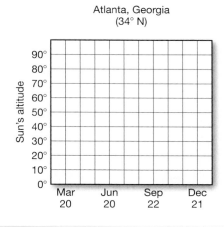
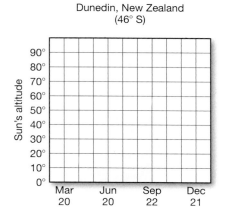

MasteringGeography™ | Looking for additional review and lab prep materials? Go to www.masteringgeography.com for pre lab videos and pre and post lab quizzes.

Name: _____ Laboratory Section: _____
Date: _____ Score/Grade: _____

Video
Exercise 7
Pre-Lab Video

Scan to view the Pre-Lab video

http://goo.gl/0SSrkb

LAB EXERCISE 7

Temperature Concepts

Temperature, a measure of sensible heat energy present in the atmosphere and other media, indicates the average **kinetic energy**, the energy of motion of individual molecules within the atmosphere. You observed air temperature as part of the "Weather Calendar" exercise in the Prologue Lab.

A variety of temperature regimes worldwide affect cultures, decision making, and resource consumption. Global temperature patterns are changing in a warming trend and are the subject of much scientific, geographic, and political interest. In Lab Exercise 7, we examine some temperature concepts and convert between Celsius and Fahrenheit. The effect of temperature in combination with wind and humidity is also presented, since wind chill and the heat index produce apparent temperatures that represent a risk in regions of North America. Lab Exercise 7 features three sections.

Key Terms and Concepts

apparent temperature (sensible temperature)
heat index
kinetic energy

sensible heat
temperature
wind chill factor

After completion of this lab, you should be able to:

1. *Differentiate* between and convert metric and English units of temperature measures.
2. *Obtain* temperature data over a 3-day period.
3. *Determine* wind-chill factors for various temperature and wind situations and heat index factors for various temperature and humidity situations.

Materials/Sources Needed
pencil
color pencils
highlighter pen
calculator

Lab Exercise and Activities

SECTION 1

Temperature Concepts, Terms, and Measurements

The amount of heat energy present in any substance is measured and expressed as its temperature. Temperature actually is a reference to kinetic energy, the speed at which atoms and molecules that make up a substance are moving. Heat that we measure with a thermometer is **sensible heat**. Changes in temperature are caused by the absorption (gain) or emission (loss) of energy. The temperature at which all motion in a substance stops is called 0° absolute temperature. Its equivalent in different temperature-measuring schemes is −273° Celsius (C), −459.4° Fahrenheit (F), or 0 Kelvin (K).

The Celsius scale (formerly called centigrade) is named after Swedish astronomer Anders Celsius (1701–1744). It divides the difference between the freezing and boiling temperatures of water into 100 degrees on a decimal scale, with freezing at 0° and boiling at 100°.

The Fahrenheit scale places the freezing point of water at 32°F (0°C, 273°K) and boiling point of water at 212°F (100°C, 373°K), with 180 F° difference between freezing and boiling. The United States remains the only major country still using the Fahrenheit scale, named after the German physicist Daniel Fahrenheit (1686–1736), who invented the alcohol and mercury thermometers and who developed this thermo-metric scale in the early 1700s. Most countries use the Celsius scale to express temperature.

The Kelvin scale, named after British physicist Lord Kelvin (born William Thomson, 1824–1907) who proposed the scale in 1848, is used in science because temperature readings are proportional to the actual kinetic energy in a material.

The formulas to use to convert between Celsius and Fahrenheit temperatures are:
To convert °C to °F:

$$°F = (°C \times 1.8) + 32$$
$$\text{or} \quad °F = 1.8°C + 32$$

To convert °F to °C:

$$°C = (°F - 32) \div 1.8$$
$$\text{or} \quad °C = (°F - 32) \times 0.556$$

When recording air temperature, the degree symbol is placed before the C or F and is read *degree Celsius* (°C) or *degree Fahrenheit* (°F)—as in "yesterday was 20°C and today is 25°C." However, when expressing the difference between two temperatures, dealing with lapse rates, temperature ranges, or units of temperature, the degree symbol is placed after the C or F and is read *Celsius degree* (C°) or *Fahrenheit degree* (F°)—as in "the high temperature yesterday was 20°C, which is 5 C° cooler than today." Another example is that ice melts at 0°C, and boils at 100°C, which means there are 100 C° between melting and boiling.

The formulas to use to convert between a number of Celsius and Fahrenheit degrees are:
To convert C° to F°:

$$F° = C° \times 1.8 \quad \text{or} \quad F° = 1.8 \, C°$$

To convert F° to C°:

$$C° = F° \div 1.8 \quad \text{or} \quad C° = F° \times 0.556$$

1. Using the given conversion formulas, complete the following temperature conversions. The first answer is provided for you in bracketed italics.

   25°C [77°F]*   30°F _____   −18°C _____   −9°F _____

   4°C _____   92°F _____   57°C _____   57°F _____

   134°F _____   59°F _____   0°C _____   15°C _____

   *(°F = ((25° × 1.8) + 32°) = (45° + 32°) = 77° F)

2. Using the given conversion formulas, complete the following degree conversions.

   5 F° _____   25 C° _____   15 F° _____   18 C° _____

   6 C° _____   90 F° _____   27 C° _____   32 F° _____

3. What temperature is it as you work on this lab exercise?

   a) Outdoor temperature?   _____ °C   _____ °F

   b) Indoor temperature?   _____ °C   _____ °F

4. Using a physical geography text, an atlas, an encyclopedia, or the Internet, answer the following. Be sure to list temperature record, date, and place.

   a) Highest natural temperature recorded on Earth (where and value)? _____

   b) Lowest natural temperature recorded on Earth (where and value)? _____

   c) Lowest natural temperature recorded for the Northern Hemisphere? _____

   d) Highest natural temperature recorded for the Southern Hemisphere? _____

Optional: Both *Geosystems* textbooks discuss the factors that produce such temperature readings (air pressure patterns, air masses, geographic location, etc.). See Chapters 5, 6, and 8 of *Geosystems* or Chapters 3, 4, and 5 of *Elemental Geosystems*.

   e) What factors produce such temperature readings in North America (air pressure patterns, air masses, geographic location, etc.)? _____

SECTION 2

Temperature Readings

Relative to atmospheric temperature, locate a properly installed thermometer either at the college or university you are attending or perhaps at home. If you do not have access to a thermometer, find another source of information for local temperature: a radio or television station, a local cable channel, the Weather Channel, a local newspaper, the National Weather Service, Environment Canada, or one of the several Internet sources available.

Perhaps you have already done this in working on your "Weather Calendar" in the Prologue Lab. The purpose of this section is to identify a reliable source of temperature information. Long after this lab course is completed, such information will benefit your lifestyle and activities. Also, this information is important to energy management of interior spaces.

1. Record the air temperature for 4 days, at approximately the same time of day, if possible. See if you can detect a trend between air temperature and other atmospheric phenomena.

Temperature	Place of observation	Time of day; date
Day 1:	Place:	Time:
Day 2:	Place:	Time:
Day 3:	Place:	Time:
Day 4:	Place:	Time:

2. Weather conditions and state of the sky at the time of your observation:

    Day 1:

    Day 2:

    Day 3:

    Day 4:

3. Do you have an outdoor thermometer available where you live? Describe the location of the thermometer from which you obtained these readings (orientation, exposure relative to compass direction, height from the ground).

4. Is temperature data available through the Geography Department at your school, or elsewhere on campus? Is there a campus weather station?

SECTION 3

The Temperatures We Feel

Our perception of temperature is described by the terms **apparent temperature** or *sensible temperature*. The combination of water vapor content, wind speed, and air temperature affect each individual's sense of comfort. High temperatures, high humidity, and low winds produce the most heat discomfort, whereas low humidity and strong winds enhance cooling sensation and effect (due to the increased evaporation of moisture from the skin). The wind chill index (given by the National Weather Service) is important to those who experience winters with freezing temperatures. The National Weather Service and the Meteorological Services of Canada (http://www.nws.noaa.gov/os/windchill/index.shtml) revised the wind chill formula and standard assumptions and introduced a new wind-chill chart during the 2001–2002 winter season. The new Wind Chill Temperature (WCT) Index is an effort to improve the accuracy of heat loss calculations.

A wind-chill chart of estimated values is presented in Figure 7.1, in both metric and English units. The **wind chill factor** indicates the enhanced rate at which body heat is lost to the air. As wind speeds increase, heat loss from the skin increases. For example, if the air temperature is −1°C (30°F) and the wind is blowing at 32 kmph (20 mph), skin temperatures will be −8°C (17°F). The **heat index** (HI) indicates the human body's reaction to air temperature and water vapor. The water vapor in air is expressed as relative humidity, a concept presented in Lab Exercise 12. For now, we can simply assume that the amount of water vapor in the air affects the evaporation rate of perspiration from the skin because the more water vapor in the air (the

Wind speed, kmph (mph)	Actual Air Temperature in °C (°F)								
	4° (40°)	21° (30°)	27° (20°)	212° (10°)	218° (0°)	223° (−10°)	229° (−20°)	234° (−30°)	240° (−40°)
Calm									
8 (5)	2° (36°)	−4° (25°)	−11° (13°)	−17° (1°)	−24° (−11°)	−30° (−22°)	−37° (−34°)	−43° (−46°)	−49° (−57°)
16 (10)	1° (34°)	−6° (21°)	−13° (9°)	−20° (−4°)	−27° (−16°)	−33° (−28°)	−41° (−41°)	−47° (−53°)	v54° (−66°)
24 (15)	0° (32°)	27° (19°)	214° (6°)	222° (−7°)	228° (−19°)	236° (−32°)	243° (−45°)	250° (−58°)	257° (−71°)
32 (20)	21° (30°)	−8° (17°)	−16° (4°)	−23° (−9°)	−30° (−22°)	−37° (−35°)	−44° (−48°)	−52° (−61°)	−59° (−74°)
40 (25)	−2° (29°)	−9° (16°)	−16° (3°)	−24° (−11°)	−31° (−24°)	−38° (−37°)	−46° (−51°)	−53° (−64°)	−61° (−78°)
48 (30)	−2° (28°)	−9° (15°)	−17° (−1°)	−24° (−12°)	−32° (−26°)	−39° (−39°)	−47° (−53°)	−55° (−67°)	−62° (−80°)
56 (35)	−2° (28°)	−10° (14°)	−18° (0°)	−26° (−14°)	−33° (−27°)	−41° (−41°)	−48° (−55°)	−56° (−69°)	−63° (−82°)
64 (40)	−3° (27°)	−11° (13°)	−18° (−1°)	−26° (−15°)	−34° (−29°)	−42° (−43°)	−49° (−57°)	−57° (−71°)	−64° (−84°)
72 (45)	−3° (26°)	−11° (12°)	−19° (−2°)	−27° (−16°)	−34° (−30°)	−42° (−44°)	−50° (−58°)	−58° (−72°)	−66° (−86°)
80 (50)	−3° (26°)	−11° (12°)	−19° (−3°)	−27° (−17°)	−35° (−31°)	−43° (−45°)	−51° (−60°)	−59° (−74°)	−67° (−88°)

Frostbite times: 30 min.   10 min.   5 min.

▲ Figure 7.1 Wind chill chart for various temperatures and wind velocities higher the humidity), the less water from perspiration the air can absorb through evaporation, thus reducing natural evaporative cooling. The heat index indicates how the air feels to an average person. Figure 7.2 is an abbreviated version of the heat index used by the NWS and now included in its daily weather summaries during appropriate months. The table beneath the graph describes the effects of heat index categories on higher-risk groups. A combination of high temperature and high humidity can severely reduce the body's natural ability to regulate internal temperature. The NWS provides a heat index forecast warning map (see http://www.nws.noaa.gov/om/heat/index.shtml); a heat index calculator is provided at http://www.crh.noaa.gov/jkl/?n=heat_index_calculator.

1. Use the wind chill chart in Figure 7.1 to determine the wind chill temperature for each of the following examples:

    °C   (°F)

   a) Wind speed: 24 kmph, air temperature: −34°C = wind-chill temp: _____
   b) Wind speed: 48 kmph, air temperature: −7°C = wind-chill temp: _____
   c) Wind speed: 8 kmph, air temperature: +4°C = wind-chill temp: _____
   d) Wind speed: 56 kmph, air temperature: −23°C = wind-chill temp: _____

2. Competitive downhill ski racers are subjected to severe wind chill, and so are average skiers and snowboarders, to a lesser degree. Assuming a downhill racer is going 80 kmph (50 mph), which is coasting on some runs, and the air temperature is −18°C (0°F), what is the wind chill the skier is feeling on any exposed skin?

   What is the skier's time to experience frostbite, given these conditions?

   What kind of outfits do these racers wear for protection?

3. Use the heat index chart in Figure 7.2 to determine the heat index temperature for each of the following examples:

                                                                                                                  °C   (°F)

   a) Air temperature: 37.8°C, relative humidity 5% = heat index temp: _____

   b) Air temperature: 32.2°C, relative humidity 80% = heat index temp: _____

   c) Air temperature: 32.2°C, relative humidity 90% = heat index temp: _____

   d) Air temperature: 43.3°C, relative humidity 10% = heat index temp: _____

4. List the heat index categories you would experience if the temperature stayed a constant 35°C (95°F) but the relative humidity dropped from 90% down to 10%.

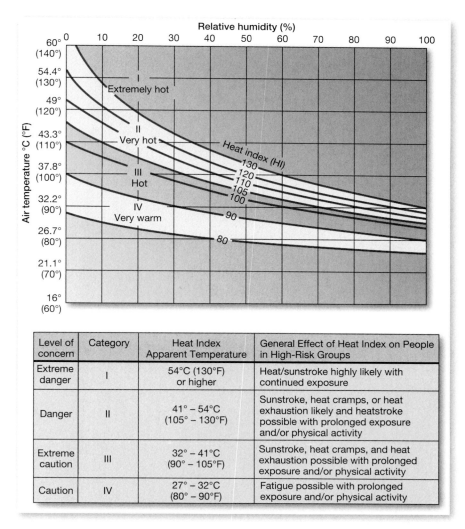

▲ Figure 7.2 Heat index for various temperatures and relative humidity levels.

MasteringGeography™ | Looking for additional review and lab prep materials? Go to www.masteringgeography.com for pre lab videos and pre and post lab quizzes.

Name: _____ Laboratory Section: _____

Date: _____ Score/Grade: _____

Video
Exercise 8
Pre-Lab Video

Scan to view the Pre-Lab video

http://goo.gl/lyol9v

LAB EXERCISE 8

Temperature Patterns

Principal controls and influences on temperature patterns include latitude, elevation, cloud cover, land–water heating differences, ocean currents and sea-surface temperatures, and general surface conditions. A variety of temperature regimes worldwide affect cultures, decision making, and resource consumption. Global temperature patterns apparently are changing in a warming trend and are the subject of much scientific, geographic, and political interest. In Lab Exercise 8 we analyze the spatial patterns of global temperatures. Lab Exercise 8 features two sections.

Key Terms and Concepts

average annual temperature
continentality
land-water heating differences
marine (maritime)

normal lapse rate
specific heat
temperature range

KEY LEARNING concepts

After completion of this lab, you should be able to:

1. *Contrast* and *compare* the temperature patterns of La Paz and Concepción, Bolivia, and *determine* the effects of elevation on temperatures.
2. *Compare* and *contrast* the temperature patterns of Vancouver, British Columbia, and Winnipeg, Manitoba, and *determine* the differing effects of marine and continental factors.
3. *Plot* and *graph* temperature data on climographs for selected stations.

Materials/Sources Needed

pencil
colored pencils
highlighter pen
calculator

Lab Exercise and Activities

SECTION 1

Elevation and Temperature

Within the troposphere, temperatures decrease with increasing altitude above Earth's surface: The **normal lapse rate** of temperature change with altitude is 6.4 C°/1000 m, or 3.5 F°/1000 ft. Worldwide, mountainous areas experience lower temperatures than do regions nearer sea level, even at similar latitudes. The consequences are that average air temperatures at higher elevations are lower, nighttime cooling increases, and the temperature range between day and night and between areas of sunlight and shadow also increases. Temperatures may decrease noticeably in the shadows and shortly after sunset. Surfaces both gain heat rapidly and lose heat rapidly to the thinner atmosphere.

Temperature data are presented in Table 8.1 for La Paz and Concepción. Both stations are approximately the same distance south of the equator but differ in elevation. La Paz is at 4103 m (13,461 ft), whereas Concepción is at 490 m (1608 ft) above sea level. The hot, humid climate of Concepción at its much lower elevation stands in marked contrast to the cool, dry climate of highland La Paz.

People living around high-elevation La Paz actually grow wheat, barley, and potatoes—crops characteristically grown in the cooler midlatitudes at lower elevations. These crops do well despite the fact that La Paz is 4103 m above sea level. The combination of *elevation* (moderating temperatures) and *equatorial location* (producing higher Sun altitude and consistent daylength) guarantee La Paz these temperature conditions, averaging about 9°C (48°F) for every month. Such moderate temperature and moisture conditions lead to the formation of more fertile soils than those found in the warmer, wetter climate of Concepción.

1. Using the temperature graphs provided in Figure 8.1, plot the data from Table 8.1 for these two cities. Use a smooth curved *line graph* to portray the temperature data. Calculate the *average* **annual temperature** and the **temperature range** (difference between the highest and lowest) for each city.

2. Why are the temperatures at La Paz more moderate in every month and so consistent overall as compared to Concepción?

3. Recall that the *normal lapse rate* of temperature change with altitude is 6.4 C°/1000 m, or 3.5 F°/1000 ft. Calculate the difference in elevation between La Paz and Concepción and calculate what the difference in their *mean (average) annual temperatures* should be, based on cooling at the normal lapse rate. Is the actual difference between the mean annual temperatures for these two cities higher, lower, or the same as that produced by calculating average normal lapse rate conditions? (Show your work.)

4. The annual march of the seasons and the passage of the subsolar point between the Tropics of Cancer and Capricorn affect these stations. Can you detect from your temperature graphs these seasonal effects? Explain.

La Paz, Bolivia

Avg. Annual Temperature (°C, °F): _____

Annual Temp. Range (C°, F°): _____

Concepción, Bolivia

Avg. Annual Temperature (°C, °F): _____

Annual Temp. Range (C°, F°): _____

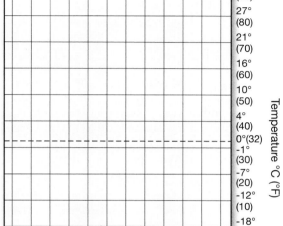
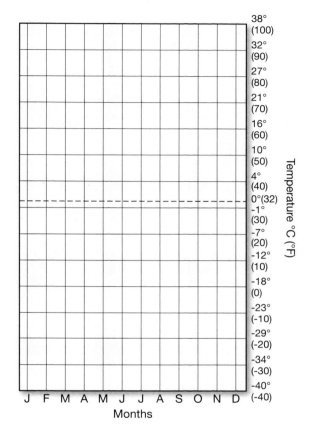

▲ Figure 8.1 Temperature graphs for La Paz and Concepción, Bolivia

La Paz, Bolivia: pop. 877,363, lat. 16.5°S, long. 68.17°W, elev. 4103 m (13,461 ft).

	Jan	Feb	Mar	Apr	May	Jun	Jul	Aug	Sep	Oct	Nov	Dec
Avg. Temp.°C	9.0	9.0	9.0	9.0	8.0	7.0	8.0	8.0	9.0	10.0	10.0	10.0
(°F)	(48.2)	(48.2)	(48.2)	(48.2)	(46.4)	(44.6)	(46.4)	(46.4)	(48.2)	(50.0)	(50.0)	(50.0)

Concepción, Bolivia: pop. 8,221, lat. 16.25°S, long. 62.05°W, elev. 490 m (1608 ft).

	Jan	Feb	Mar	Apr	May	Jun	Jul	Aug	Sep	Oct	Nov	Dec
Avg. Temp.°C	25.0	26.0	25.0	24.0	23.0	19.0	22.0	24.0	27.0	27.0	26.0	26.0
(°F)	(77.0)	(78.8)	(77.0)	(75.2)	(73.4)	(66.2)	(71.6)	(75.2)	(80.6)	(80.6)	(78.8)	(78.8)

▲Table 8.1 Temperature data for La Paz and Concepción, Bolivia

SECTION 2

Marine vs. Continental Effects

The irregular arrangement of landmasses and water bodies on Earth contributes to the overall pattern of temperature. The physical nature of the substances themselves—rock and soil vs. water—is the reason for these **land–water heating differences**. More moderate temperature patterns are associated with water bodies compared to more extreme temperatures inland.

These contrasts in temperatures are the result of the land–water temperature controls: evaporation, transmissibility, **specific heat**, movement, ocean currents, and sea-surface temperatures. The term **marine**, or *maritime*, is used to describe locations that exhibit the moderating influences of the ocean, usually along coastlines or on islands. **Continentality** refers to the condition of areas that are less affected by the sea and therefore have a greater range between maximum and minimum temperatures diurnally (daily) and yearly.

The Canadian cities of Vancouver, British Columbia, and Winnipeg, Manitoba, exemplify these marine and continental conditions. Both cities are at approximately 49°N latitude. Respectively, they are at sea level and 248 m (814 ft) elevation. However, Vancouver has a more moderate pattern of average maximum and minimum temperatures than Winnipeg. Vancouver's annual range of 11.1 C° (20.0 F°) is far below the 38.8 C° (70.0 F°) temperature range in Winnipeg. In fact, Winnipeg's continental temperature pattern is more extreme in every aspect than that of maritime Vancouver.

1. Using the data given in Table 8.2, plot the temperatures for these two cities and portray with a smooth curved *line graph* on the temperature graph in Figure 8.3. Calculate the average annual temperature and temperature range for each city.

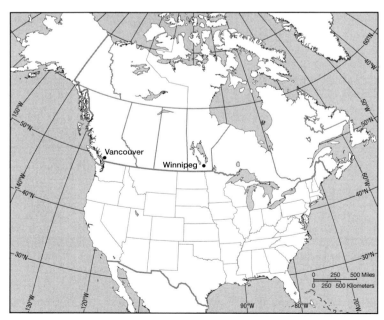

▲ Figure 8.2 Location map of Vancouver, British Columbia, and Winnipeg, Manitoba, Canada

Vancouver, British Columbia

Avg. Annual Temperature (°C, °F): _____

Annual Temp. Range (C°, F°): _____

Winnipeg, Manitoba

Avg. Annual Temperature (°C, °F): _____

Annual Temp. Range (C°, F°): _____

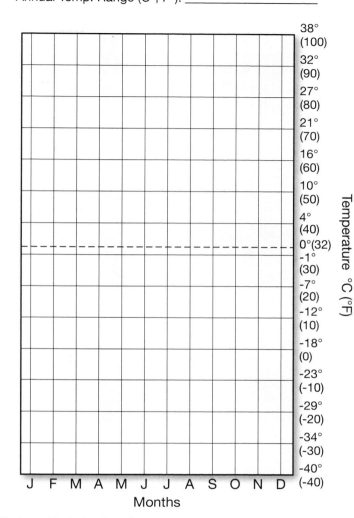

▲ Figure 8.3 Temperature graphs for Vancouver, British Columbia, and Winnipeg, Manitoba, Canada

Vancouver, British Columbia: pop. 578,041, lat. 49.18°N, long. 123.16°W, elev. sea level.

	Jan	Feb	Mar	Apr	May	Jun	Jul	Aug	Sep	Oct	Nov	Dec
Avg. Temp.°C	3.3	4.4	5.6	7.8	10.0	12.2	13.9	14.4	12.2	9.4	6.7	4.4
(°F)	(37.9)	(39.9)	(42.1)	(46.0)	(50.0)	(54.0)	(57.0)	(57.9)	(54.0)	(48.9)	(44.1)	(39.9)

Winnipeg, Manitoba: pop. 633,451, lat. 49.9°N, long. 97.23°W, elev. 248 m (813.6 ft).

	Jan	Feb	Mar	Apr	May	Jun	Jul	Aug	Sep	Oct	Nov	Dec
Avg. Temp.°C	−19.4	−16.7	−8.9	3.3	11.1	16.7	19.4	17.8	12.2	5.0	−5.6	−14.4
(°F)	(−2.9)	(1.9)	(16.0)	(37.9)	(52.0)	(62.1)	(66.9)	(64.0)	(54.0)	(41.0)	(21.9)	(6.1)

▲Table 8.2 Temperature data for Vancouver, British Columbia, and Winnipeg, Manitoba, Canada

2. Compare and contrast the marine temperature regime of Vancouver with the continental regime of Winnipeg. What significant differences do you note?

3. How many months register average temperatures below freezing in Winnipeg? How many months average below freezing in Vancouver? Explain.

Name: _____ Laboratory Section: _____

Date: _____ Score/Grade: _____

Video
Exercise 9
Pre-Lab Video

LAB EXERCISE 9

Scan to view the Pre-Lab video

http://goo.gl/5erCtb

Temperature Maps

Principal controls and influences on temperature patterns include latitude, elevation, cloud cover, land–water heating differences, ocean currents and sea-surface temperatures, and general surface conditions. A variety of temperature regimes worldwide affect cultures, decision making, and resource consumption. Global temperature patterns apparently are changing in a warming trend and are the subject of much scientific, geographic, and political interest. In Lab Exercise 9, we analyze the spatial patterns of global temperatures. Lab Exercise 9 features two sections.

Key Terms and Concepts

isotherm
thermal equator

KEY LEARNING concepts

After completion of this lab, you should be able to:

1. *Construct* isotherms on a U.S.–Canadian map and *analyze* resultant temperature patterns portrayed.
2. *Analyze* global temperature patterns by plotting temperature profiles.

Materials/Sources Needed

pencil
colored pencils
highlighter pen
calculator

SECTION 1

Working with Isotherms and Temperature Maps

Both the pattern and variety of temperature on Earth are outputs of the global energy system. Temperature maps portray the interaction of the principal temperature controls: latitude, altitude, cloud cover, and land–water heating differences. To depict the spatial aspect of data, geographers utilize *isopleths* (also called *isolines*)— lines drawn on a map connecting points of equal value, such as contour lines on a topographic map that show points of equal elevation (Lab Exercise 23).

An isopleth on temperature maps is known as an **isotherm** (similarly, pressure maps = isobars, precipitation maps = isohyets, and equal-wind velocity = isotachs). An isotherm connects points of equal temperature and presents the pattern of temperature for analysis. Geographers are concerned with the spatial distribution of temperatures, and isotherms facilitate this analysis.

In Lab Exercise 23, Section 2, you will construct contour lines to show elevation change. Isotherms, connecting equal temperature values, are drawn in much the same way. Construct the appropriate isotherms at 5 C° intervals on the weather map in Figure 9.1. The surface weather map data were recorded at 7:00 A.M. EST on October 30, 1993.

1. The 0°C (32°F) isotherm is already drawn for you (highlight it to make it distinctive). You may find it helpful to use colored pencils to "group" the temperature intervals (5 C° intervals, above and below 0°C). Use a pencil to lightly sketch the isotherms, then use a pen to darken the lines, which should be smooth curved lines. (See your geography text or the weather page of the newspaper for examples of isothermal maps.) Begin with the −5°C isotherm that is drawn for you, then draw the −10°C and −15°C isotherms. Now, continue by completing the +5°C, +10°C, +15°C (drawn for you), +20°C, and finally, +25°C isotherms (also drawn for you) across central Florida.

2. What kind of temperature pattern can be observed over the Great Plains (northern Texas, southeastern Colorado to the Dakotas). What do you think is producing this condition?

3. Describe and explain the temperature gradient over Nevada and Utah. (How rapidly are the temperatures changing from one place to another?)

4. Describe the temperature gradient over most of Texas.

5. Describe and explain the temperature gradient over the southeastern states.

6. What is the temperature range on this map? Calculate the difference between the maximum (highest) and minimum (lowest) temperature readings. In the spaces below, fill in the readings; use a map or an atlas to determine the city (station) where the data was collected.

   Maximum temperature: _____ Station (city) _____

   Minimum temperature: _____ Station (city) _____

   Range (express in C° and F°): _____

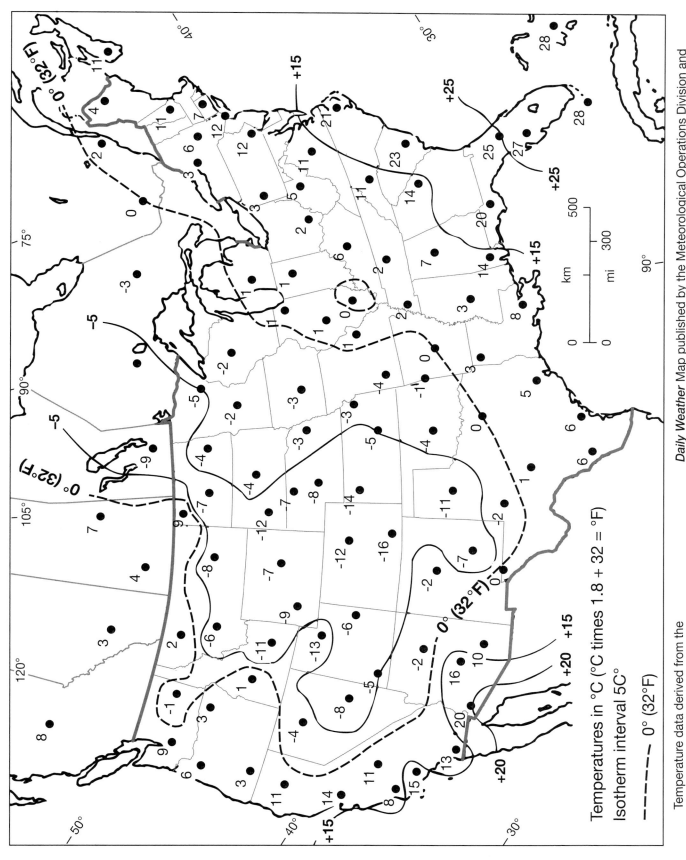

▲ Figure 9.1 Isotherm map

SECTION 2

Global Surface Temperature Map Analysis

Just as closer contour lines on a topographic map (Lab Exercise 23) mark a steeper slope, so do closer isotherms denote steepness in the temperature gradient (change across Earth's surface). In Figure 9.2a (January), note the spacing of the isotherms along the 90°W meridian between 40°–70°N latitudes (steep temperature gradient) and between 40°–70°S latitudes (gradual temperature gradient). The **thermal equator**, the location where the highest mean temperature is found is shown by a dashed line. Isotherms spaced at greater distances from one another mark a more gradual temperature change. Relate these patterns to the distribution of land and water surfaces—water is generally more moderate, land more extreme, in temperature distributions.

As you use the maps in Figures 9.2a and 9.2b, remember that these maps use a 3°C interval (example: 0°C, 3°C, 6°C, 9°C, etc.). If the data points you are plotting on the charts in Figures 9.3a, 9.3b, 9.4a, and 9.4b fall between two isotherms, you can interpolate (estimate) the temperature value to mark on the graph.

1. Temperature profiles, like topographic profiles (Lab Exercise 23), illustrate how rapidly temperatures change over Earth's surface. Use the four graphs in Figures 9.3a, 9.3b, 9.4a, and 9.4b to plot temperature data along two parallels and two meridians as noted; then complete the temperature profiles with a line graph connecting the plotted data points:

    a) January (Figure 9.2a), along 50°N parallel (already plotted and drawn for you)

    b) January (Figure 9.2a), along 90°W meridian (already plotted and drawn for you)

    c) July (Figure 9.2b), along 40°N parallel

    d) July (Figure 9.2b), along 60°E meridian

    After completing the two graphs in Figures 9.4a and 9.4b, label the parts of your plots that are over land and over water, then complete the following.

2. Using these maps and graphs, describe and explain average temperatures and temperature gradients: Where are temperatures higher/lower? Where are gradients steep or gentle? Contrast temperatures over landmasses and oceans at the same latitudes.

    a) January: _____

    b) July: _____

3. Describe and explain the differences in temperature patterns between the Northern and Southern Hemispheres in January and July. What is the general pattern of isotherms? Where are the extremes? What areas are more moderate?

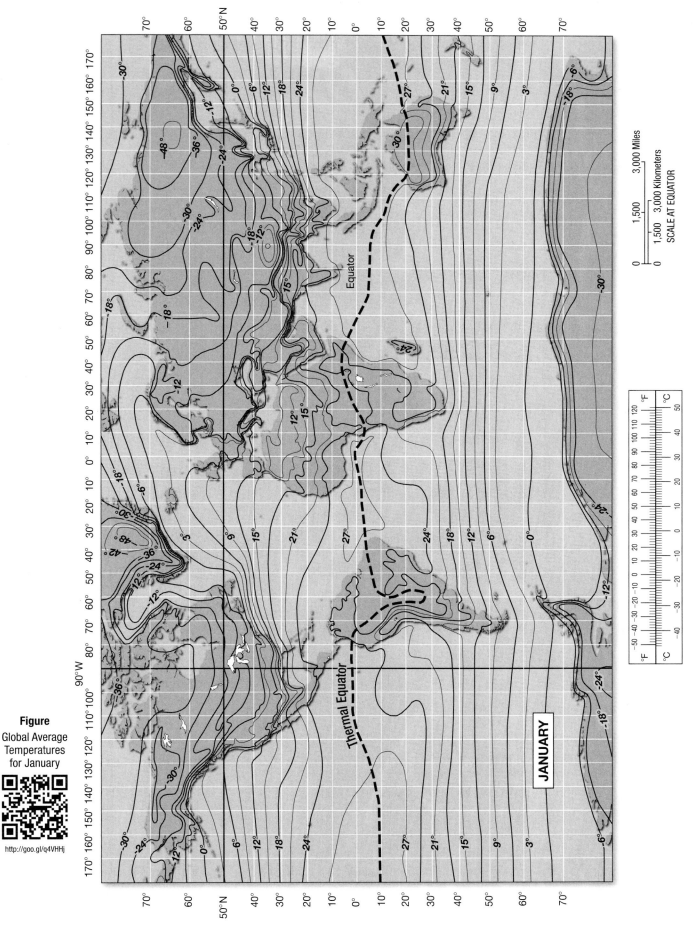

▲ Figure 9.2a Global temperatures for January

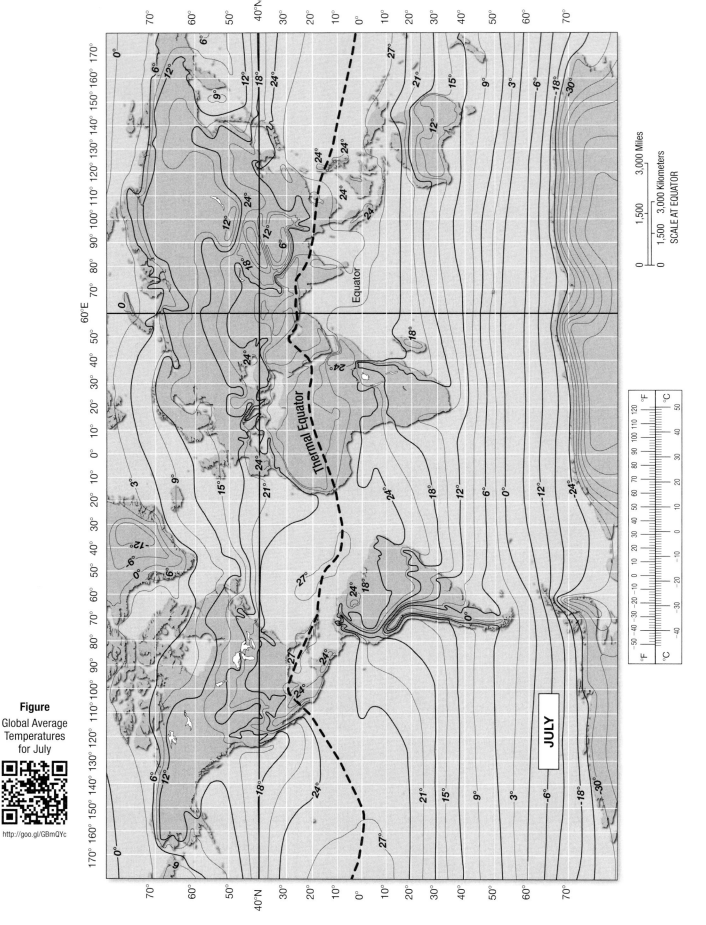

▲ Figure 9.2b Global temperatures for July (°C)

72 Lab Exercise 9

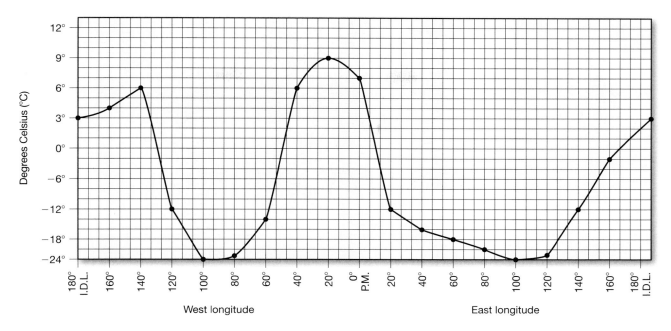

▲ Figure 9.3a Plot of average temperatures along 50°N latitude for January

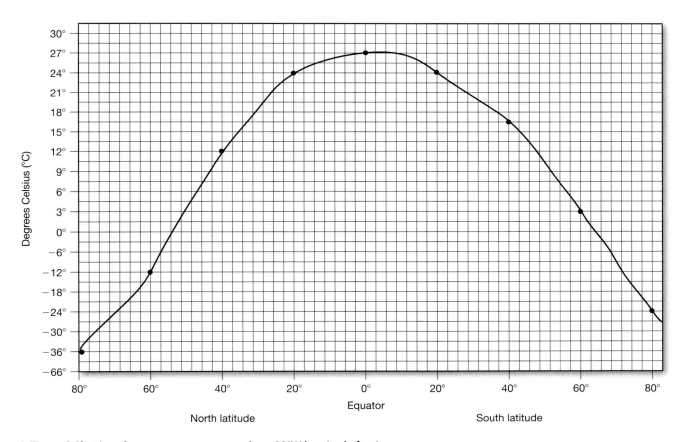

▲ Figure 9.3b Plot of average temperatures along 90°W longitude for January

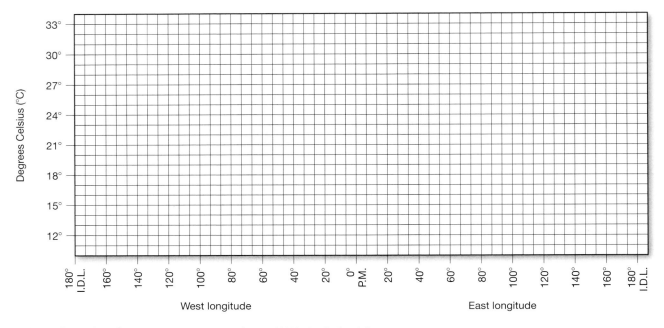

▲ Figure 9.4a Plot of average temperatures along 40°N latitude for July

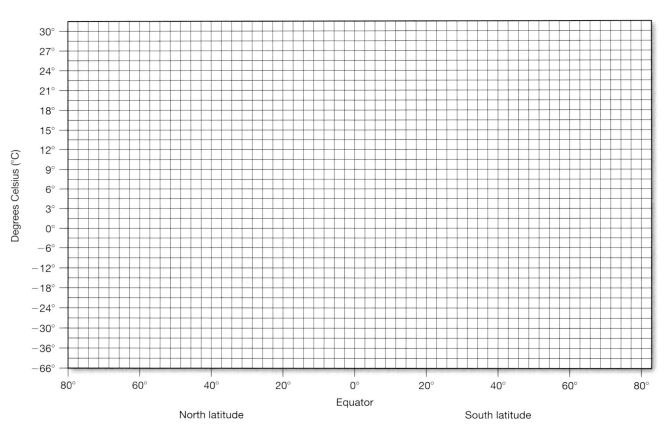

▲ Figure 9.4b Plot of average temperatures along 60°E longitude for July

Name: _____ Laboratory Section: _____

Date: _____ Score/Grade: _____

LAB EXERCISE

Scan to view the Pre-Lab video

Video
Exercise 10
Pre-Lab Video

http://goo.gl/HzD3sG

10 Earth's Atmosphere: Temperature and Pressure Profiles

The temperature of Earth's atmosphere varies greatly with altitude. The average temperature of Earth's surface is 15°C (59°F), but the temperature of the atmosphere drops to −90°C (−130°F) and climbs to over 1200°C (2200°F). These temperatures are the result of our atmosphere absorbing high-energy radiation from the Sun. The ozone layer converts ultraviolet radiation to long-wave infrared radiation, and the ionosphere absorbs gamma and x-ray radiation. While the temperature at the top of the atmosphere is over 1200°C (2200°F), there is little heat because there are so few molecules at that altitude. The gases that make up air create pressure through their motion, size, and number. This pressure is exerted on all surfaces in contact with the air. The weight of the atmosphere, or **air pressure**, exerts an average force of approximately $1 kg/cm^2$ ($14.7 lb/in^2$) at sea level. Under the acceleration of gravity, air is compressed and therefore denser near Earth's surface. The atmosphere rapidly thins with increased altitude, a decrease that is measurable because air exerts its weight as pressure. Consequently, over half the total mass of the atmosphere is compressed below 5500 m (18,000 ft), 75% is compressed below 10,700 m (35,105 ft), and 90% is compressed below 16,000 m (52,500 ft). All but 0.1% is below an altitude of 50,000 m (163,680 ft), or 50 km (31 mi).

In Lab Exercise 10 we examine Earth's atmospheric temperature and pressure patterns. Lab Exercise 10 features three sections.

Key Terms and Concepts

air pressure
aneroid barometer
barometer
mercury barometer
mesopause

mesosphere
standard atmosphere
stratopause
stratosphere
thermopause

thermosphere
tropopause
troposphere

After completion of this lab, you should be able to:

1. *Graph* altitudinal temperature changes in the atmosphere.
2. *Demonstrate* pressure trends with altitude.
3. *Explain* the standard atmosphere concept and *plot* key elements on a graph.
4. *Locate* a properly installed barometer (and, if necessary, a thermometer) and *obtain* data over a 3-day period.

Materials/Sources Needed

pencil
colored pencils

calculator
textbook

Lab Exercise and Activities

SECTION 1

Temperature Profile of the Atmosphere

Based on *temperature*, the atmosphere is divided into four distinct zones or layers: the **thermosphere**, **mesosphere**, **stratosphere**, and **troposphere**. The transition area at the top of each temperature region is named using the suffix *-pause*, which means "to cause to change," i.e., **thermopause, mesopause, stratopause**, and **tropopause.**

Temperatures in the lower atmosphere do not simply decline with altitude, for it is more complex than this and actually varies. In the upper atmosphere the temperature profile shows that temperatures rise sharply in the thermosphere, up to 1200°C (2200°F) and higher. However, we must use different concepts of "temperature" and "heat" to understand this effect. The intense radiation in this portion of the atmosphere excites individual molecules (nitrogen and oxygen) and atoms (oxygen) to high levels of vibration. This *kinetic energy*, the energy of motion, is the vibrational energy stated as "temperature." However, the density of the molecules is so low that little actual heat is produced. Heating in the lower atmosphere near Earth's surface differs because the greater number of molecules in the denser atmosphere transmit their kinetic energy as *sensible heat*, meaning that we can sense it and measure it. Figure 10.1 gives you the general trend for this temperature profile that you specifically plot in this section using data from Table 10.1.

Boundary of Atmosphere	Altitude	Temperature
Thermopause	480 km (300 mi)	1200°C (2200°F)+
Mesopause	80 km (50 mi)	−90°C (−130°F)
Stratopause	50 km (31 mi)	0°C (32°F)
Tropopause	18 km (11 mi)	−57°C (−70°F)
Surface (No. Hemis.)	Sea level	15°C (59°F)

▲Table 10.1 Standard temperature values for the atmosphere

1. Using the graph in Figure 10.1, plot the standard temperature values given in Table 10.1. (The sea level value has been done for you.) After you plot the data points, connect them with a line graph to complete the profile. Label the layers of the atmosphere and the transition areas at the top of each layer.

   Analysis and completion questions about the standard temperature profile.

2. Briefly explain why the temperature decreases as altitude increases in the troposphere (at the normal lapse rate).

3. Why do temperatures increase throughout most of the stratosphere? Specifically discuss the process that produces this warming effect.

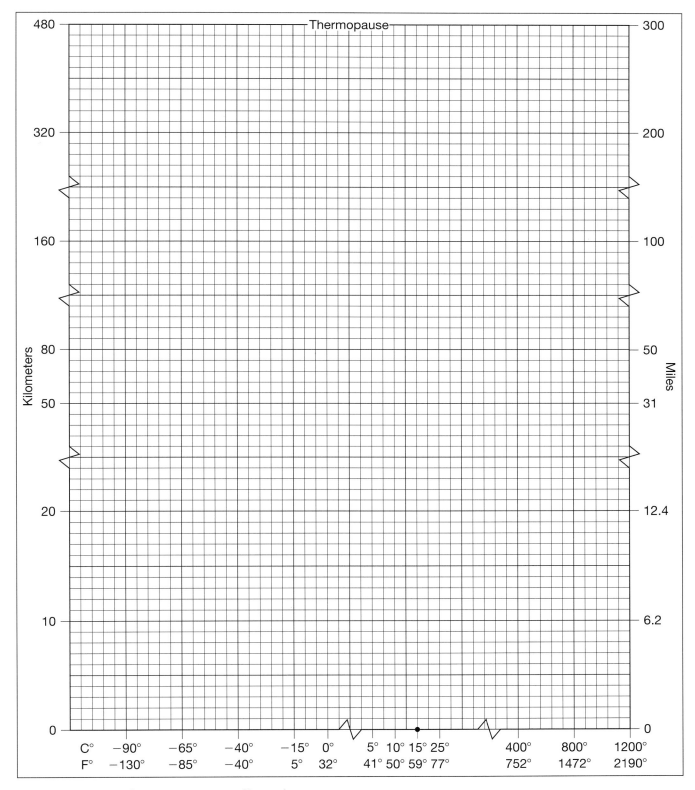

▲ Figure 10.1 Atmospheric temperature profile graph

Lab Exercise 10 77

4. Briefly explain why temperatures increase in the thermosphere.

SECTION 2

Air Pressure

Any instrument that measures air pressure is called a **barometer**. One type of barometer uses a column of mercury that is counterbalanced by the mass of surrounding air exerting an equivalent pressure on a vessel of mercury to which the column is attached. The **mercury barometer** was developed by Evangelista Torricelli in 1647. Mercury barometers are rarely used in schools because of the toxicity of mercury. A more common type of barometer, the **aneroid barometer**, is a small chamber that is partially emptied of air, sealed, and connected to a mechanism that is sensitive to changes in air pressure. As air pressure varies, the mechanism responds.

Altimetry is the measurement of altitude using air pressure. A pressure altimeter is an instrument that measures altitude based on a strict relationship between air pressure and altitude. Air pressure is measured within the altimeter by an aneroid barometer capsule, with the instrument graduated in increments of altitude. A pilot must constantly adjust the altimeter, or "zero" the altimeter, during flight as atmospheric densities change with air temperature. Another type of altimeter—a radar altimeter—sends and receives radio wavelengths between the plane and the ground to determine altitude.

Normal sea level pressure is expressed as 1013.2 mb (millibars) of mercury (a way of expressing force per square meter of surface area). At sea level standard atmospheric pressure is expressed in several ways:

- 14.7 lb/in^2 (pounds per square inch, or psi)
- 29.9213 inches of mercury (in Hg)
- 1013.250 millibars (mb)
- 101.325 kilopascals (1 kilopascal = 10 millibars)

Some convenient conversions are helpful:

- 1.0 in Hg = 33.87 mb = 25.40 mm Hg = 0.49 lb/in^2
- 1.0 mb = 0.0295 in Hg = 0.75 mm Hg = 0.0145 lb/in^2

The **standard atmosphere** for pressure in millibars and altitude in kilometers is given in Table 10.2. This is used in the activity that follows the table.

Altitude (km)	Pressure (mb)	Altitude (km)	Pressure (mb)
0.00	1013.25	10.00	264.99
0.50	954.61	12.00	193.99
1.00	898.76	14.00	141.70
1.50	845.59	16.00	103.52
2.00	795.01	18.00	75.65
2.50	746.91	20.00	55.29
3.00	701.21	25.00	25.49
4.00	616.60	30.00	11.97
5.00	540.48	35.00	5.75
6.00	472.17	40.00	2.87
7.00	411.05	50.00	0.79
8.00	356.51	60.00	0.23
9.00	308.00	70.00	0.06

▲Table 10.2 Standard atmosphere for pressure and altitude

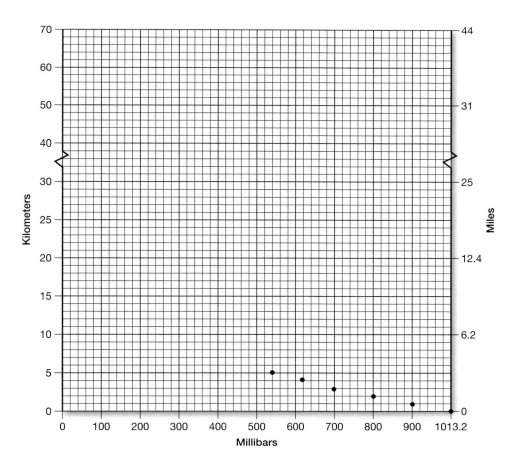

▲ Figure 10.2 Atmospheric pressure profile graph—the standard atmosphere—from the surface to 70 km.

1. Using the graph in Figure 10.2, *plot* the standard atmosphere of air pressure decrease with altitude presented in Table 10.2. After completing the plot, connect the data points with a line to complete the pressure profile of the atmosphere. The data points from 0 to 5 km are plotted for you.

2. The information in Table 10.2 allows a determination of the rate of pressure decrease with altitude, which is not at a constant rate. Remember that half of the weight of the total atmosphere occurs below 5500 m (18,000 ft); at that altitude only about half of the total atoms and molecules of atmospheric gases remain to form the mass of the atmosphere. Determine the decrease in pressure between the following altitudes. Express the difference in millibars and inches of mercury. (Conversions are presented earlier in this section.)

1 km interval difference in pressure:

0 and 1 km __[114.49]__ mb; __[3.38]__ in. Hg

2 and 3 km _____ mb; _____ in. Hg

5 and 6 km _____ mb; _____ in. Hg

8 and 9 km _____ mb; _____ in. Hg

9 and 10 km _____ mb; _____ in. Hg

10 km interval difference in pressure:

0 and 10 km __[748.26]__ mb; __[22.07]__ in. Hg

10 and 20 km _____ mb; _____ in. Hg

20 and 30 km _____ mb; _____ in. Hg

40 and 50 km _____ mb; _____ in. Hg

60 and 70 km _____ mb; _____ in. Hg

3. Using the graph you prepared in Figure 10.2, list the *approximate* answers to the following, assuming standard atmosphere conditions. The first answer is provided for you in brackets.

   a) Mount Everest's summit is 8850 m (29,035 ft) above sea level. What is the barometric pressure there, according to the standard atmosphere? [*320 mb*]

   b) Mount McKinley, 6194 m (20,320 ft); air pressure at the summit?

   c) Mount Whitney, 4418 m (14,494 ft); air pressure at the summit?

   d) Yellowstone Lake, Yellowstone N.P., 2356 m (7731 ft); air pressure?

   e) The Petronas Towers I and II, Kuala Lumpur, Malaysia, 452 m (1483 ft); air pressure?

f) In a commercial airliner taking you from San Francisco to New York at 12,000 m (39,400 ft), what percentage of atmospheric pressure is below your plane? What percentage of atmospheric pressure resides above your flight altitude?

_____

_____

4. Why does atmospheric pressure decrease so rapidly with altitude?

SECTION 3

Pressure Readings

Lab Exercise 7, Section 2 asked you to record air temperature and air pressure for 4 days. If you did not record barometric readings at that time, locate a barometer either at the college or university you are attending, or perhaps at home. Many students find that someone in the family was given an aneroid barometer years ago—and perhaps no one quite knows how to use it. If you do not have access to a barometer, find a source for information about barometric pressure (weather broadcasts, Internet sources). At the same time, you may want to record air temperature. As before, the goal is to locate and use a reliable source of barometric pressure. This will be an asset in your life, travel, and activities.

1. You started collecting data on the Weather Calendar in the Prologue Lab. As you learn more about elements of weather, you should begin to apply what you learn to the weather that you experience on a daily basis. Collecting and working with weather data will enhance your awareness and understanding of weather processes.

    For at least 4 days, record the air pressure (including the units) and air temperature at approximately the same time of day, if possible. See if you can detect a trend in air pressure changes or a relationship between air pressure and other atmospheric processes. *Record* your observations below.

Date, Pressure, and Temperature	Place of observation	Time of day	Barometer used
Day 1:	Place:	Time:	Type:
Day 2:	Place:	Time:	Type:
Day 3:	Place:	Time:	Type:
Day 4:	Place:	Time:	Type:

2. Weather conditions and state of the sky at the time of your observation:

   Day 1:

   Day 2:

   Day 3:

   Day 4:

3. What have you observed regarding the relationship between air pressure and the observed weather, including the state of the sky?

MasteringGeography™ | Looking for additional review and lab prep materials? Go to www.masteringgeography.com for pre lab videos and pre and post lab quizzes.

Name: _____ Laboratory Section: _____
Date: _____ Score/Grade: _____

LAB EXERCISE 11

Earth's Atmosphere: Pressure and Wind Patterns

Scan to view the Pre-Lab video

Video
Exercise 11
Pre-Lab Video

Differences in air pressure between one location and another produce wind, the horizontal motion of air. The driving forces that produce and determine the speed and direction of surface winds are **gravitational force**, the **pressure gradient force**, the **Coriolis force**, and **friction**. These forces, combined with the energy imbalance from the tropics to the poles, cause the circulation of Earth's atmosphere and ocean currents, including the principal pressure systems and patterns of global surface winds.

Earth's gravitational force compresses the atmosphere worldwide, and atmospheric pressure decreases with increasing altitude. The pressure gradient force causes winds by driving air from areas of higher air pressure to areas of lower pressure. High- and low-pressure areas are mainly formed by the unequal heating of Earth's surface. Cold, dense air at the poles exerts greater pressure than warm, less dense air along the equator. Coriolis force makes wind appear to be deflected in relation to Earth's rotating surface. Because Earth rotates eastward, objects appear to curve to the right in the Northern Hemisphere and to the left in the Southern Hemisphere. Friction force slows wind speeds and reduces Coriolis force close to the surface which causes winds to move across isobars at an angle. Surface friction extends to a height of about 500 m (around 1600 ft).

The primary high- and low-pressure areas appear on maps of air pressure as cells or uneven belts of similar pressure that stretch across the face of the planet, interrupted by landmasses. The primary winds flow between the primary high- and low-pressure areas. Four broad pressure areas cover the Northern Hemisphere, and a similar set exists in the Southern Hemisphere. Along the equator is a belt of low pressure called the **Intertropical Convergence Zone** or **ITCZ**. Moving away from the equator and towards the poles, there are the **subtropical high-pressure cells**; the **Hawaiian high** in the Pacific Ocean and the Azores or Bermuda high in the Atlantic Ocean. The **subpolar low-pressure cells** are more prominent in winter and include the Aleutian low in the northern Pacific and the Icelandic low in the north Atlantic.

Lab Exercise 11 features two sections.

Key Terms and Concepts

Coriolis force
friction
geostrophic winds
gravitational force
Hawaiian high
isobars

ITCZ
polar high-pressure cells
pressure gradient force
subpolar low-pressure cells
subtropical high-pressure cells

KEY LEARNING concepts

After completion of this lab, you should be able to:

1. *Identify* the four forces that produce wind and *draw* a diagram showing how they create surface and upper atmosphere wind patterns.
2. *Interpret* global pressure patterns and *construct* zonal and meridional pressure profiles.
3. *Analyze* global pressure patterns by plotting pressure profiles.

Materials/Sources Needed

pencil
colored pencils
calculator
textbook

Lab Exercise and Activities

SECTION 1

Driving Forces Within the Atmosphere and Wind Patterns

The driving forces that produce and determine the speed and direction of surface winds are gravitational force, the pressure-gradient force, the Coriolis force, and friction. Air flows from high pressure to low pressure because of the pressure-gradient force. Without the pressure-gradient force, there would be no wind. The Coriolis and friction forces act only on moving air. The strength of the pressure-gradient force is shown by the spacing of isobars. An **isobar** is an isoline that connects points of equal pressure on a weather map. Closer spacing indicates a steeper pressure gradient and a stronger pressure-gradient force. The **Coriolis force** makes wind appear to be deflected in relation to Earth's rotating surface. Because Earth rotates eastward, objects appear to curve to the right in the Northern Hemisphere and to the left in the Southern Hemisphere. The Coriolis force is zero along the equator, increases to half the maximum deflection at 30°N and 30°S latitude, and reaches maximum deflection at the poles. Without Coriolis force, winds would move along straight paths between high- and low-pressure areas. Friction force slows wind speeds and reduces Coriolis force close to the surface. Surface friction extends to a height of about 500 m (around 1600 ft). Higher-altitude winds leave the drag of surface friction behind and increase speed. This increases the Coriolis force, producing the pattern of upper-air westerly winds flowing from the subtropics to the poles. Higher-altitude winds do not flow directly from high to low but flow parallel to the isobars around the pressure areas. Such winds are **geostrophic winds** and are characteristic of upper-tropospheric circulation.

At the surface, airflow is initiated by the pressure-gradient force and modified by the Coriolis force and friction, as shown in Figures 11.1a and 11.1b. Above 500 m (1600 ft), airflow is produced by the pressure-gradient force and the Coriolis force, as shown in Figures 11.1c and 11.1d.

Pressure gradient + Coriolis + friction forces (surface winds)

Pressure gradient combined with the Coriolis force and surface friction cause wind to flow at a 45° angle to the pressure gradient. Air flows into low-pressure and turns to the left, because of deflection to the right.

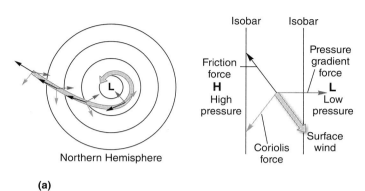

(a)

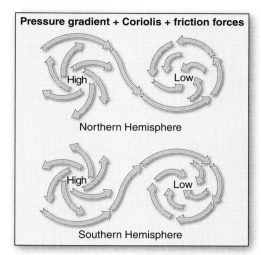

(b)

Pressure gradient + Coriolis forces (upper level winds)

Without friction, the pressure gradient and Coriolis force cause wind to flow between highs and lows parallel to isobars.

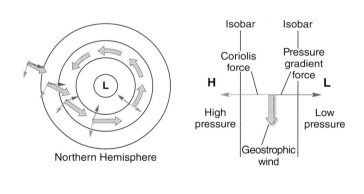

(c)

(d)

▲ Figure 11.1 Forces that produce winds

1. Figure 11.2 shows surface wind flow in the Northern and Southern Hemispheres. Place an H or an L in the center of each example to indicate what type of pressure is shown and write the name of the correct hemisphere below each of the examples. Pick one arrow from each example and draw the pressure-gradient force, Coriolis force, and friction diagram for that figure.

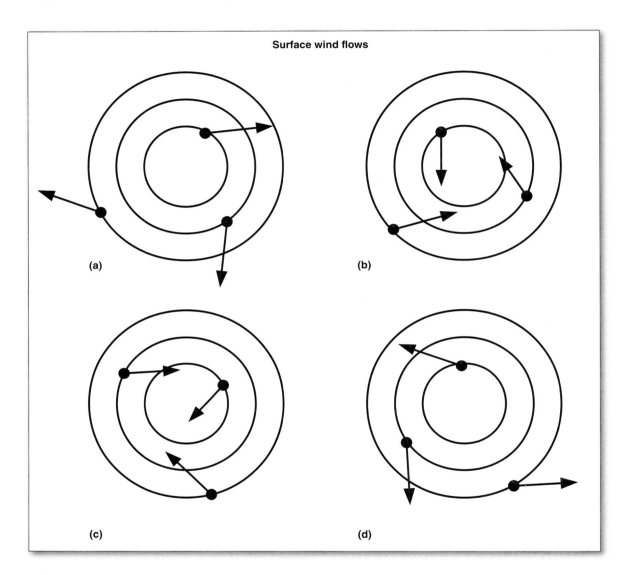

▲ Figure 11.2

2. Figure 11.3 shows geostrophic wind flows. Write the name of the correct hemisphere under each figure. Pick one arrow from each figure and draw the pressure-gradient and Coriolis force.

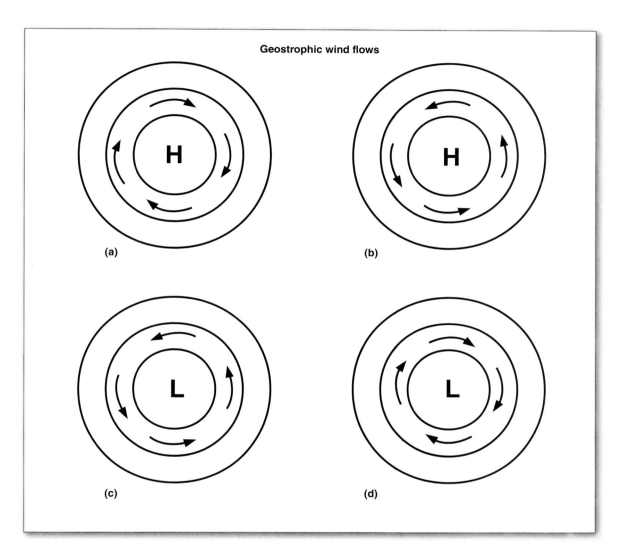

▲ Figure 11.3

3. Figure 11.4 shows the surface pressure conditions on October 18, 2007. Using a red pencil, draw at least four arrows to show the direction of wind flowing into the low-pressure center off of the Pacific Northwest. Remember that surface winds flow at a 45° angle to the pressure gradient. Do the same for the low-pressure center over Iowa. Using a blue pencil, draw at least four wind arrows to indicate the direction of wind flowing out of the high-pressure center over Nevada and Idaho. Draw at least two blue arrows to show wind flowing out of the high-pressure center off the east coast. Using an orange pencil, shade at least two regions with closely spaced isobars, indicating a steep pressure gradient. Using a green pencil, shade at least two regions with widely spaced isobars.

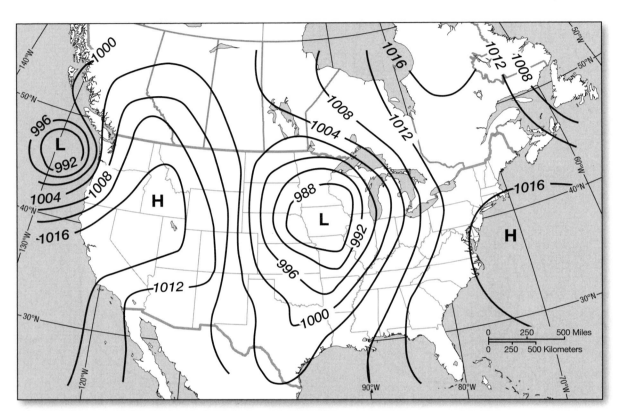

▲ Figure 11.4

4. Figure 11.5 shows the 500-mb chart for the same day. The 500-mb chart shows the position of high- and low-pressure regions and the pressure gradient. The 500-mb chart shows how many meters above sea level the 500-mb surface is rather than showing actual pressures, as the surface chart does. Draw four red arrows to show the direction of wind between the 5580 m and 5640 m height contours. Remember that wind flows parallel to the isobars in the upper atmosphere. Draw four blue arrows between the 5880 m and 5820 m contours to show the direction of wind flowing around the centers of high pressure. Using an orange pencil, shade at least two regions with closely spaced isobars, indicating a steep pressure gradient. Using a green pencil, shade at least two regions with widely spaced isobars.

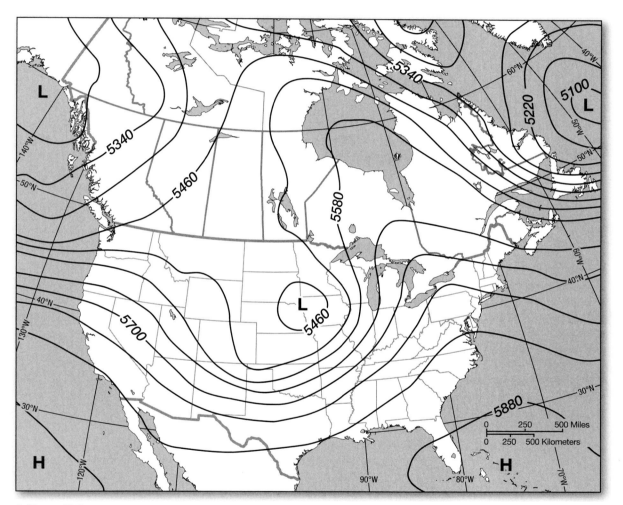

▲ Figure 11.5

SECTION 2

Air Pressure Map Analysis

High- and low-pressure areas that exist in the atmosphere principally result from unequal heating at Earth's surface and from certain dynamic forces in the atmosphere. A map of pressure patterns is created by using **isobars**, lines that connect points of equal pressure. As with other isolines that we have used—contour lines and isotherms—the distance between isobars indicates the degree of pressure difference, or pressure gradient, which is the change in atmospheric pressure over distance between areas of higher pressure and lower pressure. Isobars facilitate the spatial analysis of pressure patterns and are key to weather map preparation, interpretation, and forecasting.

Sea-level atmospheric pressure averages for January and July are portrayed in Figures 11.6 and 11.7. These maps use long-term measurements from surface stations and ship reports generally taken from the 1950s to 1970s, with some ship data going back to the last half of the nineteenth century.

Surface air pressure map analysis:

Just as closer contour lines on a topographic map mark a steeper slope and closer isotherms (Lab Exercise 9) mark steeper temperature gradients, so do closer isobars denote steepness in the pressure gradient. Isobars spaced at greater distances from one another mark a more gradual pressure gradient, one that creates a slower air flow. A steep gradient causes faster air movement from a high-pressure area to a low-pressure area.

In Figure 11.6 (January), note the spacing of the isobars along the 90°W meridian between 40°–70°N latitudes (gradual pressure gradient—weaker winds) and between 40°–70°S latitudes (steeper pressure gradient—stronger winds). As you use the maps in Figures 11.6 and 11.7, remember that these maps use a 2-mb interval (e.g., 1020 mb, 1018 mb, 1016 mb, etc.). If the data points you are plotting fall between two isobars, you can interpolate (estimate) the pressure value to mark on the graph.

1. Use the four graphs in Figures 11.8a, 11.8b, 11.9a, and 11.9b to plot pressure data along two parallels and two meridians as noted; then complete the pressure profile with a line graph connecting the plotted data points every 20°:

    a) January (Figure 11.6), along 50°N parallel (already done for you)

    b) January (Figure 11.6), along 90°W meridian

    c) July (Figure 11.7), along 40°N parallel

    d) July (Figure 11.7), along 60°E meridian

After completing the plotting of pressure data on graphs in Figures 11.8a, 11.8b, 11.9a, and 11.9b, complete the following.

2. Using these maps and your textbook, explain the patterns of average atmospheric pressure over the landmasses. Are the patterns of high and low pressure caused by thermal or dynamic forces?

    a) January: _____

    _____

    _____

    b) July: _____

    _____

    _____

3. Using Figures 11.6 and 11.7, label the primary subtropical high-pressure and low-pressure cells in the Northern Hemisphere. Draw and label the primary wind systems in the midlatitudes and tropics.

   For Figure 11.7, draw arrows indicating the monsoonal flow of air from the Indian Ocean northward into India and the Tibetan Plateau.

4. In south and central Asia, given the pressure patterns you observe on the maps, what wind patterns (direction, velocity) would you expect to find and why?

   a) January: _____

   _____

   _____

   b) July: _____

   _____

   _____

5. During the Southern Hemisphere summer, describe the pressure gradient over the Southern Ocean (South Pacific and South Atlantic Oceans). What do you think this gradient produces? Discuss with others in your lab to determine whether there is a popular term for these latitudes of such a pressure gradient.

6. Refer to the global air temperature profiles you plotted in Section 2 of Lab Exercise 9 (temperature maps in Figure 9.2a (January) and 9.2b (July), and your graphs in Figures 9.3a and 9.3b and Figures 9.4a and 9.4b). Those temperature profiles in Lab Exercise 9 and these pressure profiles you just completed for this exercise are along the same parallels and meridians.

   Comparing the two sets of graphs (Figures 9.3 and 9.4 with Figures 11.8 and 11.9), what correlation can be seen between global temperature and pressure patterns? Select a few areas that seem to illustrate a link between your graphs.

   The following pages contain:

   Figure 11.6: January world pressure map

   Figure 11.7: July world pressure map

   Figures 11.8a and 11.8b: plots for 50°N and for 90°W (11.8a and 11.8b are completed for you)

   Figures 11.9a and 11.9b: plots for 40°N and 60°E

MasteringGeography™ | Looking for additional review and lab prep materials? Go to www.masteringgeography.com for pre lab videos and pre and post lab quizzes.

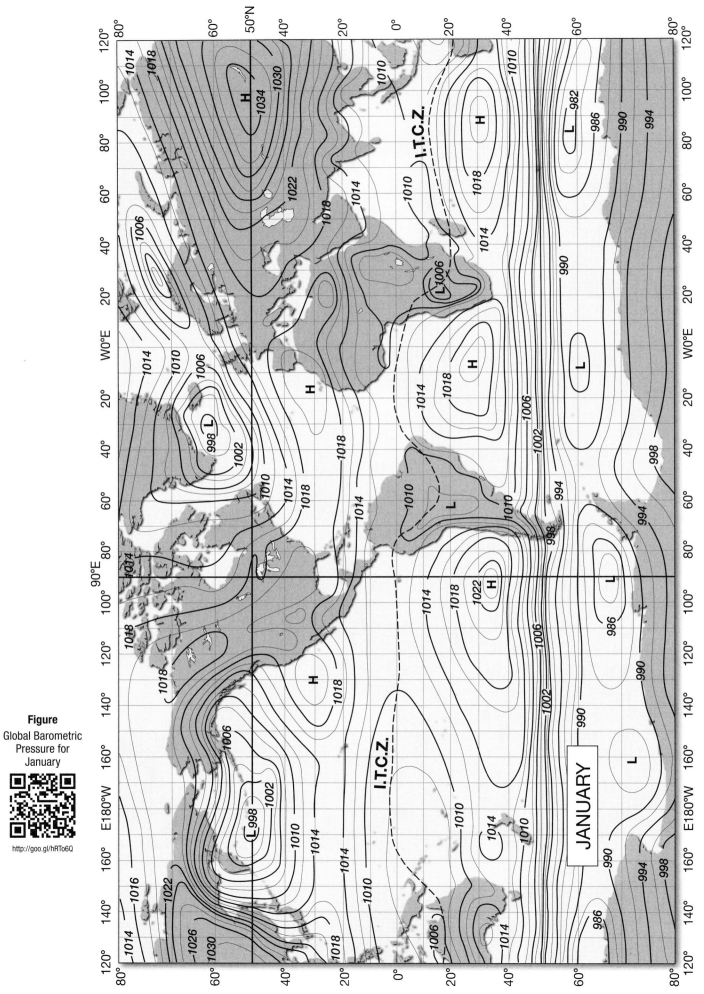

▲ Figure 11.6 Global barometric pressure for January

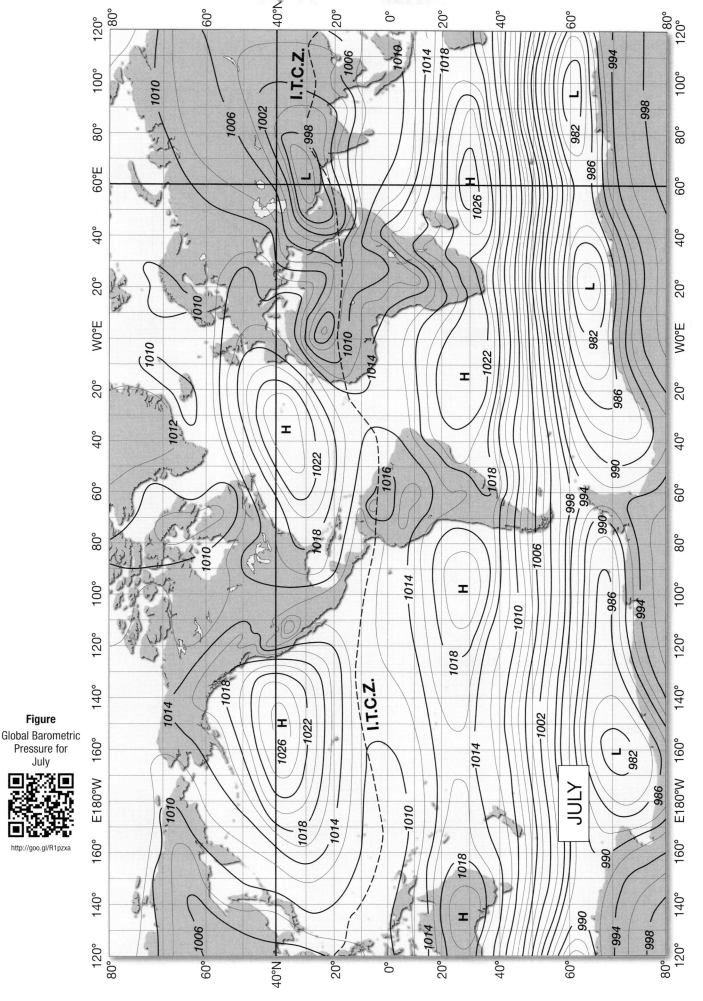

▲ Figure 11.7 Global barometric pressure for July

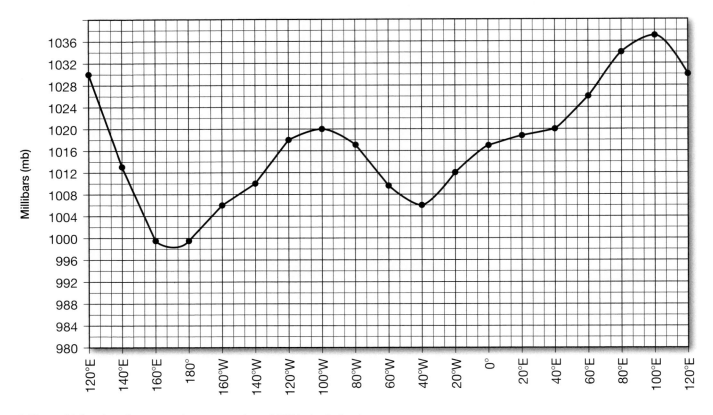

▲ Figure 11.8a Plot of average air pressure along 50°N latitude for January

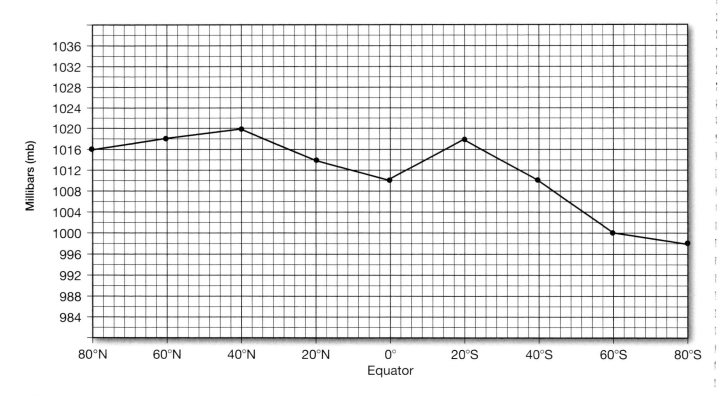

▲ Figure 11.8b Plot of average air pressure along 90°W longitude for January

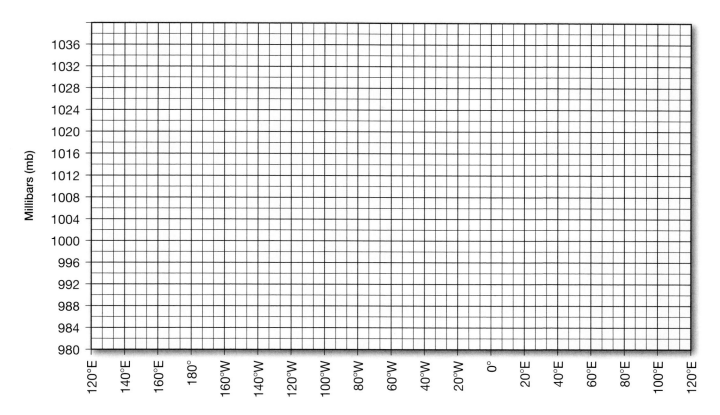

▲ Figure 11.9a Plot of average air pressure along 40°N latitude for July

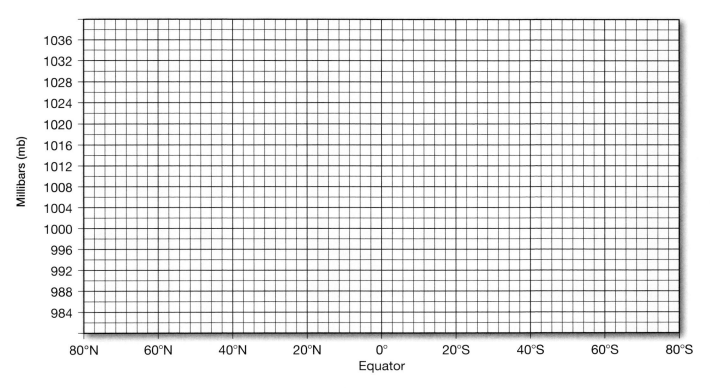

▲ Figure 11.9b Plot of average air pressure along 60°E longitude for July

Lab Exercise 11   95

Name: _____ Laboratory Section: _____

Date: _____ Score/Grade: _____

Video
Exercise 12
Pre-Lab Video

Scan to view the Pre-Lab video

LAB EXERCISE

12 Atmospheric Humidity

The powerful weather drama we witness daily in the sky is fueled by insolation as heat energy is delivered to the atmosphere by water vapor. The water vapor content of air is termed **humidity**. The capacity of air to hold water vapor is primarily a function of both the temperature of the air and of the water vapor, which are usually the same.

Warmer air has a greater capacity for holding water vapor than cooler air. To determine the energy available for weather activities, it is essential to know the water vapor content of air. Humidity is easily measurable with instruments that we use in this exercise.

Lab Exercise 12 has three sections.

Key Terms and Concepts

albedo
dew-point temperature
hair hygrometer
humidity
relative humidity

saturated
sling psychrometer
specific humidity
vapor pressure

KEY LEARNING concepts

After completion of this lab, you should be able to:

1. *Define* humidity and *describe* the several ways of expressing measurements of humidity.
2. *Demonstrate* the use of a sling psychrometer to *calculate* and *determine* relative humidity.

Materials/Sources Needed

pencils
calculator
colored pencils
hair hygrometer (if available for demonstration)
psychrometric tables (provided)
sling psychrometer

Copyright © 2015 Pearson Education, Inc. Lab Exercise 12   97

Lab Exercise and Activities

SECTION 1

Relative Humidity

Next to temperature and pressure, the most common piece of information in local weather broadcasts is **relative humidity**. Relative humidity is not a direct measurement of water vapor. Rather, it is a ratio (expressed as a percentage) of the amount of water vapor that is actually in the air (*content*) compared to the maximum water vapor the air could hold at a given temperature (*capacity*). If the air is relatively dry in comparison to its capacity, the percentage is lower; if the air is relatively moist, the percentage is higher.

Relative humidity is calculated as follows:

$$\text{Relative humidity} = \frac{\text{actual water vapor } content \text{ of the air}}{\text{maximum water vapor } capacity \text{ of the air}} \times 100$$

EXAMPLE:
An air mass at 15°C has a maximum specific humidity of 6.4 g. The maximum specific humidity (maximum capacity) of the air to hold water vapor at that temperature is 10.6 g. Therefore, the relative humidity is:

$$\text{Relative humidity} = \frac{6.4 \text{ g}}{10.6 \text{ g}} = .60 \times 100 = \mathbf{60\%}$$

Relative humidity varies due to evaporation, condensation, and temperature changes, all of which affect both the amount of water vapor present and the maximum amount of water vapor that can be present. Air is said to be **saturated**, or full, if the amount of water vapor present is equal to the maximum amount of water vapor that can be present at a given temperature. Water molecules in the atmosphere are constantly changing states between solid, liquid, and gas. If there is more heat energy is in the atmosphere, more evaporation will occur; if there is less heat energy in the atmosphere, more condensation will occur. As the temperature of the atmosphere decreases, the number of water molecules that can exist as a gas will decrease. When the number of water vapor molecules present equals the number of water vapor molecules possible at that temperature, the atmosphere is saturated with moisture. Saturated air has a relative humidity of 100%. Saturation indicates that any further addition of water vapor (change in content) or any decrease in temperature (change in capacity) will result in active condensation.

The temperature at which a given mass of air becomes saturated is termed the **dew-point temperature**. In other words, air is saturated when the dew-point temperature and the air temperature are the same. There are several ways to express humidity, relative humidity, and saturation concepts. Here we examine **specific humidity** as a way of expressing humidity. Humidity can also be expressed in millibars (mb) of **vapor pressure** since water vapor is a gas. Specific humidity is a measure of humidity expressed in grams of water vapor present per kilogram of air. Humidity can also be expressed in millibars (mb) of vapor pressure since water vapor is a gas. In terms of specific humidity, the maximum water vapor capacity of air is expressed as maximum specific humidity and is presented in Figure 12.1.

1. Using the graph provided in Figure 12.1, *plot* the maximum specific humidity data in grams and *connect* the data points with a smooth, curved line. The first several points are plotted for you.

    After finishing the maximum specific humidity graph in Figure 12.1, complete the following.

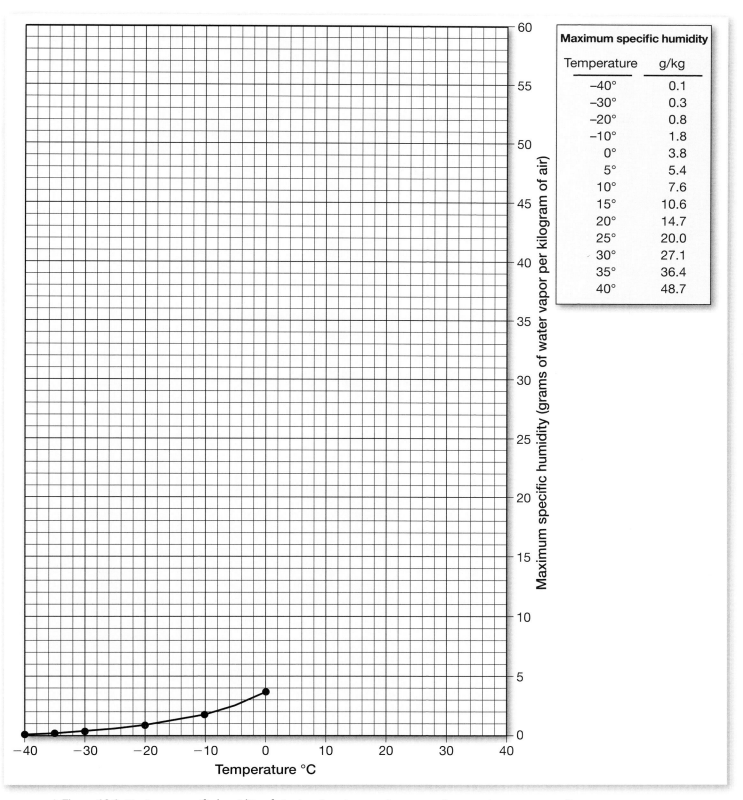

▲ Figure 12.1 Maximum specific humidity of air at various temperatures—maximum water vapor capacity of the air

2. The difference in maximum specific humidity between −30°C and −10°C is _____ mb, whereas the difference in the capacity of air to hold water vapor between +10°C and +30°C is

   _____ mb. Note the geometric relationship between increasing temperature and increasing capacity of the air to hold water vapor; for each 10°C increase in temperature, the maximum specific humidity nearly doubles.

3. Air at 10°C has a maximum specific humidity of _____ grams (*capacity*). An air mass at 10°C actually contains 3.8 grams of water vapor per kilogram of air (g/kg) (*content*). Using the equation given at the beginning of Section 1, what is the present relative humidity of this air mass?

   _____ %. Show your work:

4. What is the approximate dew-point temperature of the air mass in question #3? _____ °C? (What is the temperature for which 3.8 g/kg is the maximum capacity?) In other words, this air mass

   must cool down from its present temperature by _____ °C to achieve

   saturated conditions at the dew-point temperature. (The present temperature must cool down to the dew-point temperature for its water *content* to equal the air's water vapor *capacity*.)

5. Using the same air mass, assume that it warms to 35°C on a hot afternoon and maintains the same vapor

   pressure content of 3.8 g/kg. What is the relative humidity in the air mass at this time? _____ %. (*Hint:* Determine the capacity of 35°C air to hold water vapor, expressed as a pressure. Using the formula at the beginning of Section 1, do the simple math.) Show your work:

6. Assuming that a different air mass from the one above that has a relative humidity of 40% at a tempera-

   ture of 25°C, what is the actual humidity content, expressed as g/kg? _____ mb. Thus, by knowing relative humidity and air temperature, relative humidity provides us with an indirect method of determining the actual water vapor content of the air.

7. How is it possible to have an air mass over the hot, arid Sahara (30°N) contain an average of 14 g/kg specific humidity, whereas an air mass in the moist midlatitudes contains an average of 8 g/kg specific humidity (50°N)? (Explain this in the context of relative humidity, average temperatures in these two regions, and water vapor capacity of the air.)

8. Figure 12.1 shows that if air warms and the amount of water vapor stays the same, its relative humidity decreases. Because of this, many people who live in climates with cold winters use humidifiers in their homes to make the air more comfortable.

   a) After looking at Figure 12.1, given outside conditions of 0°C and a specific humidity of 1.8 g/kg, what is the relative humidity of the atmosphere?

   _____

   _____

b) What would the relative humidity of that air be if you brought it inside your house and warmed the air to 21°C?

9. Building engineers recommend that for optimal comfort, inside air should be 21°C, with 50% relative humidity.

   a) What is the specific humidity of air at 21°C with 50% relative humidity?

   b) How many grams of water vapor per kilogram of air would you need to add to bring the humidity up to 50% at 21°C, if the air initially had 1.8 g/kg of water vapor?

10. In question #9, you found how many grams of water vapor you would need to add to 1 kg of air at 21°C to bring it up to 50% relative humidity. Now find out how much water vapor you would need to add to make your whole house comfortable. The average American house contains 544 m^3 of air, and at 21°C, 1 m^3 of air has a mass of 1.2 kg.

    a) What is the total mass of air inside the average house at 21°C?

    b) If the air inside your house is 21°C with 1.8 g of water vapor per kilogram of air, how many grams of water would you need to add to humidify the air to 50% relative humidity?

    c) There are 1000 g of water in 1 L. Given this, how many liters of water would you need to add?

11. The EPA recommends ventilating your house so that all the air is exchanged every 3 hours, or four complete air changes per day. How many liters per day would you need to add to the air under these conditions?

12. People who live in hot, humid climates have to dehumidify their air to feel comfortable. As air cools, its relative humidity increases, making it feel muggy.

    a) What would the specific humidity of the atmosphere be if it were 32°C with 90% relative humidity?

    b) What would the dew point be?

13. a) What is the maximum specific humidity of air at 21°C?

    b) If you brought air at 32°C with 90% relative humidity into your house and cooled it down to 21°C, how many grams of water would condense per kilogram of air?

14. a) What is the specific humidity of air at 21°C and 50% relative humidity?

    b) What is the total amount of water vapor that would condense from the air if the outside air were initially 32°C, with 90% relative humidity, and it were cooled down to 21°C and dehumidified down to 50% relative humidity?

SECTION 2

Measuring Relative Humidity

The next step is to actually measure relative humidity. Relative humidity is measured with various instruments. The **hair hygrometer** uses the principle that human hair changes as much as 4% in length between 0% and 100% relative humidity. The instrument connects a standardized bundle of human hair through a mechanism to a gauge and a graph to indicate relative humidity. Remember: Relative humidity is not a direct measurement of water vapor. Once you determine the relative humidity, you can then calculate the actual water vapor content of the air.

Another instrument used to calculate relative humidity is a **sling psychrometer**. This instrument has two thermometers mounted side-by-side on a metal holder. One is called the *dry-bulb thermometer*—it simply records the *ambient* (surrounding) air temperature. The other thermometer is called the *wet-bulb thermometer*—it is set lower in the holder and has a cloth wick over the bulb that is moistened with distilled water. The psychrometer is then spun by its handle. In a weather-shelter installation, the wet-bulb thermometer can have a long cloth wick that extends to a bowl of distilled water. Instead of spinning the psychrometer, a small fan is used to move air across the dampened wick.

The rate at which water evaporates from the wick depends on the relative saturation of the surrounding air. If the air is dry, water evaporates quickly, absorbing the latent heat of evaporation from the wet-bulb thermometer, causing the temperature to drop; in other words, the wet-bulb thermometer shows a *depression* in temperature. In an area of high humidity, less water evaporates from the wick. After spinning or ventilating the psychrometer for a minute or two, the temperature on each bulb is noted and compared on a relative humidity psychrometric chart (Table 12.1) to determine relative humidity.

1. When using a sling psychrometer, what happens to the two thermometers as the psychrometer is spun? How do these changes make it possible to measure relative humidity? Explain.

Table 12.1 presents a psychrometric table that expresses dry- and wet-bulb temperature relationships in terms of *relative humidity*—the percentage of water vapor actually in the air compared to the maximum capacity that the air could hold at a given temperature. Table 12.2 presents a psychrometric table that expresses dry- and wet-bulb temperature relationships in terms of *dew-point temperatures*; the dew-point temperature is the temperature at which the actual vapor pressure and the saturation vapor pressure are equal.

When using a psychrometric table, you use the wet- and dry-bulb thermometer readings as follows:

T = Temperature of the dry bulb (air temperature)

T_w = Temperature of the wet bulb

Therefore, $T - T_w$ = Wet-bulb depression

2. Using the psychrometric charts in Tables 12.1 and 12.2, determine the relationships asked for in Table 12.3. The first line is completed for you.

Depression of the wet bulb (dry-bulb temperature minus wet-bulb temperature in °C)

Dry-bulb temperature (air temperature, °C)	0.5	1.0	1.5	2.0	2.5	3.0	3.5	4.0	4.5	5.0	7.5	10.0	12.5	15.0	17.5	20.0	22.5	25.0
−20	70	41	11															
−17.5	75	51	26	2														
−15	79	58	38	18														
−12.5	82	65	47	30	13													
−10	85	69	54	39	24	10												
−17.5	87	73	60	48	35	22	10											
−5	88	77	66	54	43	32	21	11	1									
−2.5	90	80	70	60	50	42	37	22	12	3								
0	91	82	73	65	56	47	39	31	23	15								
2.5	92	84	76	68	61	53	46	38	31	24	1							
5	93	86	78	71	65	58	51	45	38	32	1							
7.5	93	87	80	74	68	62	56	50	44	38	11							
10	94	88	82	76	71	65	60	54	49	44	19							
12.5	94	89	84	78	73	68	63	58	53	48	25	4						
15	95	90	85	80	75	70	66	61	57	52	31	12						
17.5	95	90	86	81	77	72	68	64	60	55	36	18	2					
20	95	91	87	82	78	74	70	66	62	58	40	24	8					
22.5	96	92	87	83	80	76	72	68	64	61	44	28	14	1				
25	96	92	88	84	81	77	73	70	66	63	47	32	19	7				
27.5	96	92	89	85	82	78	75	71	68	65	50	36	23	12	1			
30	96	93	89	86	82	79	76	73	70	67	52	39	27	16	6			
32.5	97	93	90	86	83	80	77	74	71	68	54	42	30	20	11	1		
35	97	93	90	87	84	81	78	75	72	69	56	44	33	23	14	6		
37.5	97	94	91	87	85	82	79	76	73	70	58	46	36	26	18	10	3	
40	97	94	91	88	85	82	79	77	74	72	59	48	38	29	21	13	6	
42.5	97	94	91	88	86	83	80	78	75	72	61	50	40	31	23	16	9	2
45	97	94	91	89	86	83	81	78	76	73	62	51	42	33	26	18	12	6
47.5	97	94	92	89	86	84	81	79	76	74	63	53	44	35	28	21	15	9
50	97	95	92	89	87	84	82	79	77	75	64	54	45	37	30	23	17	11

▲Table 12.1 Psychrometric chart of relative humidity (in percent)

Depression of the wet bulb (dry-bulb temperature minus wet-bulb temperature in °C)

Dry-bulb temperature (air temperature, °C)	0.5	1.0	1.5	2.0	2.5	3.0	3.5	4.0	4.5	5.0	7.5	10.0	12.5	15.0	17.5	20.0
−20	−25	−33														
−17.5	−21	−27	−38													
−15	−19	−23	−28													
−12.5	−15	−18	−22	−29												
−10	−12	−14	−18	−21	−27	−36										
−5	−9	−11	−14	−17	−20	−26	−34									
−2.5	−7	−8	−10	−13	−16	−19	−24	−31								
−25	−4	−6	−7	−9	−11	−14	−17	−22	−28	−41						
0	−1	−3	−4	−6	−8	−10	−12	−15	−19	−24						
2.5	1	0	−1	−3	−4	−6	−8	−10	−13	−16						
5	4	3	2	0	−1	−2	−4	−6	−8	−10	−48					
7.5	6	6	4	3	2	1	−1	−2	−4	−6	−22					
10	9	8	7	6	5	4	2	1	0	−2	−13					
12.5	12	11	10	9	8	7	6	4	3	2	−7	−28				
15	14	13	12	12	11	10	9	8	7	5	−2	−14				
17.5	17	16	15	14	13	12	12	11	10	8	2	−7	−35			
20	19	18	18	17	16	15	14	14	13	12	6	−1	−15			
22.5	22	21	20	20	19	18	17	16	16	15	10	3	−6	−38		
25	24	24	23	22	21	21	20	19	18	18	13	7	0	−14		
27.5	27	26	26	25	24	23	23	22	21	20	16	11	5	−5	−32	
30	29	29	28	27	27	26	25	25	24	23	19	14	9	2	−11	
32.5	32	31	31	30	29	29	28	27	26	26	22	18	13	7	−2	
35	34	34	33	32	32	31	31	30	29	28	25	21	16	11	4	
37.5	37	36	36	35	34	34	33	32	32	31	28	24	20	15	9	0
40	39	39	38	38	37	36	36	35	34	34	30	27	23	18	13	6
42.5	42	41	41	40	40	39	38	38	37	36	33	30	26	22	17	11
45	44	44	43	43	42	42	41	40	40	39	36	33	29	25	21	15
47.5	47	46	46	45	45	44	44	43	42	42	39	35	32	28	24	19
50	49	49	48	48	47	47	46	45	45	44	41	38	35	31	28	23

▲Table 12.2 Psychrometric chart of dew-point temperature (in °C)

(T) Dry-Bulb Temperature (°C)	(T_w) Wet-Bulb Temperature (°C)	($T - T_w$) Wet-Bulb Depression (C°)	(RH) Relative Humidity (%)	(T_{dp}) Dew-Point Temperature (°C)
−10°	−12°	[2°]	39%	−21°
5°	1°			
17.5°			86%	
25°	10°			
	30°			23°
37.5°	33°			32°

▲Table 12.3 Psychrometric relationships

SECTION 3

Measuring Relative Humidity and Temperature

At any given time, temperature and relative humidity may vary from one place to another, even over small distances. We have all experienced these changes in "micro-climates" as we have moved from one place to another. Differing local conditions can greatly influence the organisms that are present. Likewise, at a given place, the relative humidity will change over time as other atmospheric conditions change. This happens every day as air temperatures rise and fall throughout a 24-hour period.

1. Your instructor will provide you with a map of your campus (or another place, such as a park) on which a number of different sites have been indicated. These sites will vary in terms of whether they are in sunlight or shade, type of surface cover, and other variables. (As an option, your instructor will provide you with data already obtained from your campus or another place with which to complete this section of the exercise.)
   Record the data requested in Table 12.4.

Site	Site Characteristics and Albedo	(T) Dry-Bulb Temperature (°C)	(T_w) Wet-Bulb Temperature (°C)	($T - T_w$) Wet-Bulb Depression (C°)	(RH) Relative Humidity (%)
A					
B					
C					
D					

▲Table 12.4 Microclimate site characteristics and data

At each locale, record site characteristics, and from the information in Table 12.5, estimate the albedo. **Albedo**, the reflectivity of a surface, lessens absorption of Sun energy, which will, in turn, mean lower surface temperatures. The lighter the surface, the higher the albedo and lower the temperature recorded. The darker the surface, the lower the albedo and higher the temperature recorded.

Surface	Albedo (% reflected)
Fresh snow	80–95
Polluted or old snow	40–70
Ocean (Sun near horizon)	40
Ocean (Sun halfway up sky)	5
Coniferous forests	10–15
Deciduous forest (green)	15–20
Crops (green)	10–25
Bare, dry sandy soil	25–45
Bare dark soil	5–15
White sand	40
Desert	25–30
Meadow, grass	15–25
Concrete	20–30
Asphalt	5–10

▲Table 12.5 Selected albedo values of various surfaces

Following instructions given by your lab instructor, take two readings at each site with the sling psychrometer and record the temperature readings in Table 12.4. Calculate the wet-bulb depression and determine the relative humidity from Table 12.1 in Section 2.

2. What variations, if any, can you observe in the recorded temperature data for the sites?

3. What relationship, if any, can be observed between surface albedo and temperature readings at each site?

4. Describe and explain the variations you measured in the relative humidity for each site.

5. Do you feel you have identified a relation between albedo and air temperature? _____ Between air temperature and relative humidity? _____ Briefly explain:

MasteringGeography™ | Looking for additional review and lab prep materials? Go to www.masteringgeography.com for pre lab videos and pre and post lab quizzes.

Name: _____ Laboratory Section: _____
Date: _____ Score/Grade: _____

Video
Exercise 13
Pre-Lab Video

LAB EXERCISE

Scan to view the Pre-Lab video

13 Stability and Adiabatic Processes

Humidity and temperature characteristics of an air mass are key parameters in determining the stability of the air. Conditions in the environment determine whether a mass of air will lift, cool by expansion, and become saturated—unstable conditions—or whether it will resist displacement—stable conditions. This exercise examines stability aspects of air masses and adiabatic processes. Lab Exercise 13 has three sections.

Key Terms and Concepts

adiabatic
dry adiabatic rate (DAR)
environmental lapse rate
lifting condensation level
moist adiabatic rate (MAR)

normal lapse rate
orographic
rain shadow
stability

KEY LEARNING

After completion of this lab, you should be able to:

1. *Contrast* environmental lapse rates with adiabatic processes to *calculate* the stability of an air mass.
2. *Define* dew-point temperature, saturation, and lifting condensation level.
3. *Identify* the processes that lead to condensation, cloud development, and precipitation.
4. *Evaluate* orographic lifting effects and apply adiabatic heating and cooling rates.

Materials/Sources Needed

pencils
calculator
colored pencils

Copyright © 2015 Pearson Education, Inc.   Lab Exercise 13   107

Lab Exercise and Activities

SECTION 1

Atmospheric Stability

Knowing the concepts of humidity, relative humidity, dew-point temperature, and saturation allows us to take the next step and examine the stability of air masses. Temperature and moisture relationships in the environment determine whether the atmosphere is stable or unstable.

Stability refers to the tendency of a parcel of air either to remain as it is or to change its initial position by ascending or descending. An air parcel is termed *stable* if it resists displacement upward or, when disturbed, tends to return to its starting place. On the other hand, an air parcel is considered *unstable* when it continues to rise until it reaches an altitude where the surrounding air has a density similar to its own.

As noted in Lab Exercise 10, Section 1, the **normal lapse rate** is the *average* decrease in temperature with increasing altitude, a value of 6.4 C° per 1000 m, or 0.64 C° per 100 m (3.5 F° per 1000 ft). This rate of temperature change can differ greatly under varying weather conditions, and so the actual lapse rate at a particular place and time is labeled the **environmental lapse rate** (ELR). When a particular environmental lapse rate is in effect, an ascending parcel of air tends to cool by expansion, responding to the reduced pressure at higher altitudes. Descending air tends to heat by compression.

Both ascending and descending temperature changes are assumed to occur without any heat exchange between the surrounding environment and the vertically moving parcel of air. The warming and cooling rates for a parcel of expanding or compressing air are termed **adiabatic**. Adiabatic rates are measured with one of two dynamic rates, depending on moisture conditions in the parcel: **dry adiabatic rate** (DAR), 10 C° per 1000 m, or 1 C° per 100 m (5.5 F° per 1000 ft), and **moist adiabatic rate** (MAR)—sometimes also called *wet adiabatic rate* (*WAR*)—6 C° per 1000 m, or 0.6 C° per 100 m (3.3 F° per 1000 ft), or roughly 4 C° less than the DAR. *Note:* The DAR is used if the air is less than saturated; the MAR is used if the air is saturated.

Determining the degree of stability or instability involves measuring and comparing simple temperature relationships between conditions inside an air parcel and the surrounding air. This difference between an air parcel and the surrounding environment produces a buoyancy that contributes to further lifting. The following three problems involve each of the possible conditions: *stable* (if the rising air is cooler than or the same temperature as the surrounding air), *unstable* (if the rising air is warmer than the surrounding air), and *conditionally unstable* (if, in rising, the air becomes warmer than the surrounding air only after it begins cooling at the MAR). The basic temperature relationships that determine stability conditions in the atmosphere between the environmental lapse rate and dry and moist adiabatic rates are shown in Figure 13.1.

For clarity, use colored pencils to highlight each of the three lapse rates in Figure 13.1. The normal lapse rate (remember: it's an average), the DAR, and MAR are plotted according to their respective formulas. The curved set of arrows shows the overall possible range within which the actual environmental rate might fall. If the environmental lapse rate of the air surrounding a rising parcel of air is greater than the DAR (i.e., >10 C°/1000 m), the rising air will remain warmer than the surrounding environment, will be *unstable* (a), and will continue to rise. If the environmental lapse rate is less than both the DAR and the MAR (i.e., < 6 C°/1000 m), the rising column of air will be cooler than the surrounding air (which is cooling at the environmental rate) and will be *stable* (c), resisting further rising, and instead will settle back to its original position. In the circumstance that the environmental rate is between the MAR and DAR (> MAR and < DAR), the parcel of rising air will resist rising as long as the air remains unsaturated; however, if the parcel is forced to rise (over a mountain range, for example) and the air becomes saturated, it will then begin cooling at the MAR, remaining warmer than the surrounding air, and become unstable, continuing to rise (b). This is *conditionally unstable* air: unstable only under certain conditions—in this instance, being forced to rise.

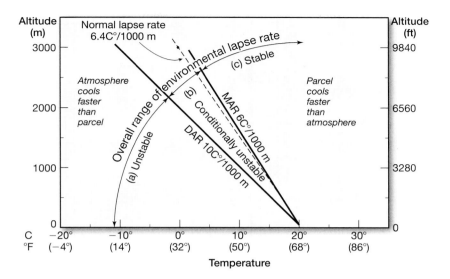

▲ Figure 13.1 The relationship between dry and moist adiabatic rates and environmental lapse rates produces three conditions of stability.

1. Describe the general relationship between the ELR and atmospheric stability. For which ELR values will the atmosphere always be stable? Unstable?

2. On the graphs provided in Figure 13.2, *plot* the following data, using colored pencils to distinguish the different rates. Begin with a surface temperature of 25°C noted on all three graphs labeled a, b, and c. Please use these colors: ELR = green pencil; DAR = red pencil; MAR = blue pencil.

   a) Plot an environmental lapse rate of 11 C° per 1000 m, DAR of 10 C° per 1000 m, MAR of 6 C° per 1000 m. This graph is completed for you.

   b) Plot an environmental lapse rate of 4.5 C° per 1000 m, DAR of 10 C° per 1000 m, MAR of 6 C° per 1000 m.

   c) Plot an environmental lapse rate of 8 C° per 1000 m, DAR of 10 C° per 1000 m, MAR of 6 C° per 1000 m.

3. Using Figure 13.1 and your graphs in Figure 13.2, note the relationship between the environmental lapse rates on each graph you made and the DAR and MAR. Determine which stability condition describes each of the three graphs you have completed. *Hint:* Compare each ELR plotted with the relations in Figure 13.2.

Stability graph (a)

Stability graph (b)

Stability graph (c)

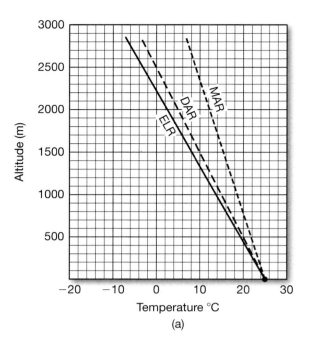
(a)

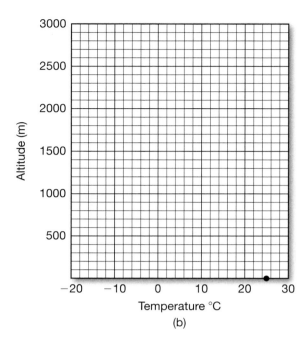
(b)

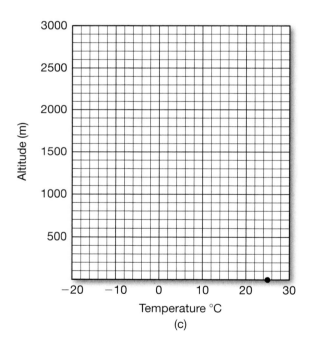
(c)

▲ Figure 13.2 Three atmospheric stability graphs

SECTION 2

Unstable and Conditionally Unstable Atmospheric Conditions

Let's work through a specific application of these stability principles. In this example, specific humidity is used as another way of describing water vapor content. Specific humidity describes the water vapor content of the air by comparing a mass of water vapor to a mass of air, expressed as grams per kilogram. Figure 13.3 presents the *maximum specific humidity* for a mass of air at various temperatures.

The graph in Figure 13.3 shows that 1 kg of air could hold a maximum specific humidity of 47 g of water vapor at 40°C (104°F), 15 g at 20°C (68°F), and 4 g at 0°C (32°F). *Locate* these three data points on the graph and use a colored pencil to *mark* them (with a dot).

If 1 kg of air at 40°C has a specific humidity of 12 g, its relative humidity is calculated as follows:

$$12 \text{ g} \div 47 \text{ g} = 0.255 \times 100 = 25.5\%$$

The actual water vapor in the air (12 g) divided by the maximum specific humidity at 40°C (47 g) gives you the relative humidity. Specific humidity is useful in describing the moisture content of large air masses that are interacting in a weather system.

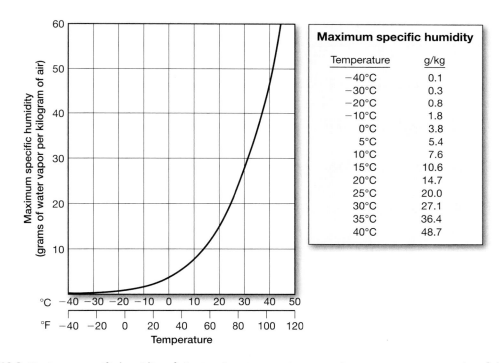

▲ Figure 13.3 Maximum specific humidity of air at various temperatures–maximum water vapor capacity of the air

1. Assume that an air parcel with a specific humidity of 5.4 g/kg and an internal temperature of 25° begins to lift. Assume an environmental lapse rate of 12 C° in the air surrounding the lifting parcel. Use standard values for the DAR and MAR given in Section 1. Using the graph provided in Figure 13.4, plot and label the environmental lapse rate from 1000 m to 3000 m. (Refer to the stability graph you plotted in Figure 13.2a.)

2. What is the dew-point temperature at which the rising air will become saturated? (See Figure 13.3 for maximum specific humidity at various temperatures.)

3. How much cooler will the parcel be when it reaches that dew-point temperature?

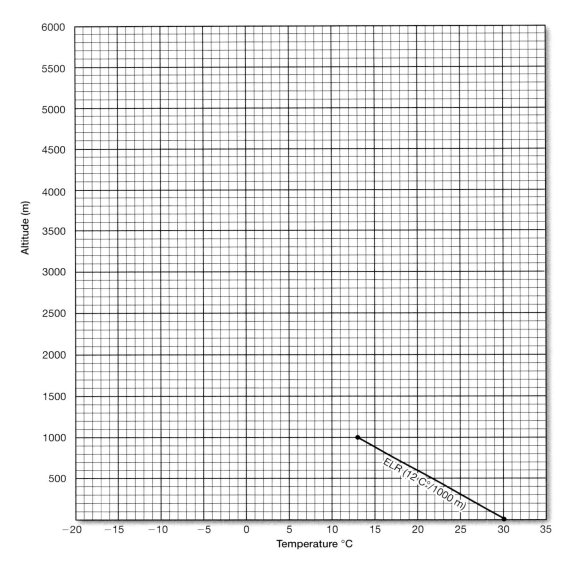

▲ Figure 13.4 Unstable atmospheric conditions and convective activity

4. Given that the parcel will cool at the DAR until it reaches that dew-point temperature, how many meters will the parcel rise before it cools to the dew-point temperature?

5. At what altitude will the parcel reach the dew-point temperature?

6. Plot the DAR on the graph for the lifting parcel of air from the surface until it cools to the dew-point temperature. Draw and label a horizontal line on the graph at the altitude that the parcel cools to the dew-point temperature and label it as the **lifting condensation level**.

7. Is the parcel of air stable or unstable at 1000 m? At 2000 m? At the altitude at which it cooled to the dew-point temperature? Explain.

112   Lab Exercise 13

8. Plot the MAR from the lifting condensation level, the altitude at which the parcel cooled to the dew-point temperature, up to 3000 m.

9. Explain why the parcel is stable or unstable at 3000 m.

10. Draw a horizontal line at the altitude at which the parcel became unstable.

11. What is the temperature of the parcel at 3000 m? What is the maximum specific humidity at that temperature? What is the difference in specific humidity in grams per kilogram between the parcel when it was on the ground and at 3000 m?

12. Assume that an air parcel with a specific humidity of 7.6 g/kg and an internal temperature of 30°C begins to lift. Assume an environmental lapse rate of 8 C° in the air surrounding the lifting parcel. Use standard values for the DAR and MAR given in Section 1. Using the graph provided in Figure 13.5, plot and label the environmental lapse rate from the surface to 6000 m. (Refer to the stability graph you plotted in Figure 13.2a.)

13. What is the dew-point temperature at which the rising air will become saturated? (See Figure 13.3 for maximum specific humidity at various temperatures.)

14. How much cooler will the parcel be when it reaches that dew-point temperature?

15. Given that the parcel will cool at the DAR until it reaches that dew-point temperature, how many meters will the parcel rise before it cools to the dew-point temperature?

16. At what altitude will the parcel reach the dew-point temperature?

17. Plot the DAR on the graph for the lifting parcel of air from the surface until it cools to the dew-point temperature. Draw and label a horizontal line on the graph, marking this lifting condensation level.

18. Is the parcel of air stable or unstable at 1000 m? At 2000 m? Explain.

19. Plot the MAR from the lifting condensation level, the altitude at which the parcel cooled to the dew-point temperature, up to 6000 m.

20. At what altitude did the parcel become unstable? Draw a horizontal line to indicate that altitude and label "unstable" above that line and "stable" below it.

21. What is the temperature of the parcel at 6000 m? What is the maximum specific humidity at that temperature? What is the difference in specific humidity in grams per kilogram between the parcel when it was on the ground and at 6000 m? How many grams of water vapor would have condensed from the parcel, in grams per kilogram?

22. If the parcel descends, how would its relative humidity change? How does this relate to the rain shadow concept?

Note: There are four principal lifting mechanisms that can force air to rise, resulting in adiabatic cooling, and perhaps reach its dew-point temperature. The same stability concepts you sketched and worked with in Figure 13.4 apply in each of these lifting situations, despite the difference in mechanism. These lifting mechanisms are:

a) *Convergent* (air flowing from different directions into an area of low pressure);

b) *Convectional* (local heating of air, a plowed field, an urban heat island, land compared to water);

c) *Orographic* (air encounters a mountain barrier and is forced to lift); and,

d) *Frontal* (conflicting air masses of different densities forcing warmer air to lift).

SECTION 3

Orographic Lifting and Rain Shadows

The physical presence of a mountain acts as a topographic barrier to migrating air masses. **Orographic lifting** (*oro* means "mountain") occurs when air is forcibly lifted upslope as it is pushed against a mountain. The lifting air cools adiabatically. An orographic barrier enhances convectional activity and causes additional lifting during the passage of weather fronts and cyclonic systems, thereby extracting more moisture from passing air masses. The wetter intercepting slope is the windward slope, as opposed to the drier leeward slope, known as the **rain shadow**. Figure 13.5 is a transect across the Sierra Nevada mountains in California. We will follow a storm across them from Fresno up to Mt. Whitney, the highest point in the continental United States, down to Lone Pine, back up to Telescope Peak, and finally down to Badwater in Death Valley, the lowest point in North America. When the parcel is descending, assume that the air is stable and descends to the surface. Use Figure 13.3 to calculate the maximum specific humidity values for each elevation.

1. What is the condensation height for this storm?

2. Fill in the answer blanks in Figure 13.5.

3. Do you live on the windward or leeward (rain shadow) side of the nearest major terrain feature? What is the nearest terrain feature that controls your precipitation?

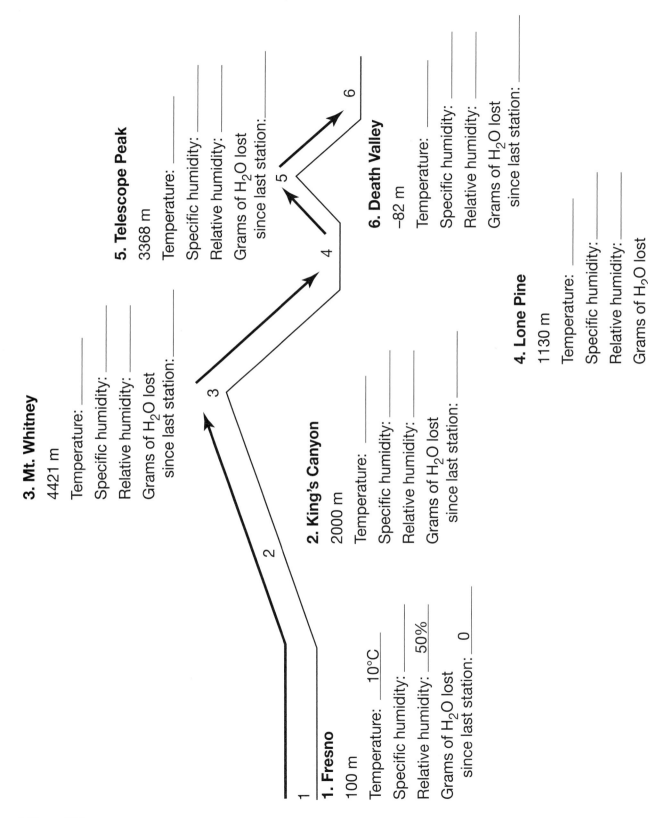

▲ Figure 13.5

The Indian Ocean is very warm and can reach temperatures of 30°C (86°F). This warm body of water, combined with low air pressure in central Asia, creates monsoonal conditions during the summer. The low air pressure pulls warm, moist air from the Indian Ocean up the slopes of the foothills and main body of the Himalayas. Cherrapunji, India, is one of the wettest places in the world and is nestled in the foothills in this region. Cherrapunji holds the record for the wettest year ever recorded—an amazing 26.47 m (86.8 ft)—although it sounds even more impressive when given in the normal rainfall reporting units of 2647 mm, or 1042 in.!

4. Take a parcel of air from the Bay of Bengal in the Indian Ocean up to Cherrapunji at 1313 m (4309 ft). Assume that its temperature was 30°C and 90% RH at 0 m (0 ft) over the Bay of Bengal. What would its initial SH be?

   a) What would its SH be at 1313 m (4,309 ft)?

   b) How many mm of H_2O would fall as rain along the way? (Assume that any moisture in excess of the maximum specific humidity will fall as rain and that 1 g of H_2O is equal to 10 mm of precipitation.)

   c) What would the parcel's condensation height be?

While moisture condenses from the lifting air mass on the windward side of the mountain, on the leeward side, the descending air mass heats by compression, and any remaining liquid water in the air evaporates. The previous questions looked at air parcels beginning their ascent while the air was warm and moist, but this question looks at an air parcel when it is finishing its descent on the leeward slope and it becomes hot and dry. As stable air descends, it is heated by compression at the dry adiabatic rate (DAR) of 10 C°/km (5.5 F° per 1000 ft). This heating causes the relative humidity to drop, causing warm dry winds. These winds go by a number of regional names, including *foehn* or *föhn* in Europe and Chinook in the Pacific North West.

5. Take a parcel of air from the top of Mt. Rainier, at 4392 m (14,410 ft), and follow it to Yakima, Washington, at 325 m (1066 ft). Assume that the initial conditions of the parcel of air are −10°C and 100% RH.

   a) What is the initial specific humidity of the parcel?

   b) What will its temperature, specific humidity, and relative humidity be in Yakima?

   c) How many degrees warmer is the parcel in Yakima than on top of Mt. Rainier?

Name: _____ Laboratory Section: _____
Date: _____   Score/Grade: _____

Video
Exercise 14
Pre-Lab Video

Scan to view the Pre-Lab video

http://goo.gl/oRG4da

LAB EXERCISE 14

Weather Maps

Weather is the short-term, day-to-day condition of the atmosphere, contrasted with **climate**, which is the long-term average (over decades) of weather conditions and extremes in a region. Weather is, at the same time, both a "snapshot" of atmospheric conditions and a technical status report of the Earth–atmosphere heat–energy budget.

We consult the broadcast media for weather information so that we can better plan our activities and travels. We turn for the weather forecast to the National Weather Service in the United States **(http://www.nws.noaa.gov)** or Canadian Meteorological Centre, a branch of the Meteorological Service of Canada (MSC) **(http://weatheroffice.ec.gc.ca/canada_e.html)** to see current satellite images and to hear weather analysis. Internationally, the World Meteorological Organization coordinates weather information (see **http://www.wmo.ch)**. Perhaps you are already working on the "Weather Calendar" assignment, gathering weather data and recording it in the proper form and with the official symbols used internationally.

In this exercise, we work with an actual weather map to better understand this important aspect of our daily lives. Lab Exercise 14 features two sections.

Key Terms and Concepts

climate
meteorology

synoptic map
weather

After completion of this lab, you should be able to:

1. *Recognize* the symbols used to depict weather conditions at stations on weather maps.
2. *Construct* an actual weather map by *drawing* isobars and fronts and *labeling* air masses, and then *portray* weather conditions experienced on that day.
3. *Analyze* an idealized weather map and *describe* current conditions at various locations.

Materials/Sources Needed
pencil
colored pencils

Lab Exercise and Activities

SECTION 1

Weather Map Symbols

A "Weather Calendar" exercise is assigned in the Prologue Lab. If you are working on the calendar exercise, then you are now familiar with the standard weather station symbol and the shorthand method that meteorologists use to describe synoptic conditions. Remember: *Weather* is the short-term condition of the atmosphere, whereas *climate* reflects long-term atmospheric conditions and extremes.

Meteorology is the scientific study of the atmosphere (*meteor* means "heavenly" or "of the atmosphere"). Embodied within this science is a study of the atmosphere's physical characteristics and motions; related chemical, physical, and geologic processes; the complex linkages of atmospheric systems; and weather forecasting.

The sample station model is shown in Figure 14.1; a simplified version of the station model appropriate for use here is presented in the Prologue Lab. Note the variety and complexity of weather conditions shown in this simple sample station model. Figure 14.1 includes symbol explanations for present weather conditions, state of the sky—sky coverage, cloud designations, air pressure changes, wind indicators and speeds—and surface front designations. Barometric pressure is presented in an abbreviated form. In this example, 025 appears, which is short for 1002.5 mb (the "10" and the decimal point are dropped); a 980 would mean 998.0 mb (the "9" and the decimal point are dropped).

Sample weather station recording using standard weather symbols:

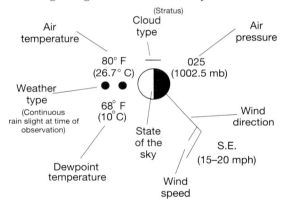

1. Select one day from your Prologue Lab "Weather Calendar" or use the current conditions outside and record the information here, in the appropriate format (shown with the model at the beginning of Section 1). Make notations in the proper form *only* for the following atmospheric conditions:

   - Air temperature (°C or °F)
   - Air pressure (millibars)
   - State of the sky (sky coverage)
   - Weather type (if any observed)
   - Wind direction (if any observed)
   - Wind speed (if any observed)
   - Cloud type (if any observed)

   Date_____

   Place_____

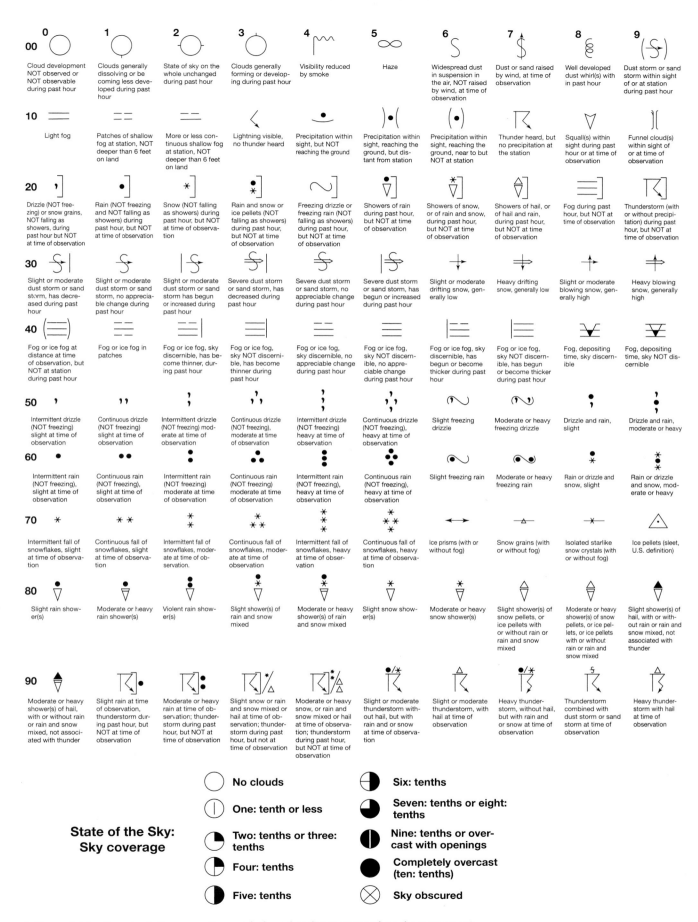

▲ Figure 14.1a Sample station reporting symbols and codes (continued on the next page)

Low altitude

Code No.	C_L	DESCRIPTION (Abridged from International Code)
1	⌒	Cu of fair weather, little vertical development and seemingly flattened
2	⌒⌒	Cu of considerable development, generally towering, with or without other Cu or Sc bases all at same level
3	⌒	Cb with tops lacking clear-cut outlines, but distinctly not cirriform or anvil-shaped; with or without Cu, Sc, or St
4	-○-	Sc formed by spreading out of Cu; Cu often present also
5	⌒	Sc not formed by spreading out of Cu
6	—	St or Fs or both, but no Fs of bad weather
7	- - -	Fs and/or Fc of bad weather (scud)
8	⌒⌒	Cu and Sc (not formed by spreading out of Cu) with bases at different levels
9	⌒	Cb having a clearly fibrous (cirriform) top, often anvil-shaped, with or without Cu, Sc, St, or scud

Middle (alto-) altitude

Code No.	C_M	DESCRIPTION (Abridged from International Code)
1	∠	Thin As (most of cloud layer semi-transparent)
2	∠	Thick As, greater part sufficiently dense to hide Sun (or Moon), or Ns
3	⌒	Thin Ac, mostly semi-transparent; cloud elements not changing much and at a single level
4	⌒	Thin Ac in patches; cloud elements continually changing and/or occurring at more than one level
5	⌒	Thin Ac in bands or in a layer gradually spreading over sky and usually thickening as a whole
6	⌒	Ac formed by the spreading out of Cu or Cb
7	⌒	Double-layered Ac, or a thick layer of Ac, not increasing; or Ac with As and/or Ns
8	M	Ac in the form of Cu-shaped tufts or Ac with turrets
9	⌒	Ac of a chaotic sky, usually at different levels; patches of dense Ci are usually present also

High altitude

Code No.	C_H	DESCRIPTION (Abridged from International Code)
1	⌒	Filaments of Ci, or mares tails, scattered and not increasing
2	⌒	Dense Ci in patches or twisted sheaves, usually not increasing, sometimes like remains of Cb; or towers or tufts
3	⌒	Dense Ci, often anvil-shaped, derived from or associated with Cb
4	⌒	Ci, often hook-shaped, gradually spreading over the sky and usually thickening as a whole
5	⌒	Ci and Cs, often in converging bands, or Cs alone; generally overspreading and growing denser; the continuous layer not reaching 45 altitude
6	⌒	Ci and Cs, often in converging bands, or Cs alone; generally overspreading and growing denser; the continuous layer exceeding 45 altitude
7	⌒	Veil of Cs covering the entire sky
8	⌒	Cs not increasing and not covering entire sky
9	⌒	Cc alone or Cc with some Ci or Cs, but the Cc being the main cirriform cloud

CLOUD ABBREVIATIONS

St—STRATUS   Fc—FRACTOSTRATUS   Sc—STRATOCUMULUS   Cu—CUMULUS   Fc—FRACTOCUMULUS   Cb—CUMULONIMBUS   Ac—ALTOCUMULUS   Ns—NIMBOSTRATUS   As—ALTOSTRATUS   Ci—CIRRUS   Cs—CIRROSTRATUS   Cc—CIRROCUMULUS

Air Pressure Change
(3 hours preceding observation)

Symbol	Description	Symbol	Description
\	Continuously falling	/	Continuously rising
_	Falling, than steady	/‾	Rising, than steady
\/	Falling before a lesser rise	√	Falling before a greater rise
/\	Rising before a greater fall	/\	Rising before a lesser fall
—	Steady		

Surface front designations

▼▼▼ Cold front
●●● Warm front
▲●▲● Occluded front
▲●▲● Stationary front

Wind Indicators and Speeds

Symbol	Knots	Miles per hour	Kilometers per hour
◎	Calm	Calm	Calm
—	1–2	1–2	1–3
	3–7	3–8	4–13
	8–12	9–14	14–19
	13–17	15–20	20–32
	18–22	21–25	33–40
	23–27	26–31	41–50
	28–32	32–37	51–60
	33–37	38–43	61–69
	38–42	44–49	70–79
	43–47	50–54	80–87
	48–52	55–60	88–96
	53–57	61–66	97–106
	58–62	67–71	107–114
	63–67	72–77	115–124
	68–72	78–83	125–134
	73–77	84–89	135–143
	103–107	119–123	192–198

▲ Figure 14.1b

Describe the weather conditions depicted for this day:

2. Record the proper symbols for each of the following listed in Figure 14.1.

   a) Visibility reduced by smoke: _____

   b) Intermittent drizzle (not freezing), heavy at time of observation: _____

   c) Ice pellets (sleet—U. S. definition): _____

   d) Fog or ice fog at a distance: _____

   e) Dust storm within sight: _____

   f) 40% sky coverage: _____

   g) Double-layered altocumulus: _____

   h) 55–60 mph (88–96 kph) winds: _____

This system of data recording is used in the preparation of the synoptic weather map such as the one you work with in Section 2.

SECTION 2

Daily Weather Map—Typical April Pattern

Daily weather maps are known as **synoptic maps**, meaning that they show atmospheric conditions at a specific time and place. The daily weather map is a key analytical tool for meteorologists. Figure 14.2 presents an adapted version of the synoptic map for a typical April morning, 7:00 A.M. EST. This exercise involves adding appropriate isobars, fronts, and air mass designations to the map. The following discussion takes you step-by-step through the completion of this map. Follow this sequence:

1. Each circle on the map represents a weather station. The number to the upper right of the station is the barometric pressure, presented in an abbreviated form used by the National Weather Service (again, reference the station model in Section 1). For example, if a *145* appears, it is short for 1014.5 mb (the "10" and the decimal point are dropped); a *980* means 998.0 mb (the "9" and the decimal point are dropped).

   a) On April 1, 1971, the center of low pressure was near Wausau, Wisconsin (see the **L** on the map), with a pressure of 994.7 mb (947 on the map). Note the wind flags around this center of low pressure. Do the wind flags at the various stations show clockwise or counterclockwise winds around the central low pressure?

   b) The high-pressure center is located near Salmon, Idaho (see the **H** on the map), with a pressure of 1033.6 mb (336 on the map). Note the pattern of air temperatures and even lower dew-point temperatures associated with a cold-air mass centered in the region of high pressure.

Daily Weather Map–April 1, 1971

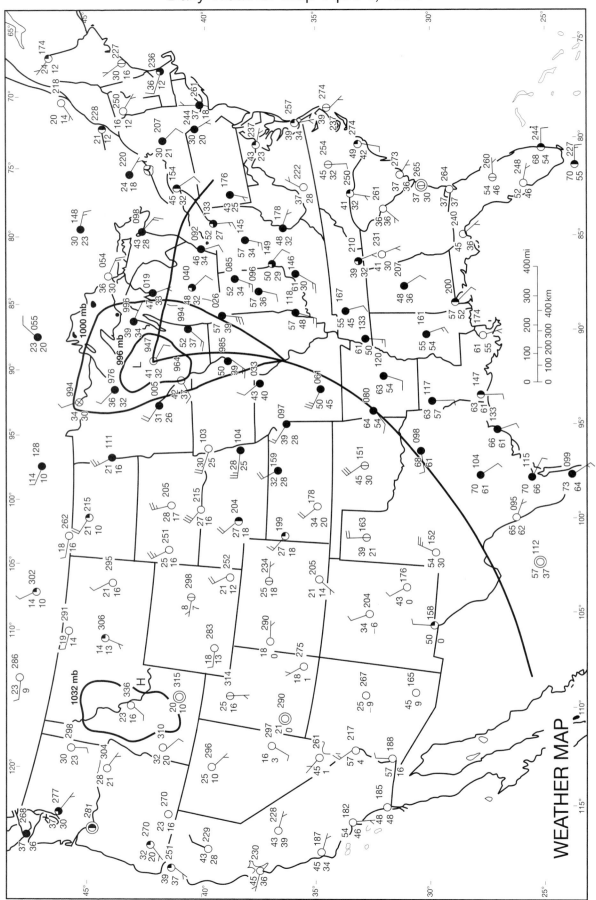

▲ Figure 14.2 Daily weather map

2. On the weather map, draw in the isobars connecting points of equal barometric pressure, at 4 mb intervals.

   a) The **996 mb** and **1000 mb** isobars around the low-pressure center are drawn for you (pressures lower than 996 go inside the isobar, higher than 996 go outside this closed isobar; pressures between 996 mb and 1000 mb fall between the two isobars).

   b) Now draw in order of increasing pressure the rest of the isobars at *4 mb intervals*: **1004, 1008, 1012, 1016, 1020, 1024,** and **1028**. The highest-value isobar of **1032 millibars** is drawn for you. Note that the 1032 mb isobar is a closed isobar that surrounds Salmon, Idaho, which is the area of highest pressure.

3. After completion of the isobars, determine the pattern of low-pressure and high-pressure systems on the map and related air masses to determine the position of the weather fronts. The southeastern portion of the country is influenced by a mild maritime tropical (mT) air mass. The high-pressure area around Idaho is under the influence of a continental polar (cP) air mass.

   a) Locate the warm front line. The warm front has a mT air mass with winds blowing from the south on one side of it, and a cP air mass with winds blowing from the east on the other side of it. The line is drawn for you, but it lacks the red semi-circles that the warm front symbol on maps. Use a red colored pencil or pen and draw in the red semi-circles on the side with the cP air mass.

   b) Locate the colf front line. The cold front has a cP air mass with winds blowing from the northwest on one side, and a mT air mass with winds blowing from the south on the other side. The line is drawn for you, but it lacks the blue triangles of the cold front symbol on maps. Use a blue colored pencil or pen and draw in blue triangles on the side with the mT air mass.

   c) Use a red colored pencil or pen and cross hatch the region that is within 600 km of the warm front or cold front and is under the mT air mass. Use a blue colored pencil and cross hatch the region within 600 km of the warm front or cold front and is under the cP air mass.

   d) Use a pencil or black marker and draw a big arrow indicating the general direction of the winds that are: the cP air mass driving the cold front; the mT air mass that is driving the warm front; and the cP air mass that is in front of the warm front.

4. On this weather map, note *air temperatures* in comparison with *dew-point temperatures*. The air temperatures and dew-point temperatures are noted to the left of each station symbol on the map.

   a) List a sampling of *air temperatures* and *dew-point temperatures* for Idaho, Wyoming, and Utah.

      Try to use specific city names (use an atlas or a map): _____

      _____

   b) Now compare these with *air temperature* and *dew-point temperatures* in Louisiana and Mississippi.

      Try to use specific city names (use an atlas or a map): _____

      _____

   Recall from Lab Exercise 12 that the dew-point temperature gives you an idea of the moisture content of the cP (dry) and mT (moist) air masses, respectively.

5. Describe the pattern of cloudiness across the map. The pattern of cloudiness is indicated by the state-of-the-sky (sky coverage) status recorded within each station symbol. The areas of frontal lifting are clearly identified by these patterns of clouds.

MasteringGeography™ | Looking for additional review and lab prep materials? Go to www.masteringgeography.com for pre lab videos and pre and post lab quizzes.

Name: _____ Laboratory Section: _____ Video
Date: _____ Score/Grade: _____ Exercise 15 Pre-Lab Video

Scan to view the Pre-Lab video
http://goo.gl/IytLBa

LAB EXERCISE 15

Midlatitude Cyclones and Hurricanes

The Sun's energy evaporates water from the tropics and sets in motion wind and ocean currents that transport heat toward the poles. Two large weather systems created by this transfer of heat are **midlatitude cyclones** and **hurricanes**. The conflict between contrasting air masses along the polar front can develop a midlatitude cyclone, also known as a **extratropical cyclone**, or *wave cyclone*. Midlatitude cyclones, which can be 1600 km (1000 mi) wide, dominate weather patterns in the middle and higher latitudes of both the Northern and Southern Hemispheres.

Tropical cyclones are another type of massive storm system. Originating entirely within tropical air masses, these are powerful manifestations of the Earth–atmosphere energy budget and can reach diameters of 1300–1600 km (800–1000 mi). Each year, approximately 45 of these tropical weather events are powerful enough to be classified as hurricanes, **typhoons**, or **cyclones**, which are different regional names for the same type of tropical storm: Hurricanes occur around North America, typhoons in the western Pacific (mainly in Japan and the Philippines), and cyclones in Indonesia, Bangladesh, Australia, and India. A full-fledged hurricane, typhoon, or cyclone has wind speeds greater than 119 kph (74 mph, or 65 knots). Hurricanes that form in the tropics are quite different from midlatitude cyclones because the air of the tropics is essentially homogeneous, with no fronts or conflicting air masses of differing temperatures. In addition, the warm air and warm seas ensure abundant water vapor and thus the necessary latent heat to fuel these storms. The warmer the ocean and atmosphere are, the more intense and powerful the storm.

The first section of this lab examines midlatitude cyclones, with an idealized weather map. The second section involves analyzing an actual weather map to gain a better understanding of the structure of these storms. The final section examines Hurricane Sandy. Students will plot the locations of Hurricane Sandy, the largest Atlantic storm ever recorded. By analyzing the relationship between hurricane behavior and sea-surface temperatures, we will examine the role of hurricanes in the Earth–atmosphere heat-energy budget.

Lab Exercise 15 has three sections.

Key Terms and Concepts

extratropical cyclone
hurricane
midlatitude cyclone

tropical cyclone
typhoon

After completion of this lab, you should be able to:

1. *Analyze* an idealized weather map and *describe* current conditions at various locations.
2. *Plot* hurricane locations, evaluate the effects of sea-surface temperatures on wind speeds and air pressure, and predict potential effects of climate change on hurricane behavior.

Materials/Sources Needed
pencil
colored pencils

Lab Exercise and Activities

SECTION 1

Idealized Weather Map Analysis

Analyze the idealized weather map in Figure 15.1 to *determine* general weather conditions at the stations noted. Using the weather symbols presented in Lab Exercise 14 and a physical geography text dealing with air masses and weather, complete the following information for six cities: wind direction and speed (if any), air pressure, air temperature, dew-point temperature, state of the sky, and weather type (if any).

1. Atlanta, Georgia—label the weather conditions you think are occurring. Describe the dominant air mass and relative humidity.

2. Columbus, Ohio—label the weather conditions you think are occurring. Describe the dominant air mass and relative humidity.

3. Springfield, IL—label the weather conditions you think are occurring. Describe the dominant air mass and relative humidity.

4. Wichita, Kansas—label the weather conditions you think are occurring. Describe the dominant air mass and relative humidity.

5. Denver, Colorado—label the weather conditions you think are occurring. Describe the dominant air mass and relative humidity.

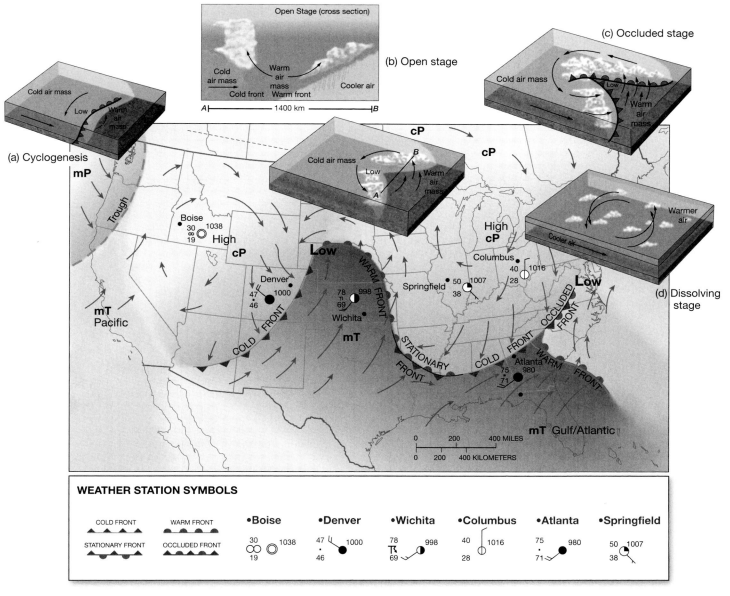

▲ Figure 15.1 Idealized weather map—assume afternoon conditions

6. Boise, Idaho—label the weather conditions you think are occurring. Describe the dominant air mass and relative humidity.

7. If you were a weather forecaster in Springfield, Illinois, and needed to prepare a forecast for broadcast covering the next 24 hours, what would you say? Begin with current conditions you listed in #3 above,

and then progress through the next 6, 12, and 24 hours. Assume that the system is moving eastward at 25 kph (15 mph) and note the scale on the map.

a) 6 hours: _____

b) 12 hours: _____

c) 24 hours: _____

d) 48 hours: _____

SECTION 2

Conflicting air masses brought together along the polar front can develop into a midlatitude cyclone. Steered by the polar jet stream, warm, tropical air collides with cold, drier air from the poles to form these storms. Air spirals into a low-pressure center, and the entire storm system rotates counterclockwise in the Northern Hemisphere; driven by the westerlies and the jet stream, the entire mass moves eastward. Figure 15.2 shows a midlatitude cyclone on October 18, 2007. Examine the weather map and complete the following questions.

1. Which stage of cyclogenesis is this storm in? Explain your answer.

2. Shade the region of the map that is experiencing an mT air mass with red cross-hatching and shade the regions with mP air masses with blue cross-hatching.

3. The lines for the warm and cold fronts are drawn for you. Identify the warm front and draw a series of red semicircles on the side of the line that the mT air mass is moving toward. Identify the cold front and draw a series of blue triangles on the side of the line that the mP air mass is moving toward.

4. Which type of front is directly to the north of the low-pressure center? Using colored pencils, draw the correct symbols for that portion of the front.

5. Consult an atlas to identify the following weather stations inside the study area box:

   a) Which station has the highest wind speed? What is the name of the city, and what is the wind speed?

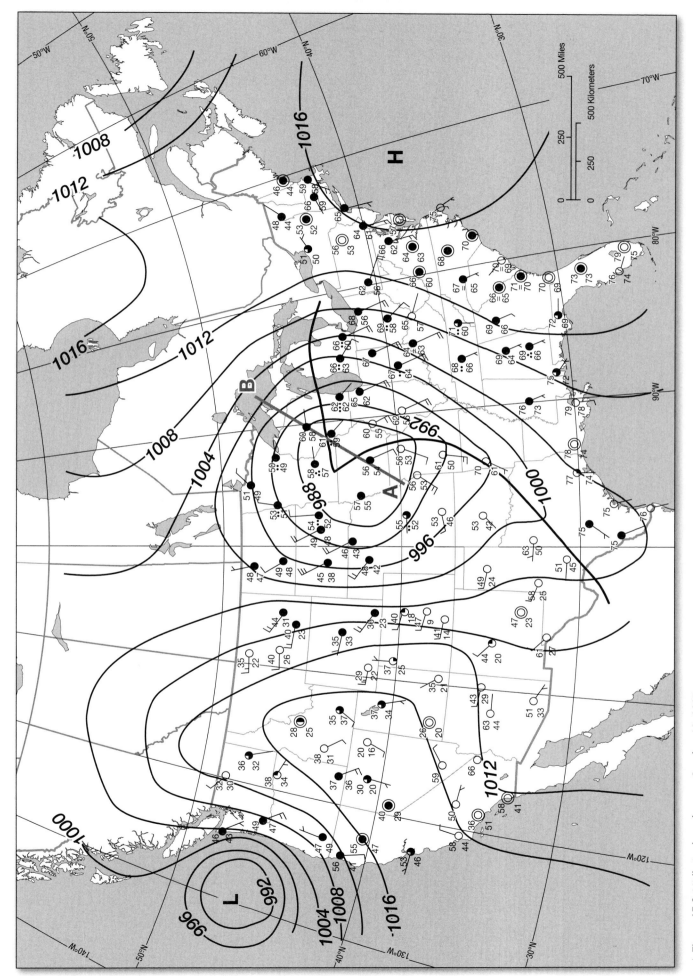

▲ Figure 15.2 Midlatitude cyclone on October 18, 2007.

b) Which station has the driest air? Look for the station with the lowest dew point. What is the name of the city, and what is the dew point? Which type of air mass is over this city?

_____

_____

c) Which station has the most humid air? Look for the station with the highest dew point. What is the name of the city, and what is the dew point? Which type of air mass is over this city?

_____

_____

6. What were the current conditions in St. Louis, Illinois, on October 18, 2007? List the temperature, dew-point temperature, wind speed and direction, and air pressure. Which type of air mass was overhead?

7. On October 19, 2007, the storm had moved 600 km (360 mi) eastward. Estimate what the conditions were in St. Louis, Illinois, on that day. List what you think the temperature, approximate dew-point temperature, and wind speed and direction were on October 19. Which type of air mass was overhead?

8. Figure 15.2 shows a view of the storm from above. There is a transect line from point A in southern Iowa to point B in Lake Superior. In the space below, draw a cross section (side view) of the cold front and the warm front that corresponds to the transect line drawn from A to B.

A   B

▲ Cross section of a midlatitude cyclone

SECTION 3

Hurricanes, Typhoons, and Cyclones

Tropical cyclones are potentially the most destructive storms experienced by humans, claiming thousands of lives each year worldwide. This is especially true when they attain the wind speeds and low-pressure readings of a full-fledged hurricane, typhoon, or cyclone (> 65 knots, > 74 mph, > 119 kph).

Record levels of activity and storm intensity occurred from 1995 to 2006. As an example of this, hurricane season usually extends to November. However, during the 2005 season, hurricane season forecasters ran through the list of names for storms and had to resort to using Greek letters to identify storms that continued well into January! Some of the other extraordinary events during this time period include:

- In March 2004, Hurricane Catarina was the first hurricane observed turning from the equator into to develop in the south Atlantic.
- In October 2005, Tropical Storm Vince became the first Atlantic tropical cyclone to strike Spain since 1842!
- In 2006, Hurricane Ioke, also referred to as Super Typhoon Ioke, strongest hurricane ever recorded in the Central Pacific developed south of Hawaii and lasted for almost 3 weeks; its final remnants caused erosion along the shores of Alaska!
- In 2007, Super Cyclone Gonu became the strongest tropical cyclone on record in the Arabian Sea. It eventually hit Oman and the Arabian Peninsula.

These four storms are examples of the increased occurrence and intensity of cyclonic systems, which are related to higher oceanic and atmospheric temperatures.

Tropical cyclones are powerful manifestations of the Earth–atmosphere energy budget. The warm air and seas where they begin provide abundant water vapor and thus the necessary **latent heat** to fuel these storms. Tropical cyclones convert heat energy from the ocean into mechanical energy in the wind; the warmer the ocean and atmosphere, the more intense the conversion and powerful the storm. For these processes, sea-surface temperatures must exceed approximately 26°C (79°F). During Hurricane Katrina, the sea-surface temperature (SST) in the Gulf of Mexico exceeded 30°C (86°F).

Warmer Atlantic waters along the 20th parallel between 20°W and 90°W, and the Caribbean and Gulf of Mexico, are fueling a greater intensity than previously experienced. A study published in *Science* stated,

> ... the trend of increasing numbers of category 4 and 5 hurricanes for the period 1970–2004 is directly linked to the trend in sea-surface temperature.... Higher SST was the only statistically significant controlling variable related to the upward trend in global hurricane strength since 1970.[*]

Statistically, the damage caused by tropical cyclones is increasing substantially as more and more development occurs along susceptible coastlines, whereas loss of life is decreasing in most parts of the world as a result of better forecasting of these storms.

The combination of more intense and numerous storms, due in part to higher SSTs, and increasing development of inappropriate areas will create the potential for future catastrophes.

Hurricane Sandy began as a typical late-season hurricane in the Caribbean Sea on October 22, 2012. Sandy passed over Jamaica as a category 1 hurricane and strengthened to a category 3 hurricane when it passed over Cuba. While over the Bahamas, it underwent complex changes, growing in diameter while weakening in intensity. It then moved northeastward, parallel to the eastern coast of the United States, and gained intensity while passing over the warmer waters of the Gulf Stream. It became a post-tropical cyclone, unofficially called "Superstorm Sandy," when it was no longer fueled by the release of latent heat of condensation close to the center of low pressure but by conflicting air masses more similar to those of a midlatitude cyclone.

By the time it dissipated on October 31, 2012, it had become the largest Atlantic basin storm in terms of diameter, claimed 286 lives in seven countries, affected 24 U.S. states, and caused almost $70 billion in damage.

---

[*]C. D. Hoyos, P. A. Agudelo, P. J. Webster, and J. A. Curry, "Deconvolution of Factors Contributing to the Increase in Global Hurricane Intensity, *Science* (April 7, 2006), v 312 no. 5770: pp. 94–97.

Hurricane Sandy

1. Plot Sandy's storm track, using the data in Figure 15.3, then fill in the SST for each of the data points in the spaces provided. The first three positions have been plotted for you.

2. What physical inputs are required to create a hurricane? Identify and describe the positive feedback loop that allows hurricanes to mature. What factors act to decrease hurricane intensity and strength?

3. Estimate the distance traveled using the scale bar on the map. How far did Sandy travel from 00 UTC on the 27th to 00 UTC on the 28th? 00 UTC on the 28th to 00 UTC on the 29th? 00 UTC on the 29th to 00 UTC on the 30th?

4. What was Sandy's maximum recorded wind speed? How fast was it traveling on that day, in kmph? What was the maximum wind speed in the right-front quadrant (wind speed combined with traveling speed)? What would the total wind speed have been in the left-rear quadrant (wind speed minus traveling speed)?

5. Mark the location on the map with the lowest mb/highest wind speed. On what day did that occur? What was the water temperature on that day?

6. How much did the central pressure drop from normal sea level pressure?

7. For hurricanes to form, the water temperature must be at least 26°C (79°F). What was the warmest water that Hurricane Sandy passed over? Coldest water? Describe the relationship between the water temperature and wind speed.

8. Plot on the graphs below the wind speed for each day and the central pressure, in mb, for each day.

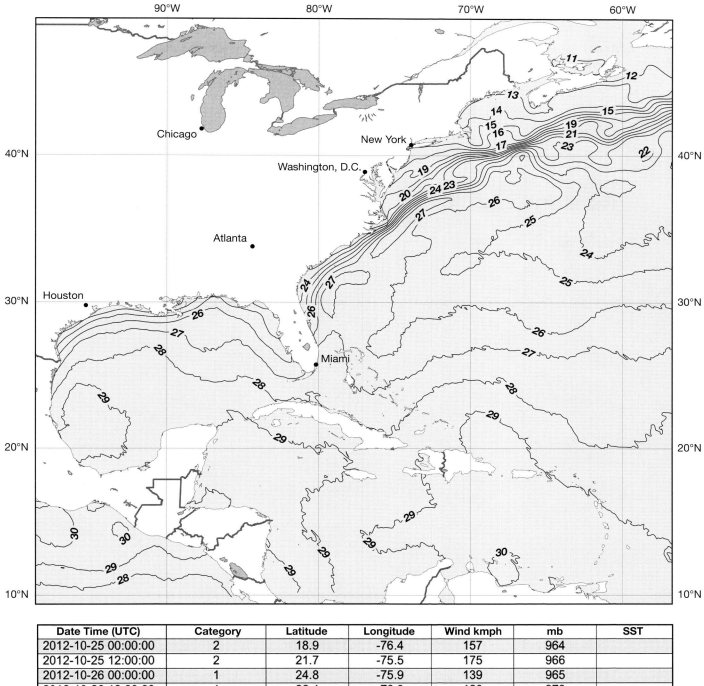

▲ Figure 15.3 Hurricane Sandy

Date Time (UTC)	Category	Latitude	Longitude	Wind kmph	mb	SST
2012-10-25 00:00:00	2	18.9	-76.4	157	964	
2012-10-25 12:00:00	2	21.7	-75.5	175	966	
2012-10-26 00:00:00	1	24.8	-75.9	139	965	
2012-10-26 12:00:00	1	26.4	-76.9	120	970	
2012-10-27 00:00:00	Tropical Storm	27.5	-77.1	111	969	
2012-10-27 12:00:00	1	28.8	-76.5	129	956	
2012-10-28 00:00:00	1	30.5	-74.7	120	960	
2012-10-28 12:00:00	1	32.0	-73.0	120	954	
2012-10-29 00:00:00	1	33.9	-71.0	129	950	
2012-10-29 12:00:00	2	36.9	-71.0	157	945	
2012-10-30 00:00:00	Extra-tropical	39.5	-74.5	129	946	
2012-10-30 12:00:00	Extra-tropical	40.1	-77.8	92	978	
2012-10-31 00:00:00	Extra-tropical	40.7	-79.8	65	992	
2012-10-31 12:00:00	Extra-tropical	41.5	-80.7	55	995	

9. Briefly summarize the relationship between SST and hurricane strength. What are the implications of this on future hurricanes? What advice would you give to coastal planners regarding future development of coastal areas on the U.S. east coast?

MasteringGeography™ | Looking for additional review and lab prep materials? Go to www.masteringgeography.com for pre lab videos and pre and post lab quizzes.

Name: _____ Laboratory Section: _____ **Video**
Date: _____ Score/Grade: _____ Exercise 16
Pre-Lab Video

LAB EXERCISE 16

Water Balance and Water Resources

Scan to view the Pre-Lab video

Because water is not always naturally available when and where it is needed, humans must rearrange water resources in time and space. The maintenance of a house plant, the distribution of local water supplies, an irrigation program on a farm, the rearrangement of river flows—all involve aspects of the water balance and water-resource management.

The water balance is a portrait of the hydrologic cycle at a specific site or area for any period of time. The water balance can be used to estimate streamflow, accurately determine irrigation quantity and timing, and show the relationship between a given supply of water and the local demand.

A water balance can be established for any defined area of Earth's surface—a continent, nation, region, or field—by calculating the total precipitation input and the total water output.

In this lab exercise, we work with a water-balance equation and accounting procedure to determine moisture conditions for two cities—Kingsport, Tennessee, and Sacramento, California. Given this data, you will prepare graphs that illustrate these water-balance relationships. Also, this lab examines the broader issues of water resources in the United States. Lab Exercise 16 features three sections.

Key Terms and Concepts

actual evapotranspiration
available water
capillary water
deficit
evaporation
evapotranspiration
field capacity
potential evapotranspiration
precipitation

soil moisture recharge
soil moisture storage
soil moisture storage change
soil moisture utilization
soil-water budget
surplus
transpiration
wilting point
withdrawal

KEY LEARNING concepts

After completion of this lab, you should be able to:

1. *Identify* and *define* the key water-balance components, as arranged in a water-balance equation.
2. *Compare* and *contrast* maps of precipitation and potential evapotranspiration in North America.
3. *Plot* water-balance components for Kingsport, Tennessee, and Sacramento, California, and *graph* the relationship between PRECIP and POTET.
4. *Explain* the dynamics of soil moisture storage changes—recharge and utilization.

Materials/Sources Needed

calculator
pencil

colored pencils

Copyright © 2015 Pearson Education, Inc. Lab Exercise 16 135

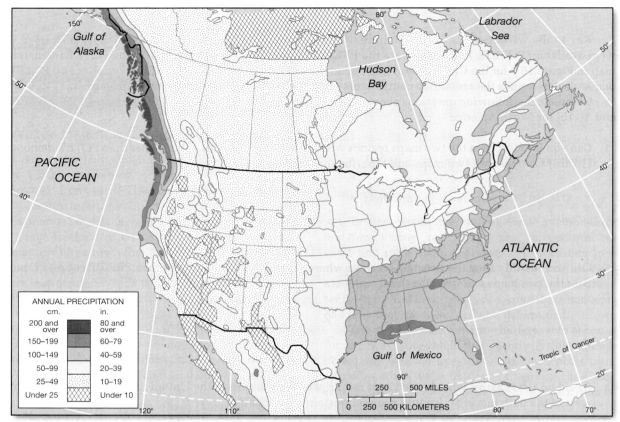

(a) Precipitation

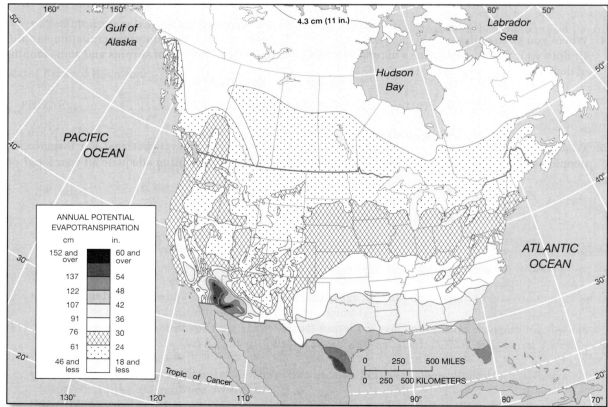

(b) Potential evapotranspiration

▲ Figures 16.1a and 16.1b **Potential evapotranspiration and precipitation for the United States and Canada**

Figure Precipitation and POTET
http://goo.gl/bx5UYw

SECTION 2

Water Budget Calculations for Kingsport, Tennessee

Soil-moisture storage is a "savings account" of water that can receive deposits and allow withdrawals as conditions change in the water balance. Soil-moisture storage (ΔSTRGE) refers to the amount of water that is stored in the soil and is accessible to plant roots. Soil is said to be at the **wilting point** when all that is left in the soil is unextractable water; the plants wilt and eventually die after a prolonged period of such moisture stress.

The soil moisture that is generally accessible to plant roots is **capillary water**, held in the soil by surface tension and hydrogen bonding between the water and the soil. Almost all capillary water is **available water** in soil moisture storage and is removable for POTET demands through the actions of plant roots and surface evaporation. After water drains from the larger pore spaces, the available water remaining for plants is termed **field capacity**, or storage capacity. This water is held in the soil by hydrogen bonding against the pull of gravity. Field capacity is specific to each soil type and is an amount that can be determined by soil surveys. If you have a garden or house plant, you no doubt are aware of how much effort is needed to maximize soils for good water retention.

Assuming a soil moisture storage capacity of 100 mm (4.0 in.) for Kingsport, Tennessee, typical of shallow-rooted plants, the months of net demand for moisture are satisfied through **soil-moisture utilization**. Various plant types send roots to different depths and therefore are exposed to varying amounts of soil moisture.

For example, shallow-rooted crops such as spinach, beans, and carrots send roots down about 65 cm (25 in.) in a silt loam, whereas deep-rooted crops such as alfalfa and shrubs exceed a depth of 125 cm (50 in.) in such a soil.

For this exercise we assume that soil moisture utilization occurs at 100%; that is, if there is a net water demand, the plants will be able to extract moisture as needed. Actually, in nature, as the available soil water is reduced by soil-moisture utilization, the plants must exert greater effort to extract the same amount of moisture. As a result, even though a small amount of water may remain in the soil, plants may be unable to exert enough pressure to utilize it. The unsatisfied demand resulting from this situation is calculated as a deficit. Avoiding such deficit inefficiencies and reduction in plant growth are the goals of a proper irrigation program, for the harder plants must work to get water, the less their yield and growth will be.

Likewise, relative to soil moisture recharge, we assume a 100% rate; if the soil moisture storage is less than field capacity, then excess moisture beyond POTET demand will go to **soil-moisture recharge**. We assume in this exercise a soil moisture recharge rate as 100% efficient, as long as the soil is below field capacity and above a temperature of −1°C (30.2°F). Under real conditions, we know that infiltration actually proceeds rapidly in the first minutes of a storm, and it slows as the upper layers of soil become saturated, even though the soil below is still dry.

Table 16.1 presents the long-term average water supply and demand conditions for the city of Kingsport in the extreme northeast corner of Tennessee. Kingsport is located at 36.5°N, 82.5°W, with an elevation of 390 m (1280 ft), and it has maintained weather records for more than half a century. The monthly values in Table 16.1 assume that PRECIP and POTET are evenly distributed throughout each month, smoothing the actual daily and hourly conditions. Kingsport experiences a humid subtropical climate with adequate moisture throughout the year. Let's first compare PRECIP with POTET by month to determine whether there is a net supply ()+ or net demand ()− for water. This is the basic bookkeeping procedure.

Questions and analysis about the water balance of Kingsport, Tennessee:

1. Calculate the positive or negative relationship between PRECIP and POTET. Record this plus or minus value in the spaces provided. Note that January and June are already figured. Complete the remainder of the water-balance table, recalling that storage capacity is 100 mm. Data are smoothed over the entire month.

	Jan	Feb	Mar	Apr	May	Jun	Jul	Aug	Sep	Oct	Nov	Dec	Total
PRECIP	97	99	97	84	104	97	132	112	66	66	66	99	1119
POTET	7	8	24	57	97	132	150	133	99	55	12	7	781
PRECIP—POTET	+90	91				−35							—
STRGE	100	100				65						100	—
ΔSTRGE	0	0				−35							
ACTET	7	8				132							
DEFIC	0	0				0							
SURPL	90	91				0							

(All quantities in millimeters)

▲Table 16.1 Kingsport water balance

2. When does Kingsport experience a net supply of water? List the months.

3. What occurs during the warm months from June through September?

4. What is the total annual PRECIP for Kingsport in mm?

5. What is the total annual POTET for Kingsport in mm?

6. Soil moisture remains at field capacity (full) through which month?

7. How much surplus is accumulated through these five months?

8. What is the net demand for water in June?

After you satisfy this demand through soil-moisture utilization, what is the remaining water in soil moisture at the end of June, to begin the month of July?

Lab Exercise 16

9. Calculate the actual evapotranspiration for each month of the year for Kingsport and note this on the table. By subtracting DEFIC from POTET, you determine the *actual evapotranspiration*, or ACTET, that takes place for each month. Under ideal moisture conditions, POTET and ACTET are about the same, so that plants do not experience a water shortage. Prolonged deficits could lead to drought conditions.

Note that given the 100% efficiency assumptions for soil-moisture utilization used in this exercise, ACTET equals POTET each month. In reality, there actually would be a slight inefficiency of soil-moisture utilization from June through September, with ACTET progressively less than the POTET, producing a 4-month deficit of 16 mm.

10. According to your calculations, do the soils of Kingsport return to field capacity (full storage) by the end of the year? _____ Are any surpluses generated in December?

    The amount?

11. What is the total ACTET for the year?

12. What is the total DEFIC for the year?

13. What is the total SURPL for the year?

14. During how many months at Kingsport does POTET exceed PRECIP?

15. According to a physical geography textbook and lab discussion, what precipitation mechanisms produce Kingsport's pattern of summer maximum precipitation and moisture throughout the year? What factors produce such consistent precipitation throughout the year?

    Winter:

    Summer:

SECTION 3

Water Budget Calculations for Sacramento, California

For comparison, let's work with a city that experiences seasonal deficits in its annual water balance. Sacramento, California (38.58° N, 121.35° W, at 11 m elevation), has a Mediterranean dry, warm summer climate. The data for Sacramento are set in Table 16.2. Please assume the same soil-moisture storage capacity of 100 mm (4.0 in.) for Sacramento as for Kingsport, typical of shallow-rooted plants. The months of net demand for moisture are satisfied through *soil-moisture utilization*, as long as the soil moisture is available. Sacramento does experience a wilting point each year.

	Jan	Feb	Mar	Apr	May	Jun	Jul	Aug	Sep	Oct	Nov	Dec	Total
PRECIP	107	74	56	36	10	3	3	3	8	23	58	76	455
POTET	15	23	40	60	85	115	140	127	99	66	30	15	815
PRECIP—POTET													—
STRGE	100					0						89	—
ΔSTRGE													
ACTET													
DEFIC													
SURPL													

(All quantities in millimeters)

▲Table 16.2 Water balance for Sacramento, California

Questions and analysis about the water balance of Sacramento, California:

1. Complete the bookkeeping procedure for Sacramento using the monthly and annual average values given in Table 16.2. Calculate PRECIP—POTET, ΔSTRGE ACTET, DEFIC, and SURPL amounts for each month, as you did for Kingsport.

2. For Sacramento, how many months does POTET exceed PRECIP?

3. According to a physical geography textbook and lab discussion, what precipitation mechanisms produce Sacramento's seasonally variable summer—drought, winter—moist precipitation pattern?

   Winter:

   Summer:

Name: _____ Laboratory Section: _____ **Video**
Date: _____ Score/Grade: _____ Exercise 17 Pre-Lab Video

Scan to view the Pre-Lab video

http://goo.gl/9zDN70

LAB EXERCISE 17

Global Climate Systems

Earth experiences an almost infinite variety of *weather*—conditions of the atmosphere at any given time and place. But if we consider the weather over many years, including its variability and extremes, a pattern emerges that constitutes **climate**. Think of climate patterns as dynamic rather than static. Climate is more than a consideration of simple averages of temperature and precipitation.

Climatology, the study of climate, is the analysis of long-term weather patterns, including extreme weather events, over time and space to find areas of similar weather statistics, identified as **climatic regions**. Observed patterns grouped into regions are at the core of climate classification.

In this lab exercise, we examine various patterns of temperature and precipitation that operate as a basis for climate **classification**. We work with a model of global climates, the classification and plotting of actual climate data, and analysis of temperature and precipitation patterns. Lab Exercise 17 features five sections (one optional).

Key Terms and Concepts

classification
climate
climatic regions
climatology

climographs
empirical classification
genetic classification

KEY LEARNING concepts

After completion of this lab, you should be able to:

1. *Identify* patterns of temperature, precipitation, and other weather elements that contribute to the climate of an area.

2. *Analyze* key climate data to *determine* the empirical (Köppen) classification and genetic (causative) factors.

3. *Plot* data (°C and cm) on climographs for analysis.

Materials/Sources Needed

colored pencils
calculator
world atlas
physical geography textbook

Lab Exercise and Activities

SECTION 1

Climatology and Climate Classification

Climatology, the study of climate, involves analysis of the patterns in time or space created by various physical factors in the environment. One type of climatic analysis involves discerning areas of similar weather statistics and grouping them into climatic regions that contain characteristic weather patterns.

Climate classifications are an effort to formalize these patterns and determine their implications to humans. Simply comparing the two principal climatic components—temperature and precipitation—reveals important relationships that are indicative of the distribution of climate types (Figure 17.1).

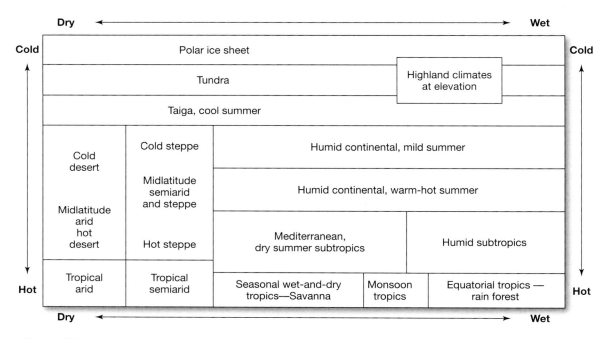

▲ Figure 17.1 Temperature and precipitation schematic showing climatic relationships

1. Mark your present location with an **X** on Figure 17.1. Place the name of this location next to the **X**. Next, mark the location (approximately) that is most characteristic of your place of birth with an **O** and record the name of that place next to the **O**.

Classification is the process of grouping data or phenomena in related classes. A classification based on causative factors—for example, the interaction of air masses—is called a **genetic classification**. An **empirical classification** is based on statistics or other data used to determine general categories. Climate classifications based on temperature and precipitation data are empirical classifications. In *Applied Physical Geography*, we use Köppen's empirical system to classify climatic regions, as well as a genetic analysis to help you appreciate the relationships between a region's location and its climate.

The principal elements of climate are insolation, temperature, pressure, air masses, and precipitation. We review them briefly here. Insolation is the energy input for the climate system, but it varies widely over Earth's surface by latitude (see Figure 6.2). The pattern of world temperatures and their annual ranges is shown in Figures 9.2a and 9.2b. Temperature variations result from a coupling of dynamic forces in the atmosphere to Earth's pattern of atmospheric pressure and resulting global wind systems (see Figures 11.6 and 11.7). Important, too, are the location and physical characteristics of air masses, those vast bodies of homogeneous air that form over oceanic and continental source regions.

Moisture is the remaining input to climate. The hydrologic cycle transfers moisture, with its tremendous latent heat energy, through Earth's climate system. Moisture patterns are important, for they are key climate control factors. Average temperatures and daylength help us approximate POTET (potential evapotranspiration), a measure of natural moisture demand. Most of Earth's desert regions, areas of permanent water deficit, are in lands dominated by subtropical high-pressure cells, with bordering lands grading to grasslands and to forests as precipitation increases. The most consistently wet climates on Earth straddle the equator in the Amazon region of South America, the Congo region of Africa, and Indonesia and Southeast Asia, all of which are influenced by equatorial low pressure and the intertropical convergence zone, or ITCZ. Simply relating the two principal climatic components—temperature and precipitation—reveals general climate types.

The basis of any empirical classification system is the choice of criteria used to draw lines between categories. The Köppen classification system, widely used for its ease of comprehension, was designed by Wladimir Köppen (1846–1940), a German climatologist and botanist. Köppen chose mean monthly temperatures, mean monthly precipitation, and total annual precipitation to devise his spatial categories and boundaries. But we must remember that boundaries really are gray areas; they are transition zones of gradual change. For our purposes of general understanding, a modified Köppen-Geiger classification is useful.

The Köppen system uses capital letters (A, B, C, D, E, plus H) to designate climatic categories from the equator to the poles. Five of the climate classifications are based on thermal criteria and one on moisture conditions as well.

> *A*—Tropical (equatorial regions)
>
> *C*—Mesothermal (Mediterranean, humid subtropical, marine west coast regions)
>
> *D*—Microthermal (humid continental, subarctic regions)
>
> *E*—Polar (polar regions)
>
> *H*—Highland (compared to lowlands at the same latitude, highlands have lower temperatures—recall the normal lapse rate—and more efficient precipitation due to lower POTET demand, justifying separation of highlands from surrounding climates)
>
> *B*—Dry, the deserts and steppes (only category based on moisture as well)

In addition to these primary types noted with capital letters, lowercase letters are used to signify certain temperature and moisture characteristics. Inside the back cover of this laboratory manual is a world map illustrating climates classified according to the Köppen system, with the addition of the main climatic influences. Table 17.1 presents a climate classification key and summary that specifies both *thermal* (A, C, D, and E) and *moisture* (B) classifications. Use this table for classifying each of the five stations presented in Lab Exercise Section 2 and two stations in Section 4. Specific formulas to calculate the *B* dry climates follow in this lab exercise in Section 4.

For example, in a *tropical rain forest Af* climate, the *A* tells us that the average coolest month is above 18°C (64.4°F, average for the month), and the *f* indicates that the weather is constantly wet (German *feucht*, for "moist"), with the driest month receiving at least 6 cm (2.4 in.) of precipitation. As you can see in the map of A climates, the *tropical rain forest Af* climate straddles the equator.

For another example, in a *Dfa* climate, the *D* means that the average warmest month is above 10°C (50°F), with at least one month falling below 0°C (32°F); the *f* says that at least 3 cm (1.2 in.) of precipitation falls during every month; and the *a* indicates a warmest summer month averaging above 22°C (71.6°F). Thus, a *Dfa* climate is a humid-continental, hot-summer climate in the microthermal D category.

Try not to get lost in the alphabet soup of this system, for it is meant to help you understand complex climates through a simplified set of symbols. To simplify the detailed criteria, a box is placed at the end of each climate category that presents Köppen's guidelines. To help keep it straight, always say the climate's descriptive name followed by its symbol: for example, *Mediterranean dry-summer Csa, tundra ET*, or *cold midlatitude steppe BSk*.

Determining primary climate classification

E — An **E climate** if warmest month is less than 10° C (50° F). If not an E then move on to B climates.

B — If precipitation supply is less than average potential evapotranspiration demand climate may be a **B climate**—go to Lab **SECTION 4** to consult formulas. If climate is more moist than B criteria then move on to A climates.

A — An **A climate** if the average coolest month is warmer than 18° C (64.4° F). If not an A then move on to C climates.

C — A **C climate** if average coldest month is below 18° C (64.4° F) but warmer than 0° C (32° F). If not an E, B, A, or C then climate is a D classification.

D — A **D climate** if average coldest month is below 0° C (32° F) and warmest month is above 10° C (50° F).

▲Table 17.1 Climate classification key

E Polar Climates

Warmest month below 10° C (50° F); always cold; ice climates.

ET - Tundra:
Warmest month 0–10° C (32–50° F); PRECIP exceeds small POTET demand; snow cover 8–10 months.

EF - Ice cap:
Warmest month between 0° C (32° F); PRECIP exceeds very small POTET demand; the polar regions.

EM - Polar marine:
All months above −7° C (20° F), warmest month above 0° C; annual temperature range <30° F (17° C).

B Dry Arid and Semiarid Climates

POTET exceeds PRECIP in all B climates. Subdivisions based on PRECIP timing and amount and mean annual temperature. Boundaries determined by formulas and graphs.

Earth's arid climates.
BWh - Hot low-latitude desert
BWk - Cold midlatitude desert
BW = PRECIP less than 1/2 POTET.
h = Mean annual temperature >18° C (64.4° F).
k = Mean annual temperature < 18° C.

Earth's semiarid climates.
BSh - Hot low-latitude steppe
BSk - Cold midlatitude steppe
BS = PRECIP MORE THAN 1/2 POTET but not equal to it.
h = Mean annual temperature >18° C.
k = Mean annual temperature < 18° C.

A Tropical Climates

Consistently warm with all months averaging above 18° C (64.4° F); annual PRECIP exceeds POTET.

Am - Tropical monsoon:
m = A marked short dry season with 1 or more months receiving less than 6 cm (2.4 in.) PRECIP; an otherwise excessively wet rainy season. ITCZ 6–12 months dominant.

Af - Tropical rain forest:
f = All months receive PRECIP in excess of 6 cm (2.4 in.).

Aw - Tropical savanna:
w = Summer wet season, winter dry season; ITCZ dominant 6 months or less, winter water-balance deficits.

C Mesothermal Climates

Warmest month above 10° C (50° F); coldest month above 0° C (32° F) but below 18° C (64.4° F); seasonal climates.

Cfb, Cfc - Marine west coast: Mild-to-cool summer.
f = Receives year-round PRECIP
b = Warmest month below 22° C (71.6° F) with 4 months above 10° C.
c = 1 to 3 months above 10° C.

Cfa, Cwa - Humid subtropical:
a = Hot summer: warmest month above 22° C (71.6° F).
f = Year-round PRECIP.
w = Winter drought, wettest summer month 10 times more PRECIP than driest winter month.

Csa, Csb - Mediterranean summer dry:
s = Pronounced summer drought with 70% of PRECIP in winter.
a = Hot summer with warmest month above 22° C (71.6° F).
b = Mild summer; warmest month below 22° C.

D Microthermal Climates

Warmest month above 10° C (50° F); coldest month below 0° C (32° F); cool temperate-to-cold conditions; snow climates. In Southern Hemisphere, only in highland climates.

Dfa, Dwa - Humid continental:
a = Hot summer; warmest month above 22° C (71.6° F).
f = Year-round PRECIP.
w = Winter drought, wettest summer month 10 times more PRECIP than driest winter month.

Dfb, Dwb - Humid continental:
b = Mild summer; warmest month below 22° C (71.6° F).
f = Year-round PRECIP.
w = Winter drought, wettest summer month 10 times more PRECIP than driest winter month.

Dfc, Dwc, Dwd - Subarctic:
Cool summers, cold winters.
f = Year-round PRECIP.
c = 1 to 4 months above 10° C.
d = Coldest month below −38° C (−36.4° F), in Siberia only.
w = Winter drought, wettest summer month 10 times more PRECIP than driest winter month.

SECTION 2

Climographs for Five Stations

Temperature patterns are influenced by several principal controls: latitudinal position, altitude, cloud cover, land–water heating differences, ocean currents and sea-surface temperatures, and general surface conditions. Review your work in Lab Exercise 8 with stations in Bolivia and Canada where you identified the effects of these controls.

This exercise features **climographs**, which show temperatures as a line graph and precipitation as a bar graph. Use the five climographs (Figures 17.2a through 17.2e) and sets of data that follow to plot mean monthly temperature (*solid-line graph*), mean monthly precipitation (*a bar graph*), and potential evapotranspiration (*a dotted line*) data for the following stations. Use Table 17.1 to determine each station's climatic classification and then determine the city that identifies each station. Using your atlas and the map inside the back cover of this laboratory manual, find and record the names of the five other cities listed below. Use the world climate map on the back cover of this laboratory manual to identify the climatic influences: high or low air pressure; types of air masses; degree of continentality; and warm and cool currents. Analyze the distribution of temperature and precipitation during the year and describe them in general terms: Is the temperature range large or small? Is it consistently warm or consistently cool? Is it dry year round, or is it wet year round? Which months are rainy, and which months are dry? Use the terrestrial biomes map inside the back cover of this laboratory manual to identify the representative biome for each region.

Selected Global Climate Stations

Section 2—Tropical, mesothermal, and microthermal climographs, six stations.

    Figure 17.2 Salvador (Bahia), Brazil: pop. 2,676,000; elev. 9 m (30 ft), completed as an example.

    Figures 17.2a through 17.2e—five stations to graph and identify (listed in alphabetical order)

        Edinburgh, Scotland: pop. 478,000; elev. 134 m (439 ft), 56°N 3.1°W

        Montreal, Quebéc, Canada: pop. 1,621,000; elev. 57 m (187 ft), 45.5°N 73.5°W

        New Orleans, Louisiana: pop. 337,000; elev. 3 m (9 ft), 30°N 90°W

        Oymyakon, Siberia, Russia: pop. 521; elev. 726 m (2382 ft), 63°N 143°E

        Sacramento, California: pop. 486,000; elev. 11 m (36 ft), 38.5°N 121.3°W

Section 3—Climate analysis

Section 4—Desert climate climographs, two stations

    Figures 17.4a and 17.4b—two stations to graph and identify (listed in no particular order)

        Reno, Nevada: pop. 221,000; elev. 1341 m (4400 ft), 39.5°N 119.5°W

        Walgett, New South Wales, Australia: pop. 2000; elev. 133 m (436 ft), 30°S 148°E

Section 5—Your climate region (optional station exercise)

Station #3:

Total annual rainfall:

Average annual temperature:

Annual temperature range:

Distribution of temperature during the year:

Distribution of precipitation during the year:

Principal atmospheric lifting mechanism(s):

Distribution of potential evapotranspiration during the year:

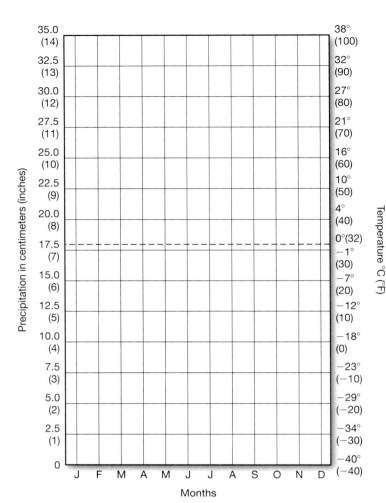

▲ Figure 17.2b

	Jan	Feb	Mar	Apr	May	Jun	Jul	Aug	Sep	Oct	Nov	Dec	Annual
Temperature °C	−47.2	−42.9	−34.2	−15.4	1.4	11.6	14.8	10.9	1.6	−16.2	−35.0	−44.0	−16.3
(°F)	−53.0	−45.2	−29.6	(4.3)	(34.5)	(52.9)	(58.6)	(51.6)	(34.9)	(2.8)	−31.0	−47.2	(2.7)
PRECIP cm	0.8	0.5	0.5	0.3	1.0	3.3	4.1	3.8	2.0	1.3	1.0	0.08	19.3
(in.)	(0.3)	(0.2)	(0.2)	(0.1)	(0.4)	(1.3)	(1.6)	(1.5)	(0.8)	(0.5)	(0.4)	(0.3)	(7.6)
POTET cm	0	0	0	0	3.0	13.5	15.7	10.9	2.3	0	0	0	45.5
(in.)	(0)	(0)	(0)	(0)	(1.2)	(5.3)	(6.2)	(4.3)	(0.9)	(0)	(0)	(0)	(17.9)

What are the main climatic influences for this station (air pressure, air mass sources, degree of continentality, temperature of ocean currents)?

Köppen climate classification symbol: ; name: ; explanation for this determination:

Representative biome (terrestrial ecosystem) characteristic of region: ; characteristic vegetation:

Station #4:

Total annual rainfall:

Average annual temperature:

Annual temperature range:

Distribution of temperature during the year:

Distribution of precipitation during the year:

Principal atmospheric lifting mechanism(s):

Distribution of potential evapotranspiration during the year:

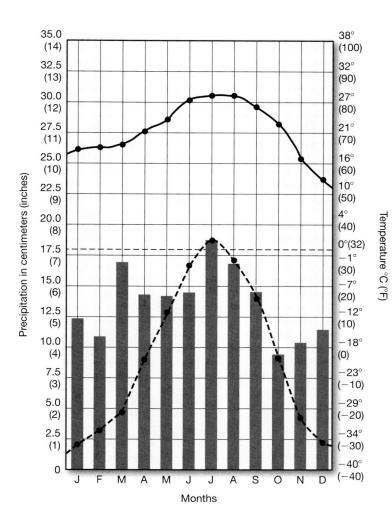

▲ Figure 17.2c

	Jan	Feb	Mar	Apr	May	Jun	Jul	Aug	Sep	Oct	Nov	Dec	Annual
Temperature °C	13.3	14.4	17.2	21.1	24.4	27.8	28.3	28.3	26.7	22.8	16.7	13.9	21.1
(°F)	(56.0)	(58.0)	(63.0)	(70.0)	(76.0)	(82.0)	(83.0)	(83.0)	(80.0)	(73.0)	(62.0)	(57.0)	(70.0)
PRECIP cm	12.2	10.7	16.8	13.7	13.7	14.2	18.0	16.3	14.7	9.4	10.2	11.7	161.3
(in.)	(4.8)	(4.2)	(6.6)	(5.4)	(5.4)	(5.6)	(7.1)	(6.4)	(5.8)	(3.7)	(4.0)	(4.6)	(63.5)
POTET cm	2.2	2.6	4.9	8.4	12.7	16.8	18.0	17.1	13.9	8.8	4.0	2.4	111.8
(in.)	(0.9)	(1.0)	(1.9)	(3.3)	(5.0)	(6.6)	(7.1)	(6.7)	(5.5)	(3.5)	(1.6)	(0.9)	(44.0)

What are the main climatic influences for this station (air pressure, air mass sources, degree of continentality, temperature of ocean currents)?

Köppen climate classification symbol: ; name: ; explanation for this determination:

Representative biome (terrestrial ecosystem) characteristic of region: ; characteristic vegetation:

Station #5:

Total annual rainfall:

Average annual temperature:

Annual temperature range:

Distribution of temperature during the year:

Distribution of precipitation during the year:

Principal atmospheric lifting mechanism(s):

Distribution of potential evapotranspiration during the year:

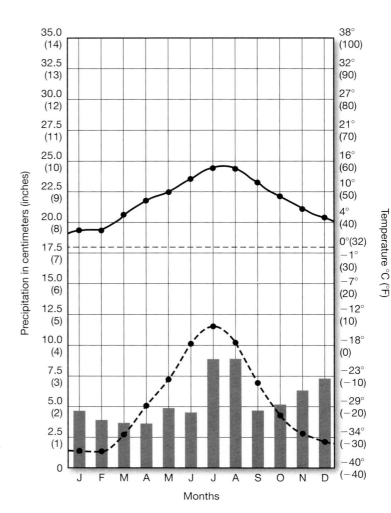

▲ Figure 17.2d

	Jan	Feb	Mar	Apr	May	Jun	Jul	Aug	Sep	Oct	Nov	Dec	Annual
Temperature °C	3.0	3.0	5.0	7.6	10.1	12.7	14.7	14.3	12.5	9.7	6.5	4.8	8.7
(°F)	(37.4)	(37.4)	(41.0)	(45.7)	(50.2)	(54.9)	(58.5)	(57.7)	(54.5)	(49.5)	(43.7)	(40.6)	(47.7)
PRECIP cm	4.8	3.6	3.3	3.3	4.8	4.6	8.9	9.1	4.8	5.1	6.1	7.4	65.8
(in.)	(1.9)	(1.4)	(1.3)	(1.3)	(1.9)	(1.8)	(3.5)	(3.6)	(1.9)	(2.0)	(2.4)	(2.9)	(25.9)
POTET cm	1.3	1.3	2.8	4.8	7.6	10.2	11.7	10.2	7.4	4.8	2.8	1.8	64.8
(in.)	(0.5)	(0.5)	(1.1)	(1.9)	(3.0)	(4.0)	(4.6)	(4.0)	(2.9)	(1.9)	(1.1)	(0.7)	(25.5)

What are the main climatic influences for this station (air pressure, air mass sources, degree of continentality, temperature of ocean currents)?

Köppen climate classification symbol: ; name: ; explanation for this determination:

Representative biome (terrestrial ecosystem) characteristic of region: ; characteristic vegetation:

152   Lab Exercise 17

Station #6:

Total annual rainfall:

Average annual temperature:

Annual temperature range:

Distribution of temperature during the year:

Distribution of precipitation during the year:

Principal atmospheric lifting mechanism(s):

Distribution of potential evapotranspiration during the year:

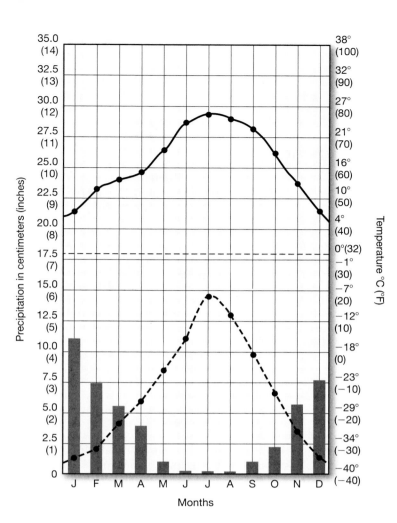

▲ Figure 17.2e

	Jan	Feb	Mar	Apr	May	Jun	Jul	Aug	Sep	Oct	Nov	Dec	Annual
Temperature °C	8.4	11.2	12.9	15.6	19.1	22.3	24.8	24.2	22.7	18.5	12.6	8.6	16.8
(°F)	(47.1)	(52.2)	(55.3)	(60.1)	(66.3)	(72.2)	(76.6)	(75.6)	(72.9)	(65.3)	(54.7)	(47.5)	(62.2)
PRECIP cm	10.7	7.4	5.6	3.6	1.0	0.3	0.3	0.3	0.8	2.3	5.8	7.6	45.5
(in.)	(4.2)	(2.9)	(2.2)	(1.4)	(0.4)	(0.1)	(0.1)	(0.1)	(0.3)	(0.9)	(2.3)	(3.0)	(17.9)
POTET cm	1.5	2.3	4.0	6.0	8.5	11.5	14.0	12.7	9.9	6.6	3.0	1.5	81.5
(in.)	(0.6)	(0.9)	(1.6)	(2.4)	(3.3)	(4.5)	(5.5)	(5.0)	(3.9)	(2.6)	(1.2)	(0.6)	(32.1)

What are the main climatic influences for this station (air pressure, air mass sources, degree of continentality, temperature of ocean currents)?

Köppen climate classification symbol: ; name: ; explanation for this determination:

Representative biome (terrestrial ecosystem) characteristic of region: ; characteristic vegetation:

SECTION 3

Climate Analysis

Temperature, precipitation, and potential evapotranspiration patterns are the elements used in determining climate classification. Factors influencing these patterns help to more fully understand the distribution of climate regions.

Use *Geosystems* to find background information for some of your responses. Answer the following questions about the completed climographs in Figures 17.2a through 17.2e.

New Orleans—Station # _____

1. What effects do latitudinal position, air mass interactions, and ocean currents have in producing the climate of New Orleans? For instance, why is the winter precipitation of New Orleans higher than in more northerly cities?

2. In examining POTET (potential evapotranspiration) values throughout the year for New Orleans and comparing them to PRECIP (precipitation) received, do you think there are moisture surpluses or deficits, given this data?

Edinburgh—Station # _____

3. Where are the several places in North America that you find this same climatic regime?

4. What controlling factors contribute to the cool summers in Edinburgh?

Sacramento—Station # _____

5. What factors produce the natural summer drought (high POTET and low PRECIP) experienced in Sacramento? (Consider air masses, pressure systems, and ocean currents.)

6. How is it possible to have extensive agricultural activity in this region, given the summer climate conditions?

Montreal—Station # _____

7. Compare the annual range of temperature (between January and July) in Montreal with your analysis of New Orleans. Why the greater range value for Montreal?

8. What contributing factors (mechanisms and conditions) produce the even distribution of precipitation throughout the year in Montreal?

Oymyakon—Station # _____

9. Briefly discuss the adaptations that you think are required to live in climates that experience these winter conditions (transportation, housing, outdoor activities).

10. Explain why this station experiences this climate. What are the contributing factors?

SECTION 4

The Desert Climates

The *B* climates are the only ones that Köppen classified according to the *amount and annual distribution of precipitation*. POTET (potential evapotranspiration) exceeds PRECIP (precipitation) in all parts of *B* climates, creating varying amounts of permanent water deficits that distinguish the different subdivisions within this climatic group. Vegetation is typically xerophytic—that is, drought resistant, waxy, hard leafed, and adapted to aridity and low transpiration loss. In general terms, the major subdivisions are the *BW deserts*, where PRECIP is *less* than one-half of POTET, and the *BS steppes*, where PRECIP is *more* than one-half of POTET.

In an effort to better approximate the dry climates, Köppen developed simple formulas, graphically presented in Figure 17.3, to determine the usefulness of rainfall, based on the season in which it falls—whether it falls principally in the summer with a dry winter, or in the winter with a dry summer, or whether the rainfall is evenly distributed throughout the year. Winter rains are the most effective because they fall at a time of lower POTET.

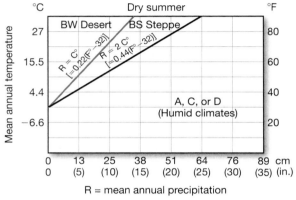

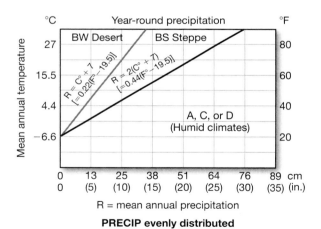

▲ Figure 17.3 *B* climate determination

As in Section 2, complete two climographs (Figures 17.4a and 17.4b). Using the climate classification key presented in Table 17.1 and the three graphs in Figure 17.3, determine the climatic classification for each station and the identity of each station—Reno, Nevada, and Walgett, New South Wales. Complete the analysis and information sheet for each of the stations.

Station #7:

Total annual rainfall:

Average annual temperature:

Annual temperature range:

Distribution of temperature during the year:

Distribution of precipitation during the year:

Principal atmospheric lifting mechanism(s):

Distribution of potential evapotranspiration during the year:

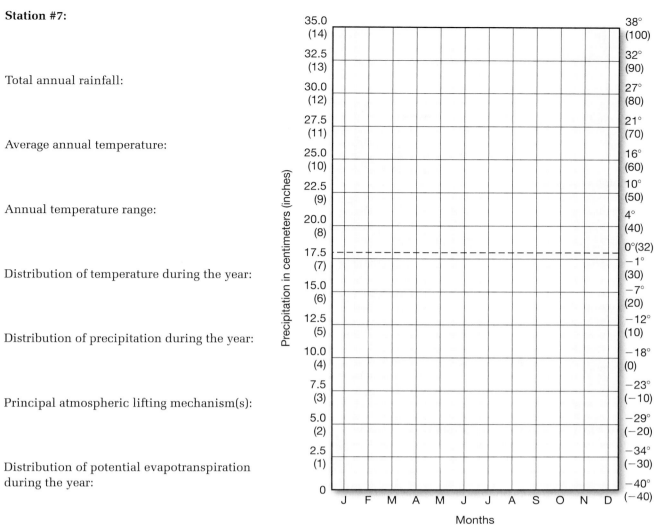

▲ Figure 17.4a

	Jan	Feb	Mar	Apr	May	Jun	Jul	Aug	Sep	Oct	Nov	Dec	Annual
Temperature °C	28.1	27.2	24.4	19.7	15.0	11.7	10.8	12.5	19.2	20.6	24.4	27.2	20.0
(°F)	(82.5)	(81.0)	(76.0)	(67.5)	(59.0)	(53.0)	(51.5)	(54.5)	(66.5)	(69.0)	(76.0)	(81.0)	(68.0)
PRECIP cm	5.3	4.8	4.1	3.0	3.8	4.1	3.3	2.8	2.5	3.0	3.8	4.3	45.0
(in.)	(2.1)	(1.9)	(1.6)	(1.2)	(1.5)	(1.6)	(1.3)	(1.1)	(1.0)	(1.2)	(1.5)	(1.7)	(17.7)
POTET cm	17.8	14.5	11.9	6.6	3.6	1.8	1.8	2.5	6.3	8.4	12.7	17.3	104.6
(in.)	(7.0)	(5.7)	(4.7)	(2.6)	(1.4)	(0.7)	(0.7)	(1.0)	(2.5)	(3.3)	(5.0)	(6.8)	(41.2)

What are the main climatic influences for this station (air pressure, air mass sources, degree of continentality, temperature of ocean currents)?

Köppen climate classification symbol: ; name: explanation for this determination:

Representative biome (terrestrial ecosystem) characteristic of region: ; characteristic vegetation:

Station #8:

Total annual rainfall:

Average annual temperature:

Annual temperature range:

Distribution of temperature during the year:

Distribution of precipitation during the year:

Principal atmospheric lifting mechanism(s):

Distribution of potential evapotranspiration during the year:

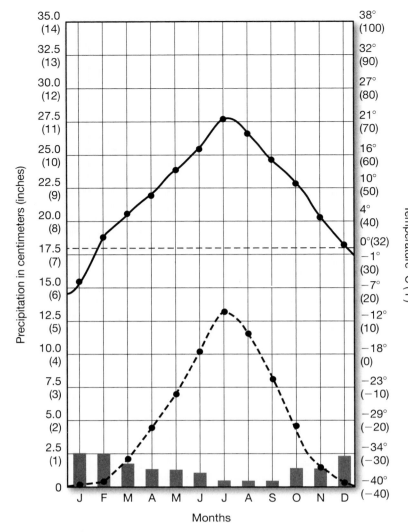

▲ Figure 17.4b

	Jan	Feb	Mar	Apr	May	Jun	Jul	Aug	Sep	Oct	Nov	Dec	Annual
Temperature °C	−0.6	2.2	5.0	8.9	12.8	16.7	21.1	19.4	15.6	10.6	4.4	0.6	10.0
(°F)	(31.0)	(36.0)	(41.0)	(48.0)	(55.0)	(62.0)	(71.7)	(67.0)	(60.0)	(51.0)	(40.0)	(33.0)	(50.0)
PRECIP cm	2.5	2.5	1.8	1.3	1.3	1.0	0.5	0.5	0.5	1.5	1.5	2.3	17.8
(in.)	(1.0)	(1.0)	(0.7)	(0.5)	(0.5)	(0.4)	(0.2)	(0.2)	(0.2)	(0.6)	(0.6)	(0.9)	(7.0)
POTET cm	0	0.7	2.2	4.0	7.1	10.1	13.0	11.7	7.8	4.3	1.7	0.2	62.8
(in.)	(0)	(0.3)	(0.9)	(1.6)	(2.8)	(4.0)	(5.1)	(4.6)	(3.1)	(1.7)	(0.7)	(0.1)	(24.7)

What are the main climatic influences for this station (air pressure, air mass sources, degree of continentality, temperature of ocean currents)?

Köppen climate classification symbol: ; name: ; explanation for this determination:

Representative biome (terrestrial ecosystem) characteristic of region: characteristic vegetation:

1. Characterize water balance conditions in **Reno, Nevada**, during June, July, and August. (Compare precipitation and potential evapotranspiration for these months.)

2. If you were working in agriculture, you might face a shortage of rainfall and need to irrigate plants. Irrigation creates a risk of causing salinization of soils in this semiarid landscape. How would you resolve water-balance deficits and potential soil problems? Explain.

3. What air mass lifting mechanism produces most of the precipitation received in Reno?

4. What air masses produce most of the precipitation received in January and February?

5. What is the annual range of temperature (between January and July) for Reno?

   What factors contribute to this value for Reno?

6. **Walgett, N.S.W.**, within a temperate short-grass plain region, is in the Southern Hemisphere (**Station #** _____). What indications of this global location can you determine from its temperature, PRECIP, and POTET data? Explain.

7. What primary economic activities are suggested by this climate regime?

SECTION 5

Your Climate Region (optional station exercise)

Now that you have examined the various climate patterns throughout the world, you can better understand the climate region in which you are located. Your instructor will give you the necessary climate data for your campus or for a nearby climate station. Use the data to complete the climograph, analysis, and information sheet (Figure 17.5). Note the climate station from Sections 2 and 4 that is most like your climate. Review the questions about the climate to better understand your local conditions.

Station:

Latitude:

Longitude:

Elevation:

Population:

Total annual rainfall:

Average annual temperature:

Annual temperature range:

Distribution of temperature during the year:

Distribution of precipitation during the year:

Principal atmospheric lifting mechanism(s):

Distribution of potential evapotranspiration during the year:

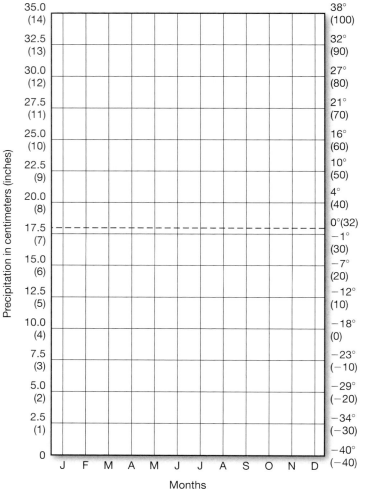

▲ Figure 17.5

Köppen climate classification symbol: name: explanation for this determination:

MasteringGeography™ | Looking for additional review and lab prep materials? Go to www.masteringgeography.com for pre lab videos and pre and post lab quizzes.

160   Lab Exercise 17   Copyright © 2015 Pearson Education, Inc.

Name: _____ Laboratory Section: _____

Date: _____ Score/Grade: _____

Video
Exercise 18
Pre-Lab Video

LAB EXERCISE

Scan to view the Pre-Lab video

http://goo.gl/8zstF6

18 Climate Change

Climate change, a significant and lasting change in weather patterns, ranging from decades to millions of years, is one of the most critical issues facing humankind in the twenty-first century. It is also an integral part of physical geography and Earth systems science. Three key elements of climate change science are the study of past climates, the measurement of current climatic changes, and the forecasting and modeling of future climate scenarios—all of which we will examine in this lab.

The physical evidence for global climate change is extensive and observable—record-breaking global average temperatures for air, land surfaces, lakes, and oceans; ice losses from mountain glaciers and the Greenland and Antarctic ice sheets; declining soil-moisture conditions and resultant effects on crop yields; changing distributions of plants and animals; and the pervasive impact of global sea-level rise, which threatens coastal populations and development worldwide. These are only a few of the many complex and far-reaching issues that climate change science must address to understand and mitigate the environmental changes ahead.

Although climate has fluctuated naturally over Earth's long history, scientists agree that modern climate change is resulting from human activities that produce greenhouse gases. The consensus is overwhelming and includes every professional scientific society, association, and council in the United States and throughout the rest of the world. The observed changes in global climate are happening at a pace much faster than that seen in the historical climate records or in the climate reconstructions that now extend millions of years into the past.

In this lab exercise, we explore the record of past climates, examine the present changes, and look at what changes the future might bring. We will analyze data from **ice cores**, analyze carbon dioxide levels and temperatures, and classify possible future climates. Lab Exercise 18 features three sections.

Key Terms and Concepts

climate change
ice core
paleoclimatology

proxy method
isotope analysis

KEY LEARNING concepts

After completion of this lab, you should be able to:

1. *Identify* patterns of CO_2 and temperature in ice core data and *evaluate* whether past climate changes have been global or hemispheric.
2. *Analyze* the relationship between CO_2 and temperature from the beginning of the Industrial Revolution to the present.
3. *Classify* future climate data using the Köppen climate classification system.

Materials/Sources Needed

color pencils
calculator

spreadsheet program (optional)
Internet access (optional)

Lab Exercises and Activities

SECTION 1

Paleoclimate

To understand present climate changes, we need to understand the climates of the past, especially climates that occurred before record-keeping with standardized instruments began almost 140 years ago. Clues to past climates are stored in a variety of environments on Earth, including gas bubbles in glacial ice, fossil marine life in ocean-bottom sediments, fossil pollen from ancient plants, growth rings in trees, and corals. Scientists extract and analyze cores to establish a chronology of environmental conditions over time periods of thousands or millions of years.

The study of Earth's past climates is the science of **paleoclimatology**, which tells us that Earth's climate has fluctuated over millions of years. To learn about past climates, scientists use **proxy methods** instead of direct measurements. A *climate proxy* is a piece of information from the natural environment that can be used to understand processes that extend back further than our present instrumentation allows. For example, a drier year may reduce the rate of tree growth, resulting in a thin tree ring. The widths of tree rings can tell us about climatic conditions that occurred thousands of years before weather record keeping began. By analyzing evidence from proxy sources, scientists are able to reconstruct climate in ways not possible using the record of firsthand scientific measurements over the past 140 years or so. Some paleoclimatology techniques yield long-term records that span hundreds of thousands to millions of years. Such records come from cores drilled into ocean-bottom sediments or into the thickest ice sheets on Earth. Once cores are extracted, layers containing fossils, air bubbles, particulates, and other materials provide information about past climates.

One basis for long-term climate reconstruction is **isotope analysis**, a technique that uses the atomic structure of chemical elements, specifically the relative amounts of their isotopes, to reveal the chemical composition and temperature of past oceans and ice masses. Oxygen-16, or "lighter" ^{16}O, is the most common isotope found in nature. Oxygen-18, or "heavier" ^{18}O, comprises only about 0.20% of total oxygen atoms. During periods of colder temperatures, more "light" oxygen is locked up in snow and ice in the polar regions, and "heavy" oxygen concentrations are highest in the oceans.

In the cold regions of the world, snow accumulates seasonally in layers, and in regions where snow is permanent on the landscape, these layers of snow eventually form glacial ice. The world's largest accumulations of glacial ice occur in Greenland and Antarctica, and it is from these ice sheets that scientists have extracted ice cores to reconstruct climate. Such ice cores, drilled thousands of meters deep into the thickest part of an ice sheet, provide a shorter but more detailed record of climate than ocean sediment cores, presently pushing the record back 800,000 years.

Within a core, any given year's accumulation consists of a layer of winter ice and a layer of summer ice, each differing in chemistry and texture. Scientists use oxygen isotope ratios to correlate these layers with environmental temperature conditions. Relatively lower ratios of $^{18}O/^{16}O$ indicate warmer conditions, and higher $^{18}O/^{16}O$ ratios indicate colder temperatures. Therefore, the oxygen isotopes in ice cores are a proxy for air temperature above the snow and ice surfaces. Ice cores also reveal information about past atmospheric composition. Within the ice layers, trapped air bubbles reveal concentrations of gases—mainly carbon dioxide and methane—indicative of past environmental conditions at the time the bubbles were sealed into the ice. By analyzing ice cores, we have reconstructed the temperatures and compositions of gases of the past 800,000 years.

Ice core projects in Greenland have produced data spanning more than 120,000 years. In Antarctica, the Dome C ice core (part of the European Project for Ice Coring in Antarctica [EPICA]), completed in 2004, reached a depth of 3270 m (10,729 ft) and produced the longest ice-core record yet: 800,000 years of Earth's past climate history. This record was correlated with a core record of 400,000 years from the nearby Vostok Station and matched with ocean sediment core records to provide scientists with a sound reconstruction of climate changes throughout this time period. In 2011, American scientists extracted an ice core from the West Antarctic Ice Sheet (WAIS) that when analyzed will reveal 30,000 years of annual climate history and 68,000 years at resolutions from annual to decadal—a higher time resolution than previous coring projects.

Analyses have shown that the greenhouse gas concentrations lag behind the temperature changes, generally by about 1000 years. This interesting relationship suggests the presence and importance of climate feedbacks.

Examine Figure 18.1. This EPICA ice core is 3270 m (10,729 ft) long. So far, scientists have analyzed the top 3189.45 m (10,462 ft), revealing the past 801,588 years of climate. The figure shows CO_2 in parts per million (ppm) and temperature anomalies in Celsius degrees relative to the average for the past 1000 years.

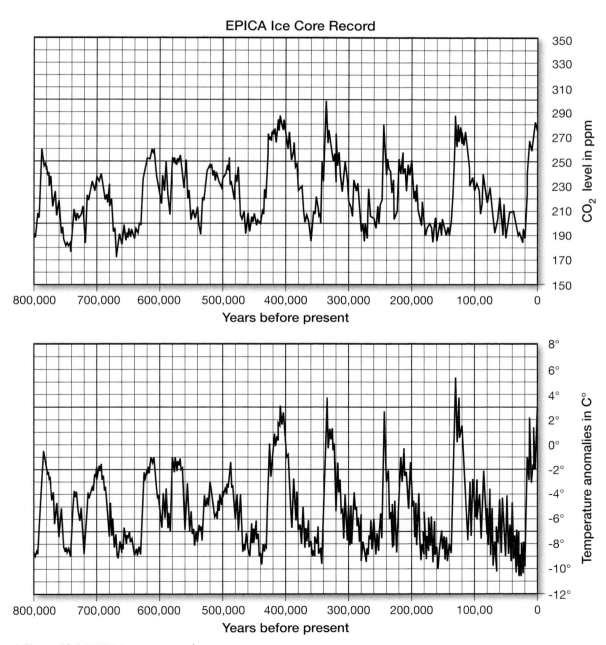

▲ Figure 18.1 EPICA ice core record

1. How many years of data are displayed from the EPICA ice core?

2. How many times has the temperature anomaly exceeded 0 C° in the past 450,000 years?

3. What are the maximum and minimum temperature anomalies recorded? What is the range of temperature anomalies over this record?

4. How many interglacials (times where the temperature deviation reached 0C° or greater) do you detect over the past 450,000 years? What is the average time between interglacials?

5. How much has CO_2 varied in the EPICA record? What were the highest and lowest values? What was the range of CO_2 values?

6. How many times has CO_2 exceeded 270 ppm in the past 450,000 years?

7. What is the period from peak to peak with regard to CO_2 and for temperature for the past 450,000 years?

8. Interglacial–glacial period climate cycles are often referred to as appearing saw-toothed, with an abrupt change followed by a more gradual change. When do we see an abrupt change: when going from interglacial to glacial times or when going from glacial to interglacial times? When do we see a gradual change?

Examine Figure 18.2. The Vostok ice core is 3310 m (10,860 ft) long. The full record extends back to 422,000 years before the present. The figure shows CO_2 in parts per million (ppm) and temperature anomalies in Celsius degrees.

1. How many years of data are displayed from the Vostok ice core?

2. How many times has the temperature anomaly exceeded 0C° in the Vostok record?

3. What are the maximum and minimum temperature anomalies recorded? What is the range of temperature anomalies in the Vostok record?

4. How many interglacials (times where the temperature deviation reached 0C° or greater) do you detect over the past 450,000 years? What is the average time between interglacials?

5. How much has CO_2 varied in the Vostok record? What were the highest and lowest values? What was the range of CO_2 values?

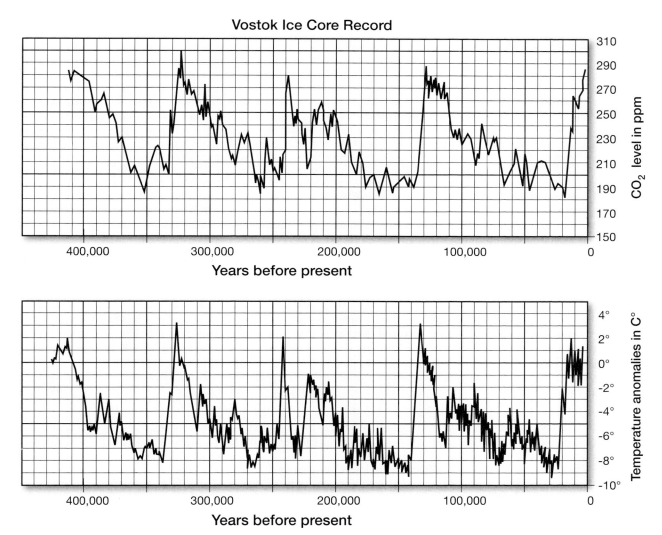

▲ Figure 18.2 Vostok ice core record

6. What is the period from peak to peak with regard to CO_2 and for temperature for the Vostok record? How does this compare with the period (time from peak to peak) of the EPICA record?

7. Interglacial–glacial period climate cycles are often referred to as appearing saw-toothed, with an abrupt change followed by a more gradual change. When do we see an abrupt change: when going from interglacial to glacial times or when going from glacial to interglacial times? When do we see a gradual change?

8. Why is it important to compare records from different locations? What other locations would you examine to study climate?

SECTION 2

Modern Climate

Carbon dioxide is a natural by-product of life processes, and it is an important component of the atmosphere. Although it accounts for only a small percentage of the atmosphere, about 0.039%, it plays a vital role in maintaining habitable global temperatures. Without CO_2 in the atmosphere, the temperature of Earth would be 15 to 33 C° cooler than it is, with Earth's average temperature around the freezing point of water. As humans have increased carbon dioxide levels in the atmosphere, global temperatures have also increased. The Dome C ice core shows that present levels of carbon dioxide, methane, and nitrous oxides are higher in the present decade than in the rest of the 800,000-year record! These are remarkable times in which we live. 2010 has tied 2005 as the warmest year ever recorded, 2000–2009 is the warmest decade ever recorded, and the years from 1998 to 2013 are the 16 warmest years ever recorded, beating records that stretch back to 1850. The Intergovernmental Panel on Climate Change (IPCC), through five major reports, including the 2014 Fifth Assessment Report (AR5), confirms that global warming is occurring. The IPCC AR5 reached a consensus, in part stating:

AR4 concluded that warming of the climate system is unequivocal. New observations, longer data sets, and more paleoclimate information give further support for this conclusion. Confidence is stronger that many changes, that are observed consistently across components of the climate system, are significant, unusual or unprecedented on time scales of decades to many hundreds of thousands of years.*

The CO_2 data for this exercise comes from the Mauna Loa observatory on the Big Island of Hawai'i and is the longest continuous direct measurement of CO_2. Dr. Charles Keeling began the project in 1958. Pay close attention to the scales of the two data series since they cover different time periods. The CO_2 units are parts per million (ppm), and the temperature values are in degrees Celsius.

Figure 18.3 shows the global annual mean temperature in degrees Celsius from 1880 to 2013, and Figure 18.4 shows annual CO_2 levels from 1958 to April 2014 as measured at the Mauna Loa observatory. Figure 18.5 shows monthly CO_2 values from 2010 to April 2014.

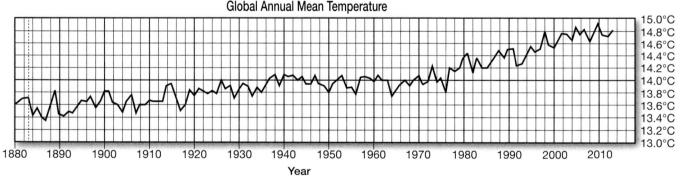

▲Figure 18.3 Global annual mean temperature

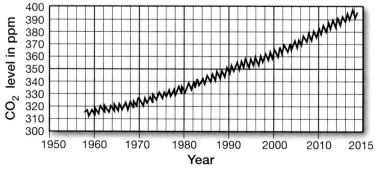

▲Figure 18.4 CO_2 1958–2014

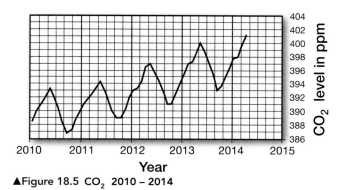

▲Figure 18.5 CO_2 2010 – 2014

1. What are the general characteristics of the CO_2 data set? What are the maximum and minimum CO_2 concentrations? When did the minimum value occur? When did the maximum value occur? What is the general trend in CO_2 values?

2. Compare the CO_2 values from 2003–2013 in Figure 18.4 with the CO_2 values on the Vostok plot in Figure 18.2. Are the values for CO_2 levels on the Vostok plot as high as they are now? If so, during which time periods were CO_2 levels as high as they are now? If not, what are the highest values for CO_2 on the Vostok plot?

3. The CO_2 values show two patterns of change: a longer pattern and a shorter pattern. How long is each oscillation? Table 18.1 shows monthly CO_2 values for 2010–2013.

4. Which months have the highest and lowest values on Figure 18.5?

5. What causes the short-term oscillation of CO_2?

6. What are the annual averages for each of the past 3 years?

7. What is the range from each of the past 3 years to the next year? Compare the range within years to the range between years. For example, compare the changes in CO_2 levels in 2010 to the change from 2010 to 2011.

8. Calculate the average annual increase as a percentage of total CO_2 and in ppm of CO_2 for 1970 to 1980 and for 2003 to 2013.

9. What is the present level of CO_2? What would it be if it were 50% greater?

10. Based on your calculation for the rate of increase from 2003 to 2013, how long do you estimate it will take for the CO_2 level to be 50% higher than the present level?

11. The Kyoto Protocol calls for an overall reduction of greenhouse gases of 5.3% below the 1990 levels. What was the average level of CO_2 in 1990? What would the level of CO_2 be if the Kyoto goal was met? How many ppm below current levels would that be? The "Group of 77" countries favor a 15% reduction below 1990 levels. What would that level be, and how far below present levels is it?

12. Dr. James Hansen of NASA has stated that we should set a goal of keeping CO_2 levels below 350 ppm. Which year did we first exceed that level? If we decreased CO_2 levels at the same rate that we have increased them over the past 10 years, how long would it take until CO_2 levels were again below 350 ppm?

Date of Eruption	Average Temp Year Before Eruption	Average Temp Year of Eruption	Average Temp 1 Year After Eruption	Average Temp 2 Years After Eruption	Average Temp 3 Years After Eruption
1883 Krakatau					
1902 Soufriere/ Pelee/ Santa Maria					
1912 Katmai					
1963 Agung					
1968 Fernandina Island					
1982 El Chichon					
1991 Pinatubo					

▲ Table 18.1 Major volcanic eruptions since 1867.

In order to examine the effects of volcanic eruptions on climate, we will look at the relationship between temperature and several major eruptions. The dashed line on Figure 18.3 at 1883 represents the eruption of Krakatau. Using colored pencils, draw vertical lines on Figure 18.3 at 1902, 1912, 1963, 1968, 1982, and 1991 to represent each of the above volcanic eruptions.

13. Complete Table 18.1 for any *three* of the above eruptions. How much of an effect do these eruptions have on global temperatures? How much do they change global temperatures? How many years does it take for the climate to return to a normal temperature after a major eruption?

SECTION 3

Future Climate

Climate regions are not static or fixed but are dynamic and respond to changes in the environment. As our climate changes, crop patterns, as well as natural habitats of plants and animals, shift to maintain preferred temperatures. According to climate models for the midlatitudes, climate regions could shift poleward by 150 to 550 km (90 to 350 mi) during this century. For example, climate conditions in Texas are projected to "migrate" northeast, as shown in Figure 18.6, so that the future climate in Illinois will resemble the current climate in east Texas.

Use the climograph (Figure 18.7) and the data that follows to plot current mean monthly temperature in blue (*solid-line graph*), projected mean monthly temperature in red (*solid-line graph*), current mean monthly precipitation in blue (*a bar graph*), projected mean monthly precipitation in red (*a bar graph*), current potential evapotranspiration in blue (*a dotted line*), and projected potential evapotranspiration in red (*a dotted line*) for the following station. You may have to color in just the outlines of the precipitation bars so you will be able to see both sets of data on the same climograph. Use Table 17.1 in Lab Exercise 17 to determine the station's current and projected climatic classifications. In addition, use the map of Earth's terrestrial biomes on the inside back cover of this manual to establish the current and projected biomes.

(a) Summer   (b) Winter

▲Figure 18.6 Migrating summer and winter Illinois climates from 2005 to 2090. These climatic projections are based on moderate CO_2 emission scenarios, which are much lower than current CO_2 emissions. After "Climate Change Projections for the United States Midwest" by D. Wuebbles and K. Hayhoe, *Mitigation Strategies for Global Change* 9: 335–363, 2004. Kluwer Academic Publishers.

1. **Chicago, IL**
   Latitude 41.81°N
   Longitude 87.68°W
   Elevation 182 m
   Population 2,714,856
   Current and projected total annual rainfall:

   Current and projected average annual temperature:

   Current and projected annual temperature range:

   Current and projected distribution of temperature during the year:

   Current and projected distribution of precipitation during the year:

   Current and projected distribution of potential evapotranspiration during the year:

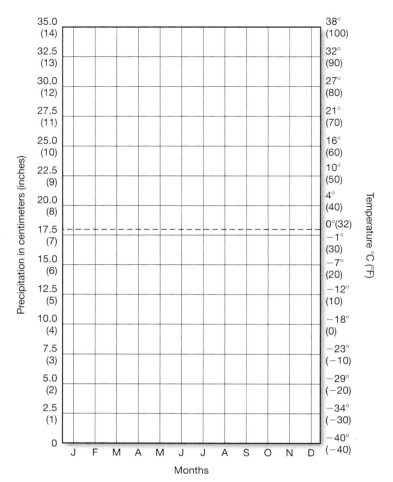

▲Figure 18.7

Current	Jan	Feb	Mar	Apr	May	Jun	Jul	Aug	Sep	Oct	Nov	Dec	Annual
Temperature °C	−6.0	−2.1	2.8	9.4	16.0	21.2	24.1	23.7	19.4	13.1	5.5	−2.3	10.4
(°F)	(21.2)	(28.2)	(37.0)	(48.9)	(60.8)	(70.2)	(75.4)	(74.7)	(66.9)	(55.6)	(41.9)	(27.9)	(50.7)
PRECIP cm	4.7	4.3	5.9	10.1	8.5	10.4	8.5	7.3	9.5	7.6	7.2	6.2	90.2
(in.)	(1.9)	(1.7)	(2.3)	(4.0)	(3.3)	(4.1)	(3.3)	(2.9)	(3.7)	(3.0)	(2.8)	(2.4)	(35.5)
POTET cm	0.0	0.0	0.9	3.9	8.4	12.8	14.7	13.3	9.0	5.0	0.8	0.0	68.8
(in.)	(0.0)	(0.0)	(0.4)	(1.5)	(3.3)	(5.0)	(5.8)	(5.2)	(3.5)	(2.0)	(0.3)	(0.0)	(27.1)

Projected	Jan	Feb	Mar	Apr	May	Jun	Jul	Aug	Sep	Oct	Nov	Dec	Annual
Temperature °C	2.9	6.9	13.6	19.1	22.6	31.0	34.1	33.1	28.2	20.0	11.8	4.6	19.0
(°F)	(37.1)	(44.3)	(56.4)	(66.4)	(72.7)	(87.8)	(93.4)	(91.6)	(82.8)	(68.0)	(53.2)	(40.3)	(66.2)
PRECIP cm	6.6	5.4	9.2	14.8	11.6	6.0	7.4	7.7	9.8	8.8	11.5	7.8	106.6
(in.)	(2.6)	(2.1)	(3.6)	(5.8)	(4.6)	(2.4)	(2.9)	(3.0)	(3.9)	(3.5)	(4.5)	(3.1)	(42.0)
POTET cm	0.0	0.5	2.9	7.1	11.5	24.3	29.5	25.6	15.9	6.5	1.7	0.2	125.7
(in.)	(0.0)	(0.2)	(1.1)	(2.8)	(4.5)	(9.6)	(11.6)	(10.1)	(6.3)	(2.6)	(0.7)	(0.1)	(49.5)

2. Current Köppen climate classification symbol: :

   Name:

   Explanation for this determination:

   Projected Köppen climate classification symbol: :

   Name:

   Explanation for this determination:

3. Compare the position of Chicago, IL, in Figure 18.5(a) to the world climate map on the back cover of this lab manual. What climate classification border is closest to Chicago, IL, under current and projected conditions? Describe its future relations to conditions on this map. Present location:

   Projected location:

4. Using the world biome map inside the back cover of this lab manual or the one in your textbook, determine:

   Current representative biome (terrestrial ecosystem) characteristic of the region: ;

   Characteristic vegetation:

   Projected representative biome (terrestrial ecosystem) characteristic of the region: ;

   Characteristic vegetation:

5. Evaluate Chicago's current annual March–May, June–August, and September–November water balances. What is the annual surplus or deficit? What are Chicago's projected annual March–May, June–August, and September–November water balances? What is the annual surplus or deficit? Which months have the greatest projected surplus? Which months have the greatest projected deficits?

6. Under moderate climate change scenarios, summer heat index temperatures in the twenty-first century are expected to rise in the American Midwest in a range between 2.7 C° and 13.8 C° (5 F° and 25 F°). A heat wave, similar to the 2003 heat wave that resulted in more than 70,000 deaths across Europe, could result in a tenfold increase in heat-related deaths in Chicago. How do you think this will affect air conditioning demand and related costs? What are your concerns, and what actions would you take if you were a Chicago public health official?

7. Go to www.globalchange.gov and find the National Climatic Assessment report for your region. Briefly summarize the impacts for your region on temperature, precipitation, heat waves, agricultural productivity, and any other significant changes. How are patterns of rainfall projected to shift? For example, will they stay the same, or will there be longer periods of no rain followed by heavy rain and flooding?

MasteringGeography™ | Looking for additional review and lab prep materials? Go to www.masteringgeography.com for pre lab videos and pre and post lab quizzes.

Name: _____ Laboratory Section: _____

Date: _____ Score/Grade: _____

Video
Exercise 19
Pre-Lab Video

LAB EXERCISE

Scan to view the Pre-Lab video

http://goo.gl/Xae7TW

19 Plate Tectonics: Global Patterns

The **plate tectonics** theory was a revolution in twentieth-century Earth science. The past few decades have seen profound breakthroughs in our understanding of how the continents and oceans evolved, why earthquakes and volcanoes occur where they do, and the reasons for the present arrangement and movement of landmasses. One task of physical geography is to explain the *spatial* implications of all this new knowledge and its effect on Earth's surface and society.

As Earth solidified after formation, heavier elements slowly gravitated toward the center, and lighter elements slowly welled upward to the surface, concentrating in the crust. Earth's interior is highly structured, with uneven heating generated by the radioactive decay of unstable elements. The results of this heating and instability are irregular patterns of moving, warping, and breaking of the crust.

In this lab exercise, we work with plate tectonics and the evolution of the present configuration of continents and ocean basins. A reality of plate tectonics is that pieces of Earth's crust are on the move, actively migrating and interacting. Plate interactions are characterized and portrayed in this lab. The motion of the Pacific plate and the formation of the Hawaiian Islands are analyzed. The ages of oceanic crust and the rates of movement of tectonic plates are calculated. Lab Exercise 19 features four sections and two optional Google Earth™ activities.

Key Terms and Concepts

continental drift
convergent boundary
divergent boundary
hot spots
mid-ocean ridges
orogenesis
Pangaea

plate tectonics
plumes
sea-floor spreading
subduction zone
transform faults
transform boundary

KEY LEARNING concepts

After completion of this lab, you should be able to:

1. *Analyze* the arrangement of Earth's crustal plates in their former positions.
2. *Contrast* types of plate boundaries and *locate* examples of each type as they are at present.
3. *Analyze* the movement of the Pacific plate and the formation of the Hawaiian-Emperor seamount chain.

Materials/Sources Needed

colored pencils
length of string
scissors

transparent tape
stereolenses

Lab Exercise and Activities

SECTION 1

Plate Tectonics

Have you ever looked at a world map and noticed that a few of the continental landmasses appear to match shape, like pieces of a jigsaw puzzle—particularly South America and Africa? The incredible reality is that the continental pieces indeed once were fitted together! Continental landmasses not only migrated to their present locations, but they continue to move at speeds up to 6 cm (2.4 in.) per year.

In 1912, German geophysicist and meteorologist Alfred Wegener publicly presented in a lecture his idea that Earth's landmasses migrate. His book *Origin of the Continents and Oceans* appeared in 1915. Wegener today is regarded as the father of this concept, called **continental drift**. Wegener postulated that all landmasses were united in one supercontinent approximately 225 million years ago, during the Triassic period. This one landmass he called **Pangaea**, meaning "all Earth," shown in Figure 19.1.

While his hypothesis would eventually be proven correct, he was unable to provide a mechanism that would explain how the continents migrated.

Wegener's proposal launched a half-century debate leading up to the adoption of plate tectonics as an all-encompassing theory of continental drift, **sea-floor spreading** along submerged **mid-oceanic ridges**, and **subduction zones**. Beginning in the 1950s and continuing through the 1970s, new evidence resulted in the acceptance of Wegner's hypothesis. The main evidence included finding magnetic reversals along spreading centers in the Atlantic, dating the ages of oceanic crust, and conducting seismic studies revealing the locations of plate boundaries. Aided by an avalanche of discoveries, the theory today is universally accepted as an accurate model of the way Earth's surface evolves.

Animation
Geography thru Geologic Time

http://goo.gl/cmX7c8

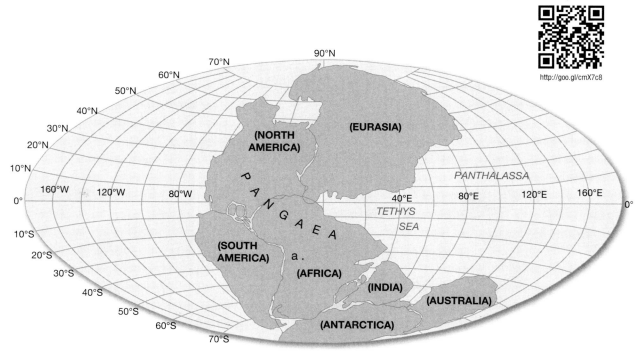

▲ Figure 19.1 Pangaea 200 to 225 million years ago

The Breakup of Pangaea

Figure 19.1 gives you an opportunity to work with an updated version of Wegener's Pangaea, 225 to 200 million years ago (Triassic–Jurassic periods). Areas of North America, North Africa, the Middle East, and Eurasia were near the equator and therefore were covered by plentiful vegetation. The landmasses focused near the South Pole and attached to the sides of Antarctica were covered by ice. We also see that Africa shared a common connection with both North and South America. The Appalachian Mountains in the eastern United States and the Atlas Mountains of northwestern Africa reflect this common ancestry; they are, in fact, portions of the same mountain range. The Atlantic Ocean did not exist. The only oceans were *Panthalassa* ("all seas"), which would become the Pacific, and the *Tethys Sea* (partly enclosed by the African and the Eurasian plates), which formed the Mediterranean Sea, with trapped portions becoming the present-day Caspian Sea.

Activity and questions about the development of the plate tectonic theory

Figure 19.1 shows Pangaea as it was 200 to 225 million years ago. Alfred Wegener used paleoclimatic evidence to support his idea of plate tectonics. For example, there are tropical regions that have been glaciated, while mid-latitude areas were once tropical swamps. For this section, you will identify locations that had very different climates in the past. For this assignment, we will assume that we are looking at mid-Permian conditions, when global temperatures were similar to current temperatures. You will have to estimate climate and biome locations using latitude and continentality to compare them to present climatic locations. For example, northern India was roughly 35°S, and on the east coast of a large landmass. A location with a similar modern climate would be coastal China today, with a humid subtropical climate.

1. For b–d below, write the letter on Figure 19.1 at the appropriate location. The first location, (a), has been marked as an example for you. You will need to use Figure 19.1 and the maps of world climates and biomes (found on the back cover and inside the back cover of this manual) to find and mark:

    a) a location that was a desert (under the subtropical high) but is now a rain forest (tropical) Done for you.

    b) a location that would have been glaciated (arctic or Antarctic) but is now tropical

    c) a location that was tropical but is now in the midlatitudes

    d) a location that was a rain forest (tropical) but is now a desert (under the subtropical high)

2. What physical clues and evidence prove that these landmasses were together 225 million years ago as Pangaea, and why do scientists think the continents are drifting (continent and sea-floor movements)?

3. If you were a petroleum geologist searching for oil deposits (derived from ancient biomass), why would you want to know the positioning of these landmasses hundreds of millions of years ago? What climate region would you look for?

4. In your opinion, what factors produced the long delay between the time of Wegener's lectures and book and the time of acceptance of his theory?

5. Why did Wegener coin the name "Pangaea"? Explain its meaning.

SECTION 2

Principal Motions of Plates and Plate Boundaries

To visualize the mechanisms that drive Earth's lithospheric plates, refer to Figure 19.2. Demonstrated in cross-section are sea-floor spreading, upwelling currents, subduction, and plate movements.

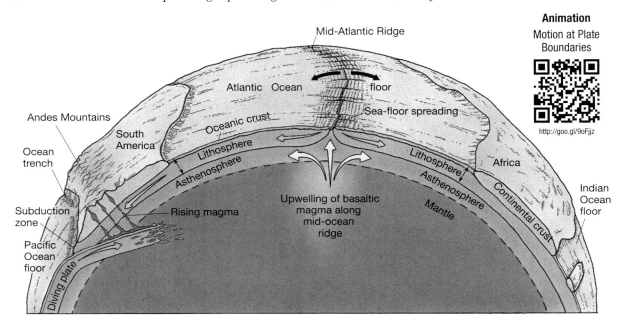

▲Figure 19.2 **The dynamics of crustal movements.**
After Peter J. Wyllie, *The Way the Earth Works.* Copyright © 1976 by John Wiley & Sons, Inc. Adapted by permission of John Wiley & Sons.

1. At the time of Pangaea, South America and Africa were attached, forming a continuous landmass. Using Figure 19.2, describe what transpired in the past 225 million years to position the two continents where they are today.

2. With respect to the movement of crustal plates in the vicinity of the Andes Mountains: Why is the oceanic plate diving beneath South America? How is this related to the **orogenesis** (or formation of mountains) of the Andes? Explain.

Plates and Plate Boundaries

Figure 19.3 shows tectonic features in modern times (the late Cenozoic Era). A spreading center extends down the center of the Atlantic and into the southern Indian Ocean. Subduction trenches are active along the west coasts of Central and South America and throughout the western Pacific Ocean basin. The northern reaches of the Indian plate have underthrust the southern mass of Asia through subduction, forming the Himalayas in the upheaval created by the collision.

Of all the major plates, the Indian traveled the farthest. From approximately 150 million years ago until the present, according to Robert Dietz and John Holden, more than half of the ocean floor was renewed. The three inset blocks demonstrate specific types of plate boundaries: **convergent boundaries**, where two plates are coming together, often leading to the subduction of one plate; **divergent boundaries**, where a plate is spreading apart, such as at sea-floor spreading, mid-ocean ridges); and **transform boundaries**, where two plates are moving past each other horizontally, for example, the San Andreas Fault. Transform boundaries occur between two plates, while **transform faults** occur within a plate, for example, the lateral motion between crests of spreading ridges produces transform faults.

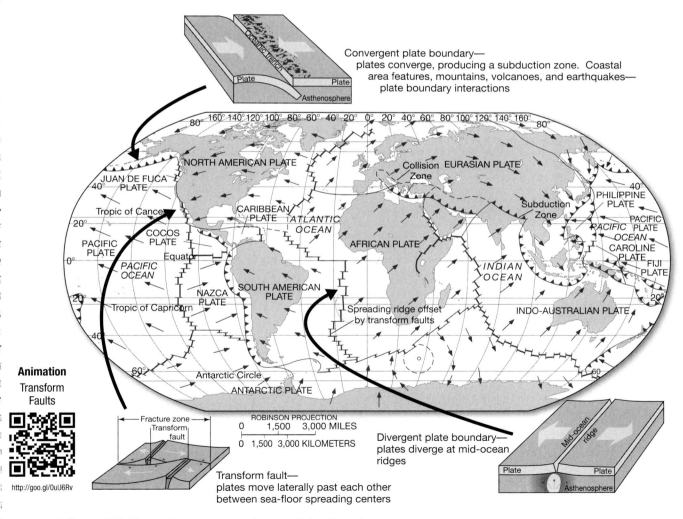

▲Figure 19.3 **The continents today and types of plate boundary interactions.**
Each arrow represents 20 million years of plate movement. The longer arrows indicate that the Pacific and Nazca plates are moving more rapidly than the Atlantic plates.

Using a physical geography textbook, such as *Geosystems*, locate the plate names, examine the different rates of plate movement, and identify types of plate boundary interactions. Using this same source, complete the following questions.

1. Using colored pencils or colored pens, trace the plate boundaries on Figure 19.3 and color them as follows: convergent boundary–yellow; divergent–red; transform–blue.

2. Of the three types of plate boundary interactions shown in Figure 19.3, characterize the following plate boundaries:

    a) On the ocean floor south of Alaska: _____

    b) Indian Ocean between the Antarctic and Indo-Australian plates: _____

    c) Beneath the Red Sea: _____

    d) Along the west coast of Central America: _____

    e) Through Iceland: _____

    f) Along the northeast coast of the Persian Gulf: _____

    g) In California between the Juan de Fuca plate and the North American Plate: _____

3. Which four plates converge on Japan?

4. Mark your location on the map in Figure 19.3. On which plate do you live? (If you live west of the San Andreas fault system in California, take care with your answer.)

5. Google Earth™ activity, San Andreas Fault, CA. For the kmz file and questions, go to mygeoscienceplace.com. Then click on the cover of *Applied Physical Geography: Geosystems in the Laboratory*.

SECTION 3

Hot Spots and the Hawaiian-Emperor Island Chain

Animation
Hot Spot
Volcano Tracks

http://goo.gl/FxVc6D

A dramatic aspect of plate tectonics is the estimated 50 to 100 **hot spots** across Earth's surface. These are individual sites of **plumes** of upwelling material from the mantle. Hot spots occur beneath both oceanic and continental crust and appear to be deeply anchored in the mantle, tending to remain fixed relative to migrating plates.

An example of an isolated hot spot is the one that has formed the Hawaiian-Emperor island chain. The Pacific plate has moved across this hot, upward-erupting plume for over 80 million years, with the resulting string of volcanic islands moving northwestward away from the hot spot. Thus, the age of each island or seamount in the chain increases northwestward from the island of Hawai'i.

The Big Island of Hawai'i actually took less than 1 million years to build to its present stature. Hawai'i Volcanoes National Park is on the island of Hawai'i, where volcanic activity is presently nearly continuous. The youngest island-to-be in the chain is still a *seamount*, a submarine mountain that does not reach the surface. It rises 3350 m (11,000 ft) from the ocean floor but is still 975 m (3200 ft) beneath the ocean surface. Even though this new island will not see the tropical Sun for many years, it is already named Lō'ihi.

The map in Figure 19.4 shows the ages of each portion or island in the chain called the Hawaiian-Emperor seamount chain.

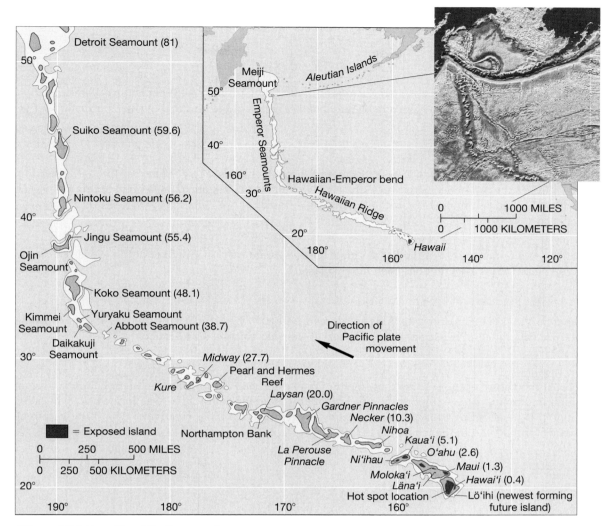

▲Figure 19.4 Hawai'i and the Emperor Seamounts.
Ages are shown in millions of years. After D. A. Clague, "Petrology and K-Ar Ages of Dredged Volcanic Rocks from the Western Hawaiian Ridge and the Southern Emperor Seamount Chain," *Geological Society of America Bulletin* 86 (1975): 991.

Using the map in Figure 19.4, complete the following questions about the Hawaiian-Emperor island chain. Questions 3 and 4 can be answered using the Google Earth™ activity Hawaii and the Emperor Seamounts. For the kmz file and questions go to mygeoscience-place.com. Then click on the cover of *Applied Physical Geography: Geosystems in the Laboratory*. Use either a piece of string and the map scale or Google Earth™ to answer the following questions.

1. What is the distance from Midway Island to Lō'ihi, in kilometers? What is the difference between their ages? What has been the rate of movement of the Pacific plate, in centimeters per year, over this time frame?

   [The distance is approximately 2400 km. Midway is 27.7 MY old. Since 1 km equals 100,000 cm, 2400 km is equal to 240,000,000 cm. Since both the distance and age are in the millions, the math to find the rate of movement is 240 (distance in millions of centimeters)/27.7 (age in millions of years) = 8.6 cm/yr. The conversion from km to centimeters will come in handy in the next section as well.]

2. What is the distance from the Detroit Seamount to the Koko Seamount, in kilometers? What is the difference between their ages? What has been the rate of movement of this region of the Pacific plate, in centimeters per year, over this time frame?

3. What is the distance from the Abbott Seamount to Lōʻihi, in kilometers? What is the difference between their ages? What has been the rate of movement of this region of the Pacific plate, in centimeters per year, over this time frame?

4. What is the distance from the Detroit Seamount through the Daikakuji Seamount to Lōʻihi, in kilometers? What is the difference between their ages? What has been the rate of movement of this region of the Pacific plate, in centimeters per year, over this time frame?

5. Has the rate of movement of the Pacific plate been constant over the lifetime of the Hawaiian-Emperor island chain? Which section has been moving faster?

6. The oldest rock that makes up Lōʻihi is an estimated 400,000 years old. Using the rate of plate movement that you calculated in question #3, how far has it traveled since its first lava cooled?

7. What is the nature of the relative movement of plates and hot spots? Briefly explain which is fixed and which is moving. Also explain how the Hawaiian-Emperor island chain was formed. Part of being a scientist is questioning everything and testing everything. Thinking creatively, what are two possible reasons for the bend in the island chain? Briefly explain how you could test these two ideas.

SECTION 4

Speeds of Plate Movement

The key to establishing the theory of continental drift was a better understanding of the sea floor. The sea floor has a remarkable feature: an interconnected worldwide mountain chain, forming a ridge some 64,000 km (40,000 mi) in extent and averaging more than 1000 km (600 mi) in width. In the early 1960s, geophysicist Harry H. Hess proposed sea-floor spreading as the mechanism that builds this mountain chain and drives continental movement. Hess said that these submarine mountain ranges were mid-ocean ridges and the direct result of upwelling flows of magma from hot areas in the upper mantle and asthenosphere and perhaps from the deeper lower mantle. Hess and Robert S. Dietz proposed that old sea floor sinks back into the Earth's mantle at deep ocean trenches and in subduction zones where plates collide. Today, scientists know that mid-ocean ridges occur at divergent plate boundaries, where plates are moving apart. As the sea floor spreads, magma rises and accumulates in magma chambers beneath the centerline of the ridge. Some of the magma rises and erupts through fractures and small volcanoes along the ridge, forming new sea floor. As the plates continue to move apart, more magma rises from below to fill the gaps. Scientists now think that this upward movement of material beneath an ocean ridge is a result of sea-floor spreading rather than the cause.

Figure 19.5 shows the ages of oceanic crust. The oldest oceanic crust shown on this map is 160–200 million years old. Not shown are parts of the eastern Mediterranean Sea, which has some of the oldest oceanic crust—about 270 million years old.

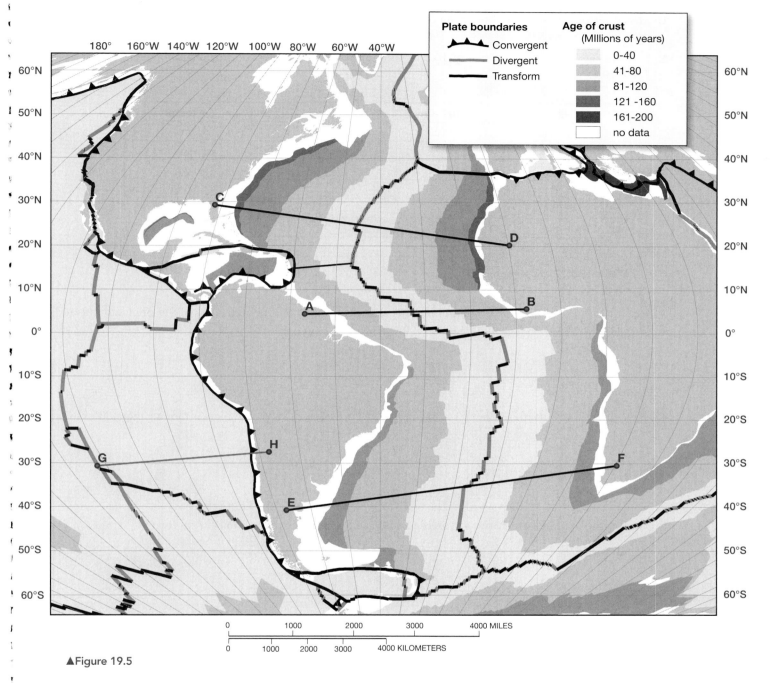

▲Figure 19.5

Figure
Ages of Oceanic Crusts

http://goo.gl/9xog30

In this section you will analyze the ages of tectonic plates and the rates of plate movement. Refer to the legend to find the ages of crust and use the scale bar to find distances. The first transect is done for you, but here are the steps. Measure from the middle of the ridge toward Point A. From the middle of the ridge to the far edge of the lightest gray band represents 40 million years (MY), to the far edge of the next gray band represents 80 MY. However, the far edge of the 80 MY band ends at the beginning of the no data section, so the full age of that band cannot be determined. To be certain of the age we are measuring, we will just measure to the far edge of the 0-40 MY section. Note that the far edges of the oldest bands are uncertain. To obtain an age that we are certain of, measure to the far edge of the oldest full band on that transect. From the ridge to Point C, the transect line crosses the 0-40 MY, 41-80 MY, 81-120 MY, and 121-160 MY bands. The oldest edge of the 121-160 MY band ends in the no data section, so we are unable to accurately measure the length to the end of the 200 MY band and instead we will measure to the edge of the 121-160 MY band.

Copyright © 2015 Pearson Education, Inc.   Lab Exercise 19   181

1. Measure along the transect lines from the center of the ridge to the edge of the oldest full band and enter that data into Table 19.1.

2. Enter the ages of the crust along each transect and calculate the speed of movement for each section in Table 19.1.

Leg	Dist. km	Age MY	Speed cm/yr
A to Ridge	1100	40	[2.8 cm/yr]
B to Ridge			
C to Ridge			
D to Ridge			
E to Ridge			
F to Ridge			
H to Edge			

▲Table 19.1 Plate Movement

3. Are the three sections in the Atlantic moving at the same speed? Are both sides of the ridge moving at the same speed? Which section is moving the fastest?

4. Compare the speed of the Atlantic sections with the Pacific transect. Which plate is moving more quickly?

5. What types of plate boundaries are on the Pacific and Atlantic shores of South America?

MasteringGeography™ | Looking for additional review and lab prep materials? Go to www.masteringgeography.com for pre lab videos and pre and post lab quizzes.

Name: _____ Laboratory Section: _____

Date: _____ Score/Grade: _____

Video
Exercise 20
Pre-Lab Video

LAB EXERCISE 20

Scan to view the Pre-Lab video

Plate Tectonics: Faulting and Volcanism

Earth's endogenic systems produce flows of heat and material toward the surface to form crust. Ongoing processes produce continental landscapes and oceanic sea-floor crust, sometimes in dramatic episodes. Earth's physical systems move to the front pages whenever an earthquake strikes or after a few of the 50 yearly volcanic eruptions threaten a city, although most of these occurrences are in remote locations or beneath the ocean surface.

We examine processes that construct Earth's surface and create world structural regions. Tectonic processes deform, recycle, and reshape Earth's crust. These processes occur sometimes in dramatic episodes but most often in slow, deliberate motions that build the landscape. Continental crust has been forming throughout most of Earth's 4.6-billion-year existence.

The arrangement of continents and oceans, the origin of mountain ranges, the topography of the land and the seafloor, and the locations of earthquake and volcanic activity are all evidence of our dynamic Earth. Principal seismic and volcanic zones occur along plate boundaries and hot-spot locations, thus linking plate tectonics to the devastation of major earthquakes and local threats of major volcanic eruptions and their attendant potential global climatic impact. We examine the patterns of movement of the San Andreas fault and the morphology of current and ancient volcanic features. We conclude by studying the relationship between volcanic location, magma composition, and volcanic shape.

Lab 20 features four sections and three optional Google Earth™ activities.

Key Terms and Concepts

asthenosphere
cinder cone
crater
composite volcano
dikes
effusive eruption
elastic-rebound theory

explosive eruption
flood basalts
lava
pyroclastics
plug dome
shield volcano
volcano

After completion of this lab, you should be able to:

1. *Identify* typical landscape features associated with faulting.
2. *Calculate* rates of movement of the San Andreas fault.
3. *Relate* volcanic profiles to volcanic location and chemical composition of magma.
4. *Analyze* the Kilauea Caldera, Hawai'i, and Shiprock, New Mexico, volcanic landscapes.

Materials/Sources Needed

colored pencils
length of string
scissors

transparent tape
stereolenses

Lab Exercise and Activities

SECTION 1

The San Andreas Fault

Crustal plates do not glide smoothly past one another. Instead, tremendous friction exists along plate boundaries. The specific mechanics of how a fault breaks remain under study, but **elastic-rebound theory** describes the basic process. Generally, two sides along a fault appear to be locked by friction, resisting any movement despite the powerful forces acting on the adjoining pieces of crust. Stress continues to build strain along the *fault plane* surfaces, storing elastic energy like a wound-up spring. When the strain buildup finally exceeds the frictional lock, the two sides of the fault plane lurch into new positions, moving from centimeters to several meters, and release enormous amounts of seismic energy into the surrounding crust. This energy radiates throughout the planet, diminishing with distance but sufficient to register on instruments worldwide.

Pinnacles National Monument and the Neenach formation formed 23 million years ago as a single volcano on the San Andreas fault (Figure 20.1). They were together when they formed but have been split apart by the movement of the San Andreas fault.

1. How many kilometers separate Pinnacles and southern portion of the Neenach formation? What is the average speed of movement of this section of the San Andreas fault over the past 23 million years, in centimeters per year?

2. The average rate of motion across the San Andreas fault zone during the past 3 million years is 5.6 cm/yr. How far apart are San Francisco and Palm Springs, in kilometers? How many million years will pass until they are sister cities across a new bay?

The 1906 San Francisco earthquake resulted in a maximum 8.5m (28 ft) of movement along the San Andreas Fault around Point Reyes.

3. How many years of stress were relieved by the 1906 earthquake, if the speed of movement has been 5.6 cm/yr?

4. If that section of the fault has not moved since 1906, how many centimeters of stress have accumulated, assuming a speed of movement of 5.6 cm/yr?

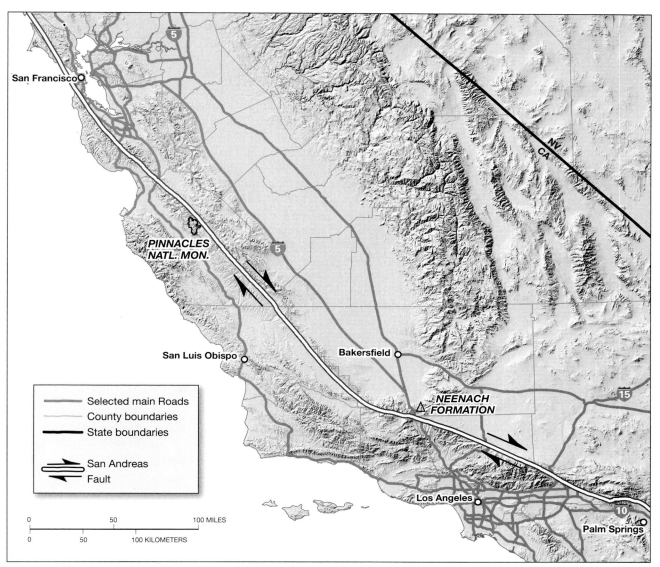

▲Figure 20.1 San Andreas Fault, Pinnacles National Monument, and the Neenach Volcano

Animation
Volcano Types

http://goo.gl/HrYBxb

SECTION 2

Volcanoes

Volcanic eruptions across the globe remind us of Earth's internal energy. Ongoing eruption activity matches patterns of plate tectonic activity. More than 1300 identifiable land volcanic cones and mountains exist on Earth, although fewer than 600 are active. In an average year, about 50 volcanoes erupt worldwide, varying from modest activity to major explosions.

A **volcano** forms at the end of a central vent or pipe that rises from the asthenosphere and upper mantle through the crust into a volcanic mountain. A **crater**, or circular surface depression, usually forms at or near the summit. Magma rises and collects in a magma chamber deep below the volcano until conditions are right for an eruption. **Lava** (molten rock), gases, and **pyroclastics**, or *tephra* (pulverized rock and clastic materials of various sizes ejected violently during an eruption), pass through the vent to the surface and build the volcanic landform.

The fact that lava can occur in many different textures and forms accounts for the varied behavior of volcanoes and the different landforms they build. In this section, we look at three volcanic landforms and their origins: shield volcanoes, cinder cones, composite volcanoes, and plug domes.

Volcanoes are of two general types, based on the chemistry and the viscosity of the magma involved. An **effusive eruption** produces a **shield volcano** (such as Kílauea in Hawai'i) and extensive deposits of **flood basalts**, or plateau basalts. Magma of higher viscosity leads to an **explosive eruption** (such as Mt. St. Helens), producing a **composite volcano,** made of multiple flows of lava and ash. The chemical content of the magma is related to the location of the volcano and determines its eruptive behavior and overall shape.

Effusive eruptions of low-viscosity fluid basalt flows create shield volcanoes. These volcanoes are gently sloped, gradually rising from the surrounding landscape to a summit crater. The shape is similar in outline to a shield of armor lying face up on the ground, hence the name *shield volcano*. The upper slopes are gently sloped around 5° because the magma is quite hot. However, it cools as it travels, so the lower slopes are a steeper 10°. A related effusive feature is a **cinder cone**, a small, cone-shaped hill usually less than 450 m (1500 ft) high, with a truncated top formed from cinders that accumulate during moderately explosive eruptions. Cinder cones are made of pyroclastic material and *scoria* (cindery rock, full of air bubbles) during an initial phase of an effusive eruption. Because they are made of loose material, their slope is quite steep, often with a 34° *angle of repose*. The angle of repose is the slope of a body of loose material.

Volcanic activity near subduction zones produces the infamous explosive volcanoes. Magma produced by the melting of subducted oceanic plate and other materials is thicker (more viscous) than magma that forms effusive volcanoes. It is 50%–75% silica and high in aluminum; consequently, it tends to block the magma conduit. The blockage traps and compresses gases, causing pressure to build for a possible *explosive eruption*, such as the 1980 eruption of Mt. St. Helens.

Unlike the volcanoes in Hawai'i Volcanoes National Park, where tourists gather at observation platforms or hike out near the cooling lava flows to watch the relatively calm effusive eruptions, these explosive volcanoes do not invite close inspection and can erupt dangerously with little warning.

Plug dome volcanoes are formed from magma that is high in silica, and very thick. The lava from these volcanoes is too stiff to flow; instead it extrudes from the ground, like toothpaste. A plug dome volcano often has a solid core of cooled lava with flanks made of loose pyroclastics.

Some volcanoes are formed from multiple eruptions—some of ash and some of lava. The term composite volcano describes these explosively formed mountains. They are sometimes called *stratovolcanoes* because they are built up in alternating layers of ash, rock, and lava, but shield volcanoes also can exhibit a stratified structure, so *composite* is the preferred term. Composite volcanoes tend to have steep sides, are more conical in shape than shield volcanoes, and therefore are also known as *composite cones.* If a single summit vent erupts repeatedly, a remarkable symmetry may develop as the mountain grows in size, as demonstrated by Mt. Rainier, Mt. Fuji, and Mt. St. Helens before it erupted in 1980. These volcanoes have a steeper upper section with slopes of up to 30° and lower flanks made of eroded materials with slopes of 5° to 10°.

In this section, you will be calculating the slope and relief of three volcanoes in order to better understand the relationship between their shape and the chemical content of their magma. You do not need to measure distances and elevations along the entire transect. For example, for Figure 20.2a, you could just measure the distance and elevation from the 6400' contour to the 6800' contour.

1. Calculate the relief for each volcano by subtracting the lower elevation on the transect line from the higher elevation. Enter the relief into Table 20.1.

2. Calculate the length along each transect between the contour lines you used to measure the relief, using the scale bar. You can also calculate the length of each transect by measuring the transect with a ruler and using the written scale to convert it to feet. Enter the length into Table 20.1.

3. Calculate the slope in percentage by dividing the relief by the length. Enter the slope into Table 20.1.

Figure	Relief	Length	Slope
A			
B			
C			

▲Table 20.1 Volcano Type

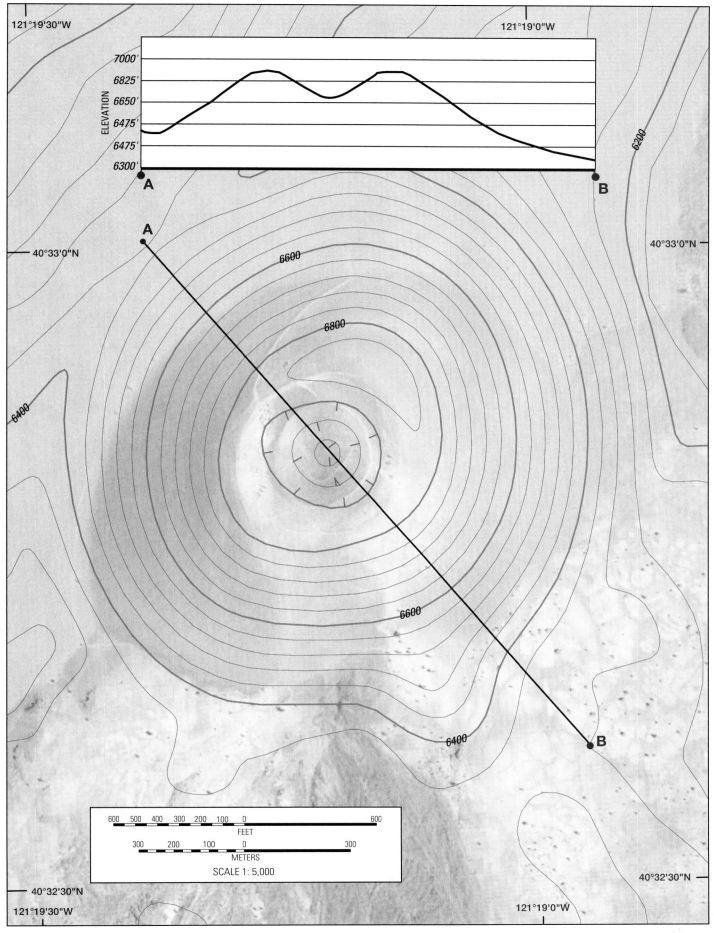

▲Figure 20.2a

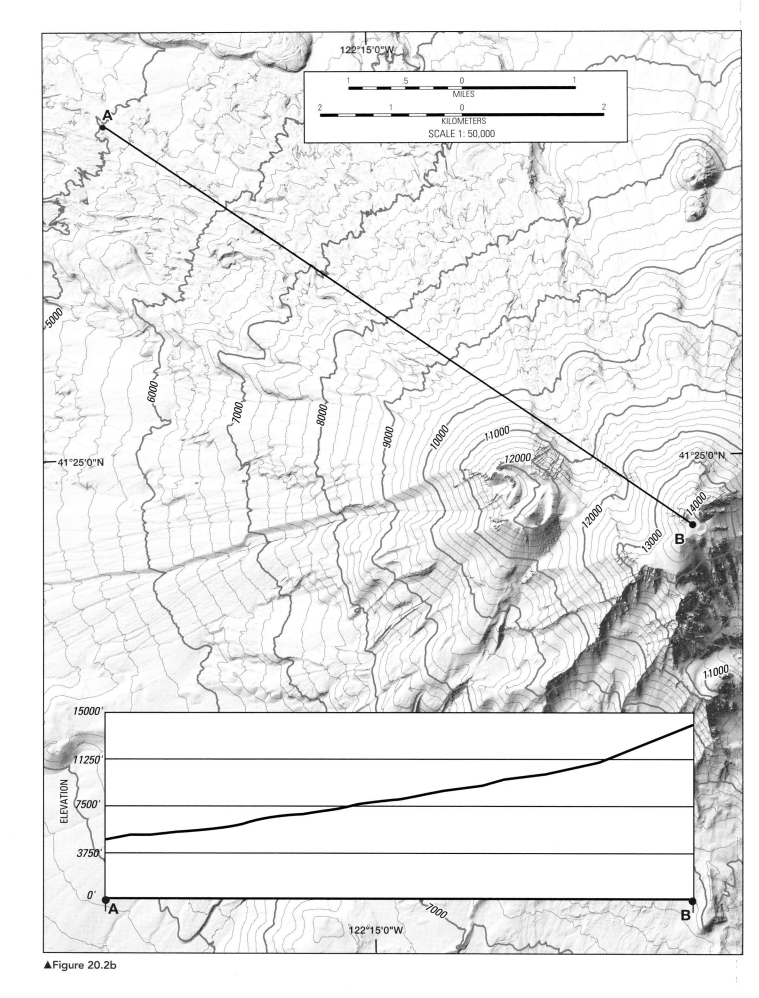
▲Figure 20.2b

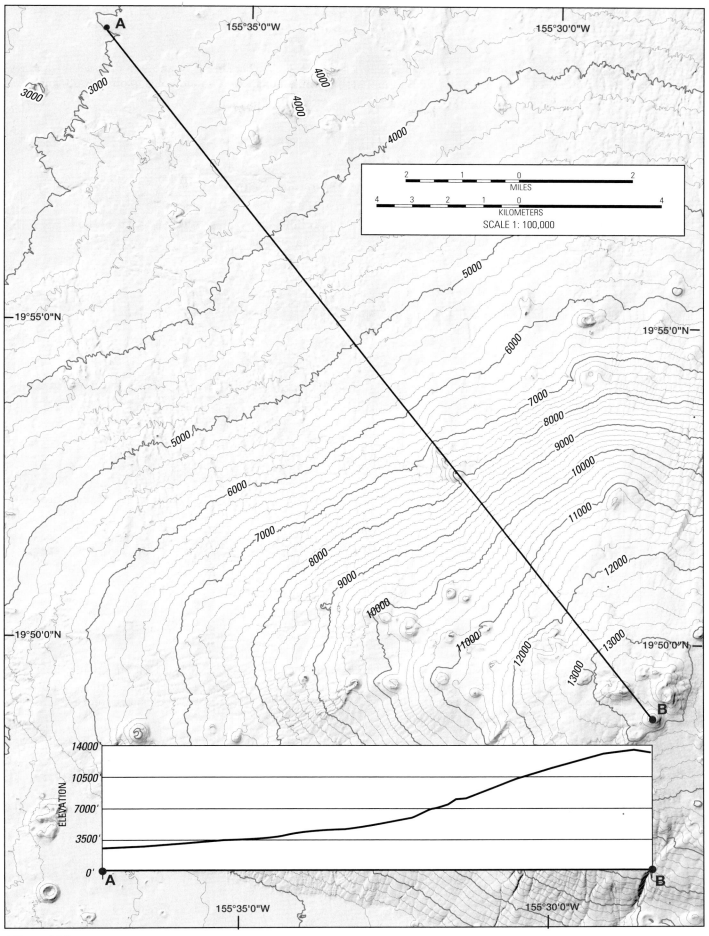
▲Figure 20.2c

4. Compare the results you recorded in Table 20.1 with Table 20.2. The three volcanoes in this section are the Cinder Cone at Lassen Volcanic National Monument; Mauna Kea, a shield volcano on Hawai'i; and Mount Shasta, a composite volcano in California.

Volcano Type	Slope and Relief	Tectonic Setting	Eruptive Behavior
Shield volcano	2°–10° (3%–18%), up to 10 km	Hot spot	Effusive
Cinder cone	34° (67%), up to 300 m	Rifting, hot spot, subduction	Lava fountain
Composite	Up to 30° (57%), over 4 km	Subduction	Explosive

▲Table 20.2   Volcano Type

5. Which volcano is shown in Figure 20.4a? Explain your conclusion. Which tectonic setting is the source of its magma?

6. Which volcano is shown in Figure 20.4b? Explain your conclusion. Which tectonic setting is the source of its magma?

7. Which volcano is shown in Figure 20.4c? Explain your conclusion. Which tectonic setting is the source of its magma?

8. Which volcano is composed of the most fluid lava? The least fluid lava? Explain.

9. In the boxes on the next pages, draw an artistic rendering of a cross-section of the tectonic situation that has created each of these volcanoes. For example, you could include a side view of tectonic rifting if one of the volcanoes had been made of flood basalt flows, or a hot spot if one of the volcanoes were a shield volcano.

▲ Figure 20.3 Setting for volcano in Figure 20.2a

▲ Figure 20.4 Setting for volcano in Figure 20.2b

▲ Figure 20.5 Setting for volcano in Figure 20.2c

SECTION 3

Kílauea Caldera in Hawai'i

Virtual Tour
Kilauea
Caldera

http://goo.gl/o48qGe

On the island of Hawai'i the continuing Kílauea eruption represents the longest eruptive episode in Hawaiian history, active since 1983! During 1989–1990, lava flows from Kílauea actually consumed several visitor buildings in the Hawai'i Volcanoes National Park and homes in Kalapana, a nearby town. Kílauea has produced more lava than any other vent on Earth. Characteristic of a hot spot, eruptions emanating from the **asthenosphere** are low-viscosity magmas that are very fluid and yield a dark, basaltic rock (less than 50% silica and rich in iron and magnesium). Gases readily escape from this magma because of its texture, causing an effusive eruption that pours out on the surface, with relatively small explosions and little tephra. These are the relatively gentle eruptions that produce enormous volumes of lava annually on the sea floor and in places like Hawai'i. However, dramatic fountains of basaltic lava sometimes shoot upward, powered by jets of rapidly expanding gases.

An effusive eruption may come from a single vent or from a flank eruption through side vents in surrounding slopes. If such vents form a linear opening, they are called *fissures* and sometimes form a dramatic curtain of fire during eruptions. In Iceland, active fissures are spread throughout the plateau landscape. In Hawai'i, rift zones capable of erupting tend to converge on the central crater, or vent. The interior of such a crater is often a sunken caldera, which may fill with low-viscosity magma during an eruption, forming a molten lake, which may then overflow lava downslope to the sea.

Mauna Loa is an enormous shield volcano rising 5800 m (19,029 ft) from the sea floor to the surface and a further 4169 m (13,678 ft) above sea level—a total height of 9969 m (32,707 ft)! It is 18 km by 9 km (11 mi by 5.5 mi). **Topographic Map #1** illustrates a portion of the Kílauea Volcano that has formed on the southeast slope of Mauna Loa. The topographic map was prepared in 1981.

Questions about and analysis of **Topographic Map #1** and the Kílauea Caldera.

1. What is the scale of this map?

   What is the contour interval?

2. What are the north-south length and east-west width of Kílauea Crater?

   in miles:

   in kilometers:

3. What is the vertical relief (the difference between the highest and lowest elevations) between the volcano observatory at Uwekahuna Bluff on the rim of the crater and the floor of the crater at the 3524-foot bench mark? Note the tick marks on the contour lines.

4. Within the crater is a "fire pit" named Halemaumau. This pit has varied in size during eruptions. What is the relief from the rim of Halemaumau to the 3412-foot benchmark on the floor of the pit?

5. Record the dated lava flows noted on the map in the Kílauea Crater, from oldest to youngest:

Virtual Tour
Shiprock

http://goo.gl/o48qGe

SECTION 4

Shiprock Volcanic Neck

Shiprock is an interesting igneous rock formation in northwestern New Mexico. Magma (molten rock) flowed through a conduit to the surface, forming a volcanic cone. When eruptions ceased, the volcanic material eroded away, leaving the jagged lava plume standing starkly above the surrounding landscape—a volcanic neck of basalt. Also, magma flowed along linear cracks from the conduit, radiating outward to form *dikes*. Millions of years of erosion exposed this volcanic neck and radiating ridges composed of fine-grained basalt. To the Navajo Nation, the sacred 518 m (1700 ft) tall formation is the "rock with wings."

Using the topographic map in Figure 20.6 and the stereopair photos in Figure 20.7, examine this volcanic feature. (Refer to the section in the Prologue Lab on the use of these photos and stereolenses.) As you work with this image, complete the following.

1. Based only on the appearance of the surrounding landscape, does Shiprock appear to be composed of different rock or similar rock materials than surrounding rocks? Note the observations that led you to your opinion.

2. Is Shiprock intrusive igneous or extrusive igneous, in your opinion? Explain.

3. How many radiating dikes do you see?

4. Describe the weathering processes, physical and chemical, that produced this volcanic formation.

5. Even though northwestern New Mexico is a desert, do you see any indication of water at work as an agent of erosion and transportation? Explain your observations from the photo stereopairs.

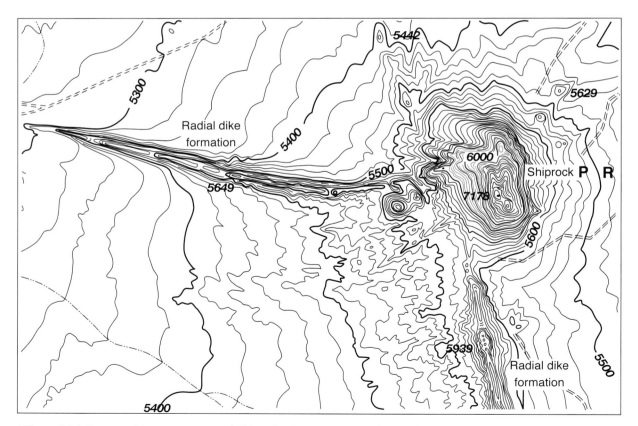

▲Figure 20.6 Topographic map segment of Shiprock, New Mexico. North is to the right, west to the top. USGS.

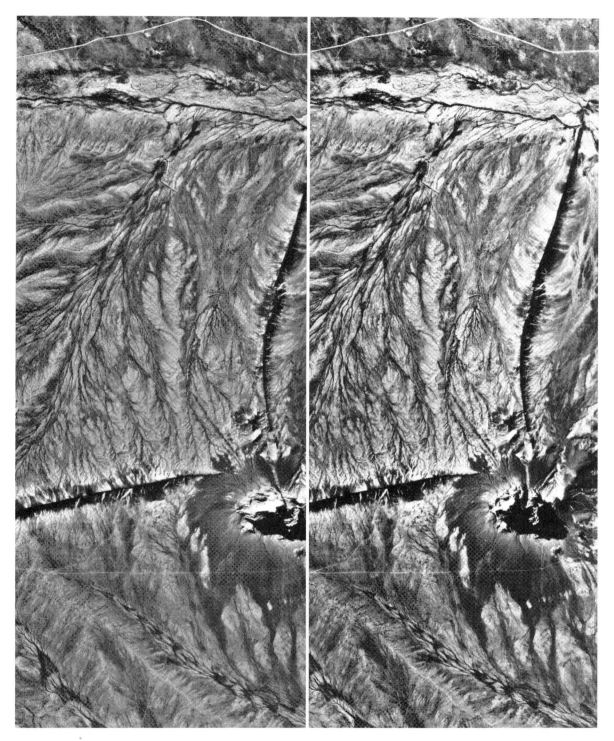

▲Figure 20.7 Photo stereopair of Shiprock, New Mexico. North is to the right, west to the top. Photos by NAPP, USGS, Eros Data Center.

Name: _____ Laboratory Section: _____

Date: _____ Score/Grade: _____

Video
Exercise 21
Pre-Lab Video

LAB EXERCISE

Scan to view the Pre-Lab video

21 The Rock Cycle and Rock Identification

A **rock** is usually an assemblage of **minerals** bound together (such as granite, a rock containing three minerals) but can also be a mass composed of a single mineral (such as rock salt), an undifferentiated material (such as the noncrystalline glassy obsidian), or even solid organic material (such as coal). A mineral is an inorganic natural solid compound having a specific chemical formula and usually a crystalline structure. Each mineral has its own characteristic color, texture, crystal shape, and density. There are more than 4200 known minerals, and of these about 30 are very common. Scientists have classified thousands of different rocks, all of which can be grouped into three types, based on the process that formed them.

Once we understand how a rock was formed, we gain insight into the environmental and tectonic settings present when it formed. **Igneous rocks** are rocks that solidified and crystallized from hot molten magma or lava, and we can tell whether they cooled quickly at the surface or slowly deep under the surface. Igneous rocks can also tell us if their magma was made of continental sediment or oceanic crust before melting. **Sedimentary rocks** are formed by the compaction, cementation, or hardening of sediments derived from other rocks; they tell us whether a region was a high-energy beach, a shallow inland sea, or a deep ocean trench. **Metamorphic rocks** are formed from existing igneous, sedimentary, or even metamorphic rock that has been transformed by going through profound chemical changes under pressure and increased temperature. Metamorphic rocks can tell us about the tectonic activity of a region, as well as about the conditions that formed the rock before it was transformed into a metamorphic rock. Lab Exercise 21 has two sections and a rock chart after the topographic maps.

Key Terms and Concepts

extrusive igneous rock
felsic
igneous rock
intrusive igneous rock
mafic

metamorphic rock
mineral
rock
rock cycle
sedimentary rock

After completion of this lab, you should be able to:

1. *Diagram* the rock cycle.
2. *Describe* the three types of rock.
3. *Identify* common rocks.

Materials/Sources Needed

magnifying glass
rock chart from back of manual

rock hand samples

Copyright © 2015 Pearson Education, Inc. Lab Exercise 21 197

Lab Exercise and Activities

SECTION 1

The Rock Cycle

The **rock cycle** is the name for the continuous alteration of Earth materials from one rock type to another (see Figure 21.1). For example, igneous rock formed from magma may break down into sediment by weathering and erosion and then lithify into sedimentary rock. This rock may subsequently become buried and exposed to pressure and heat deep within Earth, forming metamorphic rock. This may, in turn, break down and become a sedimentary rock again. Igneous rock may also take a shortcut through that cycle by directly becoming metamorphic rock.

As the arrows in Figure 21.1 indicate, there are many pathways through the rock cycle. Any rock type can become any other rock type in only one step. Rocks can also cycle through the same process; for example, an igneous rock can melt and become a new igneous rock. Two cyclic systems drive the rock cycle. At and above Earth's surface, the hydrologic cycle, fueled by solar energy, drives the exogenic processes. Below Earth's surface and within the crust, the tectonic cycle, powered by internal heat, drives the endogenic processes.

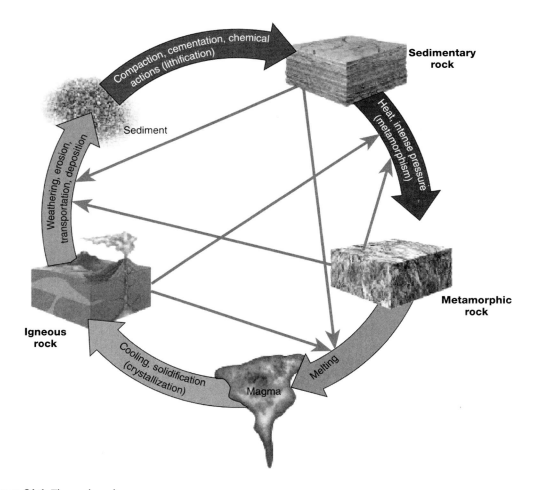

▲ Figure 21.1 The rock cycle.
In this schematic of the relationships among igneous, sedimentary, and metamorphic processes, the thin arrows indicate "shortcuts"—such as when igneous rock is melted and becomes metamorphic rock without first going through a sedimentary stage.

To identify a rock, one of the first steps is to determine which group it belongs to. Each rock has characteristics that allow it to be identified as igneous, sedimentary, or metamorphic. Relatively hard igneous rocks are generally made of randomly oriented, interlocking crystals of two or more minerals. Some have obvious physical characteristics, such as a dark, glassy appearance or many holes left by gas pockets that identify them as igneous rocks.

Sedimentary rocks are typically made of weathered material such as silt, sand, or pebbles. Some sedimentary rock is composed of fossils or plant debris, but other sedimentary rock is formed when minerals precipitate out of water, forming interlocking crystals. These rocks can be differentiated from crystalline igneous rocks because they are composed of readily identifiable minerals such as halite, calcite, or gypsum.

Metamorphic rocks are formed by the alteration of other rocks by heat and pressure. Some metamorphic rocks have a foliated appearance, caused by the parallel alignment of flat or elongated crystals, while others are made of stretched or deformed pebbles.

1. Place a check mark in the rock type column that corresponds to the process that could form or alter that rock.

Process	Igneous	Sedimentary	Metamorphic
Crystals left by evaporation of water			
Lithification of sediment			
Fossils present			
Melting of rock			
Folding of rock			
Foliated			
Heating but not melting			
Cooling of lava or magma			
Cementation of grains			
Compaction of sediment			

▲Table 21.1 Rock types and processes.

2. Identify a location in the United States where each step of the rock cycle is occurring. Describe what is happening at each location to create that rock type.

3. Describe a sequence in which a sedimentary rock forms and is modified to become a metamorphic rock.

SECTION 2

Rock Identification

To understand and classify rocks from a scientific viewpoint, we must begin with minerals, which are the building blocks of rocks. Identifying the minerals within a rock can help to identify that rock.

A rock is composed of a mineral or several minerals bound together, or solid organic material. The thousands of different rocks can be sorted according to three types, depending on the processes that formed them: igneous, sedimentary, and metamorphic.

Igneous Rocks

Approximately 90% of Earth's crust is made of igneous rocks, although sedimentary rocks, soil, or oceans frequently cover them.

Magma either intrudes into crustal rocks, cooling and hardening below the surface to form **intrusive igneous rock**, or it extrudes onto the surface as lava and cools to form **extrusive igneous rock**. The location and rate of cooling determine whether the rock is made of coarser (larger) or finer (smaller) crystals. Magma cools slowly below the surface, which allows more time for crystals to grow, resulting in coarse-grained rocks such as granite, which can be exposed to the surface by later tectonic uplift. Lava cools quickly at the surface and forms finer-grained rocks, such as basalt, the most common extrusive igneous rock. Basalt makes up 71% of the ocean floor, and it is actively forming the Big Island of Hawaii. If cooling is so rapid that crystals cannot form, the result is a glassy rock such as obsidian, or volcanic glass. Pumice does not have a crystal structure, and it forms when the bubbles from escaping gases create a frothy texture in the lava. Pumice is full of small holes, is light in weight, and is low enough in density to float in water.

Igneous rocks are classified according to their texture and composition. The same magma that produces coarse-grained granite when it cools beneath the surface can form fine-grained rhyolite when it cools above the surface. The mineral composition of a rock, especially the relative amount of silica (SiO_2), provides information about the source of the magma that formed it and affects its physical characteristics. **Felsic** igneous rocks, such as granite, are high in silicate minerals, such as *fel*dspar and quartz (pure *si*lica). Rocks formed from felsic minerals generally are lighter in color, less dense than those from mafic minerals, have lower melting points, and are less fluid when molten.

Mafic igneous rocks, such as basalt, are derived from *ma*gnesium and iron (the Latin word for iron is *fe*rrum). Compared to felsic magma, mafic rocks are lower in silica, higher in magnesium and iron, have high melting points, and are much more fluid when molten. Rocks formed from mafic minerals are darker in color and of greater density than those from felsic minerals. Ultramafic rocks have the lowest silica content; examples are peridotite (less than 45% silica) and the rocks of the mantle.

Sedimentary Rocks

Sedimentary rock can be formed from loose grains or fragments called *clasts* that are cemented together. Clasts come from several sources: the weathering and erosion of existing rock (the origin of the sand that forms sandstone), shells on the ocean floor (which make up one form of limestone), organic matter from ancient plants (which form coal), and the precipitation of minerals from water solution (the origin of the calcium carbonate, $CaCO_3$, that forms chemical limestone, although most limestone is biogenic). Sedimentary rocks are divided into several categories—clastic, biochemical, organic, and chemical—based on their origin.

Clastic Sedimentary Rocks

The formation of clastic sedimentary rock involves several processes. Weathering and erosion disintegrate and dissolve existing rock into clasts, which are carried across the landscape by gravity, water, wind, and ice. The clasts are now called *sediment*. Transport occurs from "higher-energy" sites, where the carrying medium has the energy to pick up and move the material, to "lower-energy" sites, where the sediment is deposited. Deposition is the process whereby sediment settles out of the transporting medium and results in material dropped along river channels, on beaches, and on ocean bottoms, where it is eventually buried. Lithification occurs as loose sediment is hardened into solid rock. This process involves compaction of buried sediments as the weight of overlying material squeezes out the water and air between clasts and cementation by minerals, which fill any remaining spaces and fuse the clasts—principally quartz, feldspar, and clay minerals—into a coherent mass. The type of cement varies in different environments. Calcium carbonate ($CaCO_3$) is the most common cement, followed by iron oxides and silica. Drying and heating can also unite particles. The different sediments that make up sedimentary rock range in size from boulders to gravel to sand to microscopic clay particles. After lithification, these size classes, combined with their composition, sorting, and cement characteristics, determine the common sedimentary rock types. For example, pebbles and gravels become conglomerate, silt-sized particles become siltstone or mudstone, and clay-sized particles become shale.

Biochemical, Organic, and Chemical Sedimentary Rocks

Some sedimentary rocks are formed from calcium carbonate from the shells of organisms (a biochemical process) or from dissolved minerals that precipitate out of water (a chemical process) and build up to form rock. Chemical precipitation is the formation of a separate solid substance from a solution, such as when water evaporates and leaves behind a residue of salts. The most common chemical sedimentary rock is limestone, and the most common form of limestone is biochemical limestone from marine organic origins. When calcium carbonate shelled organisms die, the solid material builds up on the ocean floor and lithifies to become limestone. The salt deposits left by evaporating water can build up to form evaporites, a type of chemical sedimentary rock. Examples of these evaporites are found in Utah on the Bonneville Salt Flats, created when an ancient lake evaporated. Both clastic and chemical sedimentary rocks are deposited in layered strata that form an important record of past ages. Using the principle of superposition, scientists use the stratigraphy (the ordering of layers), thickness, and spatial distribution of strata to determine the relative age and origin of the rocks. The strata also indicate the region's climatic history, since each layer was formed under different environmental conditions.

Metamorphic Rocks

Igneous, sedimentary, or even metamorphic rock may be transformed into a metamorphic rock by going through profound physical or chemical changes under pressure and increased temperature. The name metamorphic comes from a Greek word meaning "to change form." Metamorphic rocks generally are more compact than the original rock and therefore are harder and more resistant to weathering and erosion. Some of the processes that can cause metamorphism are heating, pressure, heating and pressure together, and hydrothermal metamorphism—such as along subduction zones and mid-ocean ridges. When heat is applied to rock, the atoms within the minerals may break their chemical bonds, move, and form new bonds, leading to new groupings of minerals. When pressure is applied to rock, its structure may change as atoms become packed more closely. When rock is subject to both heat and pressure at depth, the original mineral assemblage becomes unstable and changes. Finally, rocks may be compressed by overlying weight and subject to shear when one part of the mass moves sideways relative to another part. These processes change the shape of the rock, leading to changes in the mineral alignments within. Metamorphic rock may form from igneous rocks as the lithospheric plates shift, especially when one plate is thrust beneath another. Contact metamorphism occurs when magma rising within the crust "cooks" adjacent rock; this type of metamorphism results from heat alone. Regional metamorphism occurs when a large areal extent of rock is subject to metamorphism. This can occur when sediments collect in broad depressions, if their own weight creates enough pressure in the bottommost layers to transform the sediments into metamorphic rock. Regional metamorphism also occurs as lithospheric plates collide and mountain building occurs. Metamorphic rocks have textures that are foliated or nonfoliated, depending on the arrangement of minerals after metamorphism. Foliated rock has a banded or layered appearance, demonstrating the alignment of minerals, which may appear as wavy striations (streaks or lines) in the new rock. Nonfoliated rocks do not exhibit this alignment.

1. What characteristics indicate whether a rock is an intrusive or extrusive igneous rock?

2. What characteristics would indicate whether a rock is a clastic or nonclastic sedimentary rock?

3. Explain how a metamorphic rock could develop a foliated appearance or a nonfoliated appearance.

Use the following rock identification keys, Figures 21.2, 21.3 and 21.4, and the images of rock samples in the back of this manual to identify the rock samples that your instructor provides. Fill out Tables 21.2, 21.3, and 21.4 as you analyze the rocks.

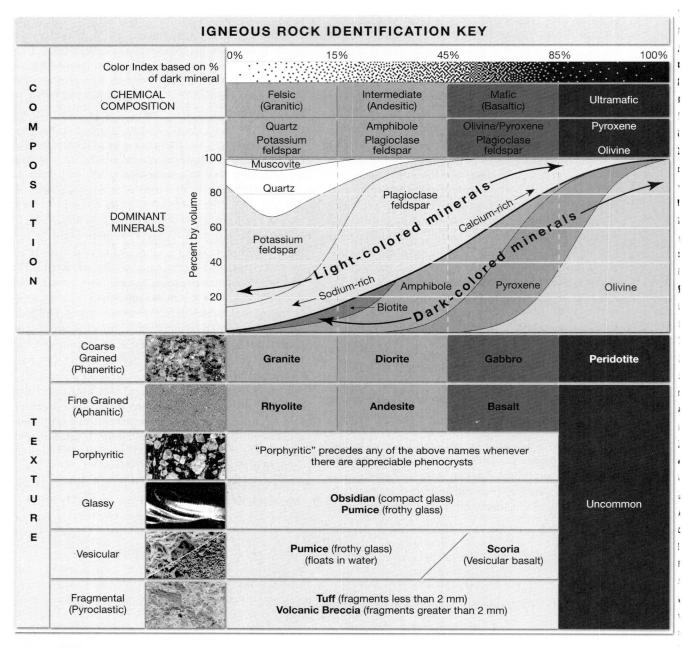

▲ Figure 21.2

SEDIMENTARY ROCK IDENTIFICATION KEY

Detrital Sedimentary Rocks

Clastic Texture (particle size)	Distinctive Properties	Rock Name	
Gravel Coarse (over 2 mm)	Rounded rock or mineral fragments, typically poorly sorted	Conglomerate	
	Angular rock or mineral fragments, typically poorly sorted	Breccia	
Sand Medium (1/16 to 2 mm)	Quartz grains, typically rounded, well sorted	Quartz Sandstone	Sandstone
	At least 25% feldspar, typically poorly sorted, angular fragments	Arkose	Sandstone
	Mixture of sand and mud, typically poorly sorted	Graywacke	Sandstone
Silt/mud Fine (1/16 to 1/256 mm)	Mostly silt-size quartz and clay, blocky, gritty	Siltstone	Mudstone
Clay/mud Very fine (less than 1/256 mm)	Mostly clay, splits into layers, may contain fossils	Shale	Mudstone
	Mostly clay, crumbles easily	Claystone	Mudstone

Chemical Sedimentary Rocks

Composition	Distinctive Properties	Rock Name	
Calcite, $CaCO_3$ (effervesces in HCl)	Fine to coarse crystalline	Crystalline Limestone	Limestone
	No visible grains, may exhibit conchoidal fracture	Micrite	Limestone
	Visible shells and shell fragments loosely cemented	Coquina	Limestone
	Various size shells and shell fragments, well cemented	Fossiliferous Limestone	Limestone
	Microscopic shells and clay, soft	Chalk	Limestone
	Faint layering, may contain cavities or pores	Travertine	Limestone
Quartz, SiO_2	Microcrystalline, may exhibit conchoidal fracture, will scratch glass	Chert (light colored)	
		Flint (dark colored)	
		Agate (banded)	
Gypsum $CaSO_4 \cdot 2H_2O$	Fine to coarse crystalline, soft	Rock Gypsum	
Halite, NaCl	Fine to coarse crystalline, tastes salty	Rock Salt	
Altered plant fragments	Black brittle organic rock, may be layered	Bituminous Coal	

▲ Figure 21.3

METAMORPHIC ROCK IDENTIFICATION KEY

Distinctive Properties	Texture	Grain Size	Rock Name	Parent Rock
Excellent rock cleavage, smooth dull surfaces	Foliated	Very fine	Slate	Shale, or siltstone
Breaks along wavy surfaces, glossy sheen	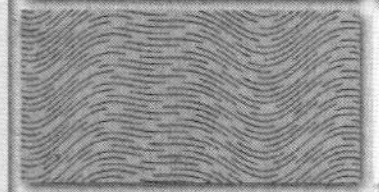 Foliated	Fine	Phyllite	Shale, slate, or siltstone
Micas dominate, breaks along scaly foliation	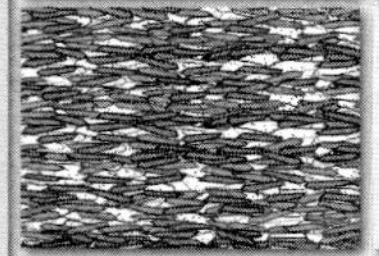 Foliated	Medium to Coarse	Schist (Mica schist, Chlorite schist, Talc schist, Garnet mica schist)	Shale, slate, phyllite, or siltstone
Compositional banding due to segregation of dark and light minerals	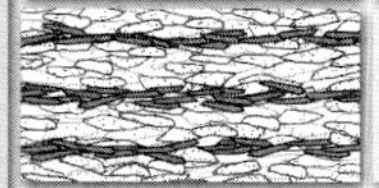 Foliated	Medium to Coarse	Gneiss	Shale, schist, granite, or volcanic rocks
Banded, deformed rock with zones of light and dark colored crystalline minerals	 Foliated	Medium to Coarse	Migmatite	Shale, granite, or volcanic rocks
Interlocking calcite or dolomite crystals nearly the same size, soft, reacts to HCl	Nonfoliated	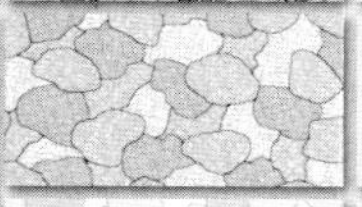 Medium to coarse	Marble	Limestone, dolostone
Fused quartz grains, massive, very hard	Nonfoliated	Medium to coarse	Quartzite	Quartz sandstone
Round or stretched pebbles that have a preferred orientation	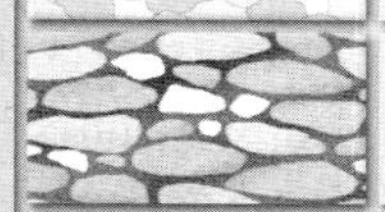 Nonfoliated	Coarse-grained	Metaconglomerate	Quartz-rich conglomerate
Shiny black rock that may exhibit conchoidal fracture	Nonfoliated	Fine	Anthracite	Bituminous coal
Usually, dark massive rock with dull luster	Nonfoliated	Fine	Hornfels	Any rock type
Very fine grained, typically dull with a greenish color, may contain asbestos fibers	Nonfoliated	Fine	Serpentinite	Basalt or ultramafic rocks
Broken fragments in a haphazard arrangement	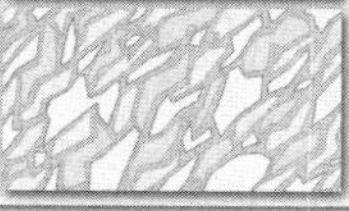 Nonfoliated	Medium to very coarse	Fault breccia	Any rock type

Increasing Metamorphism (applies to foliated rocks from Slate through Migmatite)

▲ Figure 21.4

IGNEOUS ROCKS

Sample Number	Texture	Color	Mafic or Felsic	Rock Name
1				
2				
3				
4				
5				
6				
7				
8				
9				
10				

▲Table 21.2 Igneous rocks

SEDIMENTARY ROCKS

Sample Number	Texture or Grain Size	Clastic or Chemical	Rock Name
1			
2			
3			
4			
5			
6			
7			
8			
9			
10			

▲Table 21.3 Sedimentary rocks

METAMORPHIC ROCKS

Sample Number	Grain Size	Foliated or Nonfoliated	Rock Name	Parent Rock
1				
2				
3				
4				
5				
6				
7				
8				
9				
10				

▲Table 21.4 Metamorphic rocks

MasteringGeography™ | Looking for additional review and lab prep materials? Go to www.masteringgeography.com for pre lab videos and pre and post lab quizzes.

Name: _____ Laboratory Section: _____

Date: _____ Score/Grade: _____

Video
Exercise 22
Pre-Lab Video

Scan to view the Pre-Lab video

http://goo.gl/vOD3Pe

LAB EXERCISE 22

Recurrence Intervals for Natural Events

Geographers compile spatial data in order to recognize patterns among natural events, make predictions, and encourage preventive or protective actions in an attempt to mitigate the impact of various environmental hazards. Using statistical data to determine the probability of recurrence of a natural event is the focus of this lab exercise. Probabilities are important in guiding protective policies, planning, and hazard zoning, if allowed by the usually sluggish political process. In this lab exercise, we focus on stream processes and the recurring floods that endanger society.

Hydrology is the science of water, its global circulation, distribution, and properties, specifically water at and below Earth's surface. For hydrology links on the web, see http://nwis.waterdata.usgs.gov/nwis or http://water.usgs.gov. A **recurrence interval,** or *return period*, is the average time period within which a given event will be equaled or exceeded once. This exercise examines annual streamflow data to establish recurrence intervals for flooding and to see how this can be used for improved planning and disaster management. Lab Exercise 22 has three sections.

Key Terms and Concepts

discharge
flood
flood frequency curve
flood probability
gage height (stage)

hydrology
rating curve
recurrence interval (return period)
streamflow

After completion of this lab, you should be able to:

1. *Define* stream discharge and gage height and *describe* the relationship between them.
2. *Use* selected hydrographic data to construct rating curves to *illustrate* streamflow characteristics.
3. *Define* recurrence intervals and use hydrographic data to *calculate* and *graph* them.
4. *Use* rating curves and flood frequency curves to *assess* risk potential and *make* land-use decisions based on flood potential.

Materials/Sources Needed
pencils
calculator
colored pencils
ruler

Lab Exercise and Activities

SECTION 1

Streamflow, Discharge, and Gage Height

Stream channels vary in *width* and *depth*. The streams that flow in them vary in *velocity* and in the *sediment load* they carry. All these factors increase with increasing **discharge**, or the stream's volume of flow per unit time. Discharge is calculated by multiplying the velocity of a stream by its width and depth for a specific cross section of the channel, as stated in the simple expression:

$$Q = wdv$$

where Q = discharge, w = channel width, d = channel depth (measured above an arbitrary datum called the stage), and v = steam velocity. As Q increases, some combination of channel width, depth, and stream velocity increases. Discharge is expressed in cubic meters per second (cms, or m^3/sec) or cubic feet per second (cfs, or ft^3/sec). Given the interplay of channel width and depth and stream velocity, the cross section of a stream varies over time, especially during heavy floods. Flood avoidance or management requires extensive measurement of streamflow characteristics and discharge (flow per unit of time).

Table 22.1 contains data from 1984 to 2013 for the Red River of the North, collected at the gaging station at Grand Forks, North Dakota. (This is only a portion of the data table, which actually contains records since 1882, covering 131 years overall.) The maximum peak discharge and corresponding **gage height** in meters are shown for the selected years. **Stage** is also called stream stage or gage height and is the height of the water surface, in meters, above an established elevation where the stage is zero. This zero elevation is based on a reference elevation datum, either the National Geodetic Vertical Datum of 1929 (NGVD29) or the later North American Vertical Datum of 1988 (NAVD88).

Notes for Table 22.1:
Station name: Red River of the North at Grand Forks, Grand Forks County, ND
Station number: 05082500
Latitude: 47.93°N
Longitude: 97.03°W
Basin name: Sandhill-Wilson

Drainage area (square kilometers): 77,959 km^2

Contributing drainage area (square kilometers): 68,117 km^2
Gage datum elevation: 126 m (above NGVD 29)
Base discharge (cms): 237 cms

Gage heights are given in meters above gage datum elevation.
Discharge is listed in the table in cubic meters per second.
Peak flow data were retrieved from the USGS National Water Information System (NWIS).

Using the data provided in Table 22.1, answer the following questions and completion items.

1. **a)** The highest discharge recorded for the listed years: _____

   b) Year recorded: _____

   c) The lowest discharge recorded for the listed years: _____

   d) Year recorded: _____

2. **a)** Average discharge (1984–2013): _____

   b) Range of discharge (1984–2013): _____

1	2	3	4	5
Peak Discharge Date	Discharge (cms)	Gage Height at Peak (m)	Rank	Recurrence Interval
April 2, 1984	904	11.3	16	1.94
May 19, 1985	498	7.9	22	1.41
April 2, 1986	893	11.3	17	1.82
March 29, 1987	490	10.1	24	1.29
April 5, 1988	238	6.4	27	11.5
April 13, 1989	1,109	13.2		
April 5, 1990	141	5.4	29	1.07
July 8, 1991	136	5.4	30	1.03
March 12, 1992	224	7.1	28	1.11
August 3, 1993	734	11.1	21	1.48
July 12, 1994	750	10.5	20	1.55
March 31, 1995	974	12.1	14	2.21
April 21, 1996	1,635	14.0		
April 18, 1997	3,836	15.9		
May 21, 1998	832	10.7	19	1.63
March 31, 1999	1,400	13.4		
June 26, 2000	882	11.3	18	1.72
April 14, 2001	1,618	13.7		
July 13, 2002	1,064	11.6	12	2.58
June 28, 2003	476	7.4	25	1.24
April 1, 2004	960	11.7	15	2.07
June 18, 2005	1,072	12.2	11	2.82
April 6, 2006	2,038	14.6		
June 22, 2007	988	11.8	13	2.38
June 16, 2007	496	8.1	23	1.35
April 1, 2009	2,148	15.0		
March 20, 2010	2,761	14.0		
April 14, 2011	2,450	15.2		
March 19, 2012	409	7.2	26	1.19
April 30, 2013	1,232	12.5		

▲Table 22.1 Peak annual discharge (in cubic meters per second) and gage height (in meters above gage datum elevation) for the Red River of the North at Grand Forks, North Dakota, 1984–2013. (USGS National Water Information Service, NWIS, http://nwis.waterdata.usgs.gov/usa/nwis/peak)

3. a) The highest gage at peak discharge: _____

   b) Year recorded: _____

   c) The lowest gage at peak discharge: _____

   d) Year recorded: _____

4. Scan the *Discharge* (column 2) and *Gage Height at Peak* (column 3) data columns. As gage height at peak (stream depth) gets higher, what usually happens to the discharge?

5. In general, there is a direct relationship between discharge and width, depth (gage height), and velocity. A stream's **rating curve** relates stream discharge to the gage height. Using the data in Table 22.1 and the logarithmic graphing paper in Figure 22.1, *plot* the gage height against discharge and draw a smooth-curved line through the plotted points to complete the rating curve. (*Note*: The numbers on the x-axis (discharge) are multiples of 10. Therefore, the 2012 discharge of 409 cms would be plotted along the x-axis at 41, as can be seen on the graph.) The highest discharge year and years 11–30 have been plotted for you.

6. a) We can use the graph in Figure 22.1 to read data from the chart. Use the rating curve to determine the gage height associated with the following discharges. (*Hint:* Use colored pencils to mark them on the graph first.)

   100 cms ____[3.6 m]____

   500 cms _____

   1000 cms _____

   3000 cms _____

   b) Determine the discharge associated with a gage height of

   6 meters ____[200 cms]____

   9 meters _____

   12 meters _____

   15 meters _____

7. a) As gage height at peak (stream depth) gets higher, what usually happens to the discharge?

   b) Compare the data in Table 22.1 for the following pairs of years: 1989 and 2013; 1995 and 2002. What do you observe in the gage height and discharge relationship in these years?

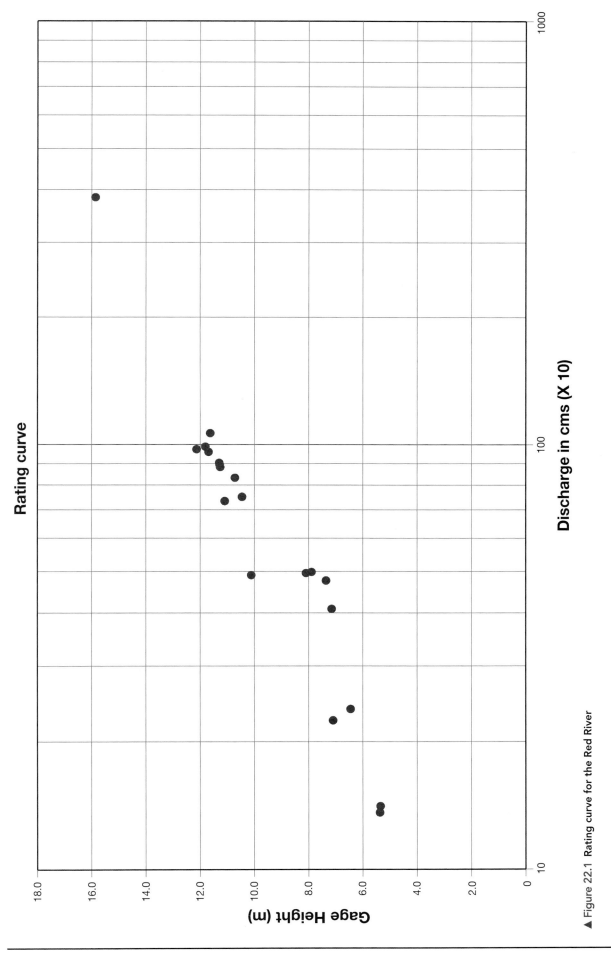

▲ Figure 22.1 Rating curve for the Red River

Lab Exercise 22   211

SECTION 2

Floods and Recurrence Intervals

Flood patterns in a drainage basin are as complex as the weather, for floods and weather are equally variable, and both include a level of unpredictability. Measuring and analyzing the behavior of each large watershed and stream enables engineers and concerned parties to develop the best possible flood-management strategy. The key to flood avoidance or management is to possess extensive measurements of **streamflow**, a stream's discharge, and how it performs during a precipitation event.

A **flood** is a high water level that overflows the natural river bank along any portion of a stream. Both floods and the floodplains they might occupy are rated statistically for the expected time intervals between floods.

A recurrence interval, or return period, is the average time period within which a given event will be equaled or exceeded once. A 10-year flood is the highest flood level (gage height) that is likely to occur once every 10 years; a 50-year flood is the highest that is likely to occur once every 50 years. A 10-year flood has a 10% likelihood of being equaled or exceeded in any one year and is likely to occur about 10 times in each century. It is important to note that these probabilities estimate *random recurrence*. Several decades may pass without a 10-year flood, or floods of such magnitude may occur two or three years in a row. The higher the return interval, the more catastrophic the event is likely to be and the less likely the probability of the event occurring in any given year.

Initial reports on the Midwest floods of 1993 and the Red River floods of 1997 included various references to 300-year and 500-year floods. Data now show that the 1993 Midwest floods easily exceeded a 1000-year **flood probability**. Let us now calculate the recurrence interval for the 1997 Red River floods.

1. Refer once again to Table 22.1 and at the far right locate columns 4 and 5, *Rank* and *Recurrence Interval*, which have not been used as yet. Rank the floods (using the gage height data in column 3), with the rank of 1 for the largest flood and continue in descending order to the smallest, entering your ranking numbers in column 5.

   The recurrence interval (*RI*) is calculated using the following formula:

$$RI = \frac{n+1}{m}$$

   Where

   RI = recurrence interval (in years)
   n = number of years of record
   m = rank of the flood

   For example, the flood with a rank of 1 and 30 years of records would have a recurrence interval calculated as follows:

$$RI = \frac{30+1}{1} = 31$$

2. Calculate the rank and recurrence intervals for the gage heights and enter your results in column 5 of Table 22.1. Some (namely, flood rank 11–30) have been done for you on the table to get started.

3. On the logarithmic paper in Figure 22.2, *plot* the recurrence (return) interval against the discharge and draw a straight line through the plotted points (so that the line fits with approximately half of the dots above and half below the line) to complete the **flood frequency curve**. (*Note:* The discharge is now on the y-axis and is once again multiplied by 10.) The sample points for flood ranks 1 and 11–30 have been plotted, but the line cannot be drawn until all points are plotted.

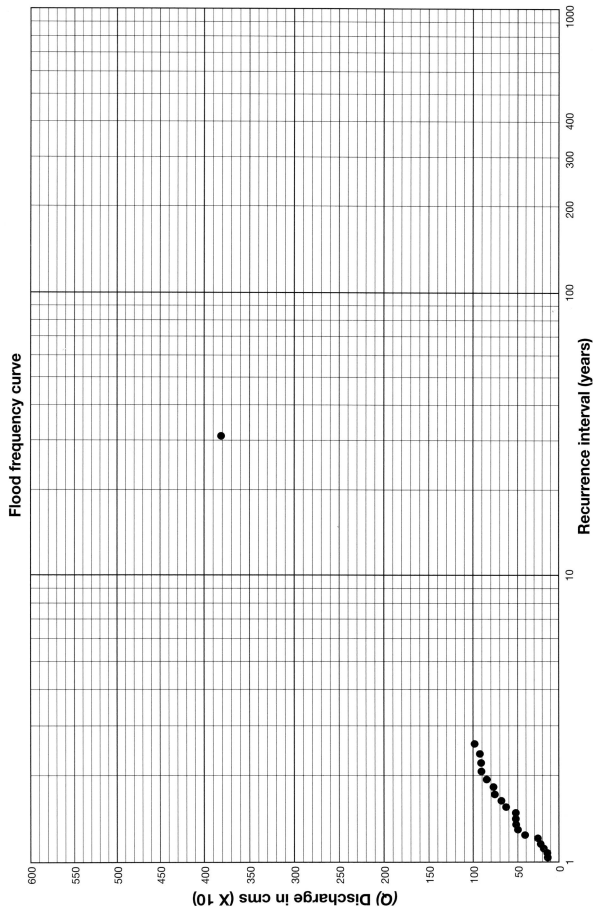

▲ Figure 22.2 Flood frequency curve for the Red River

Once again you can use the graphs you created to obtain information about hypothetical discharges and recurrence intervals.

4. a) Using the flood frequency curve in Figure 22.2 (and colored pencils to mark on graph), estimate the recurrence interval of a flood of the following magnitudes:

   200 cms _____

   1500 cms ___[4 years]___

   3000 cms _____

   b) Estimate the discharge of the

   50-year flood _____

   100-year flood _____

   500-year flood _____

5. a) Use Figure 22.2 along with Figure 22.1 to determine the gage height of these flood frequencies. (Use the flood frequency curve in Figure 22.2 to obtain the discharge, and then use that discharge on the rating curve to determine the associated gage height.)

   20-year flood _____

   300-year flood _____

   400-year flood _____

   b) Now use both graphs (Figure 22.1 first, and then 22.2) to determine the recurrence interval of the gage height at

   10 meters _____

   12 meters _____

   15 meters _____

6. How many times in a century would you expect the Red River to crest at a gage height of 13 meters?

   What is the probability of a flood of that magnitude in any given year?

7. According to the flood frequency curve, what is the return interval for a flood the magnitude of the spring 1997 flood?

SECTION 3

Probability, Prediction, and Risk Assessment

Throughout history, civilizations have settled floodplains and deltas, especially since the agricultural revolution of 10,000 years ago, when the fertility of floodplain soils was discovered. We build artificial levees to keep rivers within their channels and construct dams to hold back and regulate streamflow. How high must the levees and dams be? What type of zoning of the floodplain is required to safeguard settlements? Hydrographic data are compiled and the gage height and flood frequency curves are plotted in order to determine flood risk potential.

Neuse River 1999 Floods

The abuse and misuse of river floodplains brought catastrophe to North Carolina in 1999. In short succession during September and October, Hurricanes Dennis, Floyd, and Irene delivered several feet of precipitation to the state, each storm falling on already saturated ground. With the soil storage at capacity, much of the surplus entered streams and rivers as runoff. About 50,000 people were left homeless and at least 50 died, while more than 4000 homes were lost and an equal number were badly damaged. The economic estimate for the ongoing disaster exceeded $10 billion.

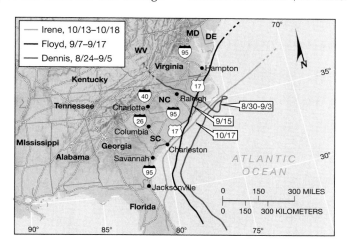

▲ Figure 22.3 **North Carolina floodplain disaster.**
Three hurricanes deluged North Carolina with several feet of rain during September and October 1999, Hurricane Floyd being the worst.

Hogs, in factory farms, outnumber humans in North Carolina. More than 10 million hogs, each producing 2 tons of waste per year, were located in about 3000 agricultural factories. These generally unregulated operations collect almost 20 million tons of manure into nearly 500 open lagoons, many set on river floodplains. The hurricane downpour flushed out these waste lagoons, spewing hundreds of millions of gallons of untreated sewage into wetlands, streams, and eventually Pamlico Sound and the ocean—a spreading "dead zone." Add to this waste hundreds of thousands of hog, poultry, and other livestock carcasses, industrial toxins, floodplain junkyard oil, and municipal waste, and you have an environmental catastrophe.

Notes for Table 22.2:
Station name: Neuse River at Kinston, Lenoir County, North Carolina
Station number: 02089500
Latitude: 35.25°N
Longitude: 75.6°W

Drainage area (square kilometers): 6,454 km^2
Gage datum elevation: 3.3 m (above NGVD)

1. Using Table 22.2, complete the rating curve on Figure 22.4 for the Neuse River at Kinston. The sample points for flood ranks 1 and 11–30 have been plotted for you.

2. Calculate the rest of the recurrence intervals for the gage heights and enter your results in column 5 of Table 22.2.

3. On the logarithmic paper in Figure 22.5, *plot* the recurrence (return) interval against the discharge and draw a straight line through the plotted points (so that the line fits with approximately half of the dots above and half below the line) to complete the flood frequency curve. (*Note:* The discharge is now on the y-axis and is once again multiplied by 10.)

1	2	3	4	5
Peak Discharge Date	Discharge (cms)	Gage Height at Peak (m)	Rank	Recurrence Interval
March 28, 1983	426	5.7	6	5.17
March 31, 1984	305	5.1	14	2.21
February 20, 1985	253	4.7	20	1.55
December 8, 1985	148	3.7	28	1.11
March 9, 1987	521	6.1	3	
April 24, 1988	104	3.0	30	1.03
May 11, 1989	398	5.6	7	4.43
April 8, 1990	273	4.8	19	1.63
August 6, 1991	239	4.6	21	1.48
August 20, 1992	300	5.0	15	2.07
January 16, 1993	339	5.2	10	3.10
March 11, 1994	330	5.2	13	2.38
February 26, 1995	353	5.5	9	3.44
September 17, 1996	759	7.1	2	
February 23, 1997	288	5.1	17	1.82
March 17, 1998	468	6.1	4	
September 22, 1999	1016	8.4	1	
October 25, 1999	451	6.0	5	
April 9, 2001	218	4.5	24	1.29
January 31, 2002	196	4.3	25	1.24
April 7, 2013	336	5.4	11	2.82
August 24, 2004	218	4.6	23	1.35
March 24, 2005	187	4.2	27	1.15
June 24, 2006	297	5.1	16	1.94
November 30, 2006	367	5.5	8	3.88
April 12, 2008	194	4.3	26	1.19
March 8, 2009	225	4.5	22	1.41
February 14, 2010	336	5.3	12	2.58
October 7, 2010	274	4.9	18	1.72
March 29, 2012	127	3.3	29	1.07

▲Table 22.2 Peak annual discharge and gage height for the Neuse River at Kinston, NC, 1980–2009

4. The Neuse River reaches **flood stage** at a gage height of 4.26 m, moderate flood stage at a gage height of 5.48 m, and major flood stage at a gage height of 6.4 m. Flood stage indicates that the river has topped its banks, moderate flooding requires some evacuation and road closures, and major flooding requires evacuation of people and livestock, as well as major road closures.

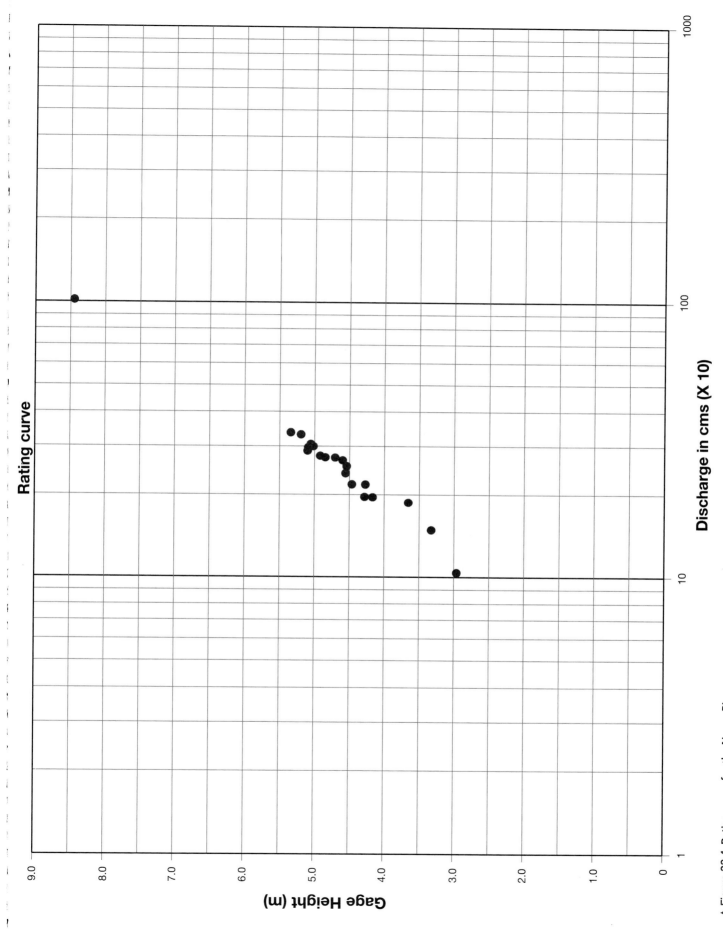

▲ Figure 22.4 Rating curve for the Neuse River

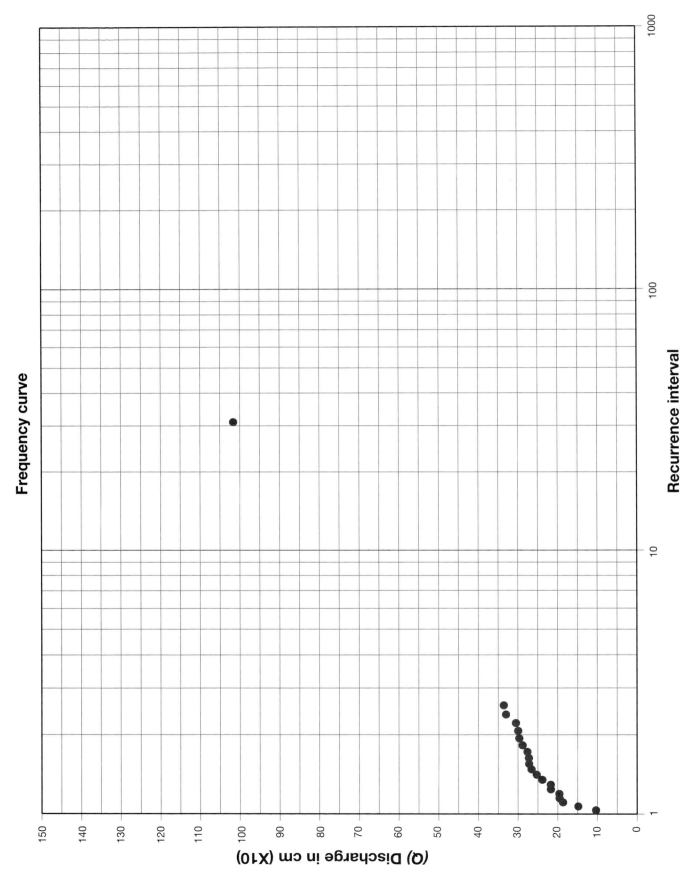

▲ Figure 22.5 Flood frequency curve for the Neuse River

a) Using the flood frequency curve from Figure 22.5 (and colored pencils to mark on the graph), estimate the recurrence interval of a flood of the following magnitudes:

196 cms ___[1.24 years]___

350 cms _____

518 cms _____

b) Estimate the discharge of the

50-year flood _____

100-year flood _____

500-year flood _____

5. a) The highest flow recorded on the Neuse River was 1092 cms, recorded in July 1919. What is the recurrence interval for such a flood, given 86 years of records? _____

b) Based on the rating curve, what would the gage height be for a discharge of 1092 cms? _____

6. What is the recurrence interval for the 1999 flood event?

Given this interval, what recommendations would you make for managing current and future land use in the Neuse River floodplain?

Red River, Grand Forks

The levees along the Red River at Grand Forks were 15.5 m (50.8 ft) high. In one 1997 event the gage height at peak discharge measured 15.86 m (52 ft), resulting from unusual icing conditions upstream, similar to a "dam break." Four days later the river rose even higher, to its peak gage height for the year, measuring 16.56 m (54.4 ft). Within a few days the river topped the levees twice, causing over $1 billion in damage!

Engineers have raised the Grand Forks levees to 18.2 m (60 ft), including an added lift of 0.91 m (3 ft) to allow for wave action at times of flooding. This was a $403 million project, shared by the citizens of Grand Forks. Ironically, other cities at similar risk along the river have not been as farsighted. This uneven mitigation effort leads to friction and problems between cities and governments when flood events occur.

Rivers have been important in the geography of human history, influencing where settlements were built, where livelihoods were made, and where borders were drawn. But to what extent should the general public continue to subsidize these risky sites? Rating curves and flood frequency curves can help us make wiser decisions about floodplain settlement and hazard zoning specifications.

7. After reading the above introduction, refer to question #4 in Section 2.

a) What is the recurrence interval for a flood the magnitude of the 1997 flood, based on your plotted line? ___[90 years]___

b) According to your charts, what is the recurrence interval for Grand Forks' peak gage height for that year?

c) What are the probabilities of floods of these magnitudes occurring in any given year?

April 18, 1997 (15.86 m) _____

April 22, 1997 (16.56 m) _____

d) What would be the recurrence interval for a flood to reach the top of the new levee? _____

e) How high above the gage datum would you want to locate a new house in Grand Forks to make certain it would be safe from a 30-year flood? _____

MasteringGeography™ | Looking for additional review and lab prep materials? Go to www.masteringgeography.com for pre lab videos and pre and post lab quizzes.

Name: _____ Laboratory Section: _____

Date: _____ Score/Grade: _____

Video
Exercise 23
Pre-Lab Video

LAB EXERCISE

Scan to view the Pre-Lab video

http://goo.gl/Y7mNBn

23 Contours and Topographic Maps

The character and general configuration of Earth's surface is called **topography**. An important diagnostic tool for landform analysis is the **topographic map**, a map that shows physical landscape features and elevation. While specific application and analysis of topographic maps relating to various types of landforms will be dealt with in later exercises, you have already explored the basic concepts of topographic maps when you analyzed temperature and pressure patterns. **Isolines**, lines connecting points of equal value, are seen on topographic maps as **contour lines**, lines connecting points of equal elevation. These isolines are conceptually similar to the isotherms (equal temperatures) and isobars (equal atmospheric pressure) that you used in Lab Exercises 9 and 11, as well as isohyets (equal precipitation). Therefore, a familiarity with reading and interpreting isolines will be useful at this point. A general discussion of topographic maps and contour lines is the focus of Lab Exercise 23. Lab Exercise 23 features four sections and two optional Google Earth™ activities. These optional questions allow you to compare the topographic maps and your work in the lab manual with Google Earth™ imagery and topographic maps draped over the landscape.

Key Terms and Concepts

Alber's projection
contour interval
contour lines
geographic index number
index contours
isogonic map
isolines
local relief
map view
planimetric map

profile view
relief
stereoscopic contour map
topography
topographic map
topographic profile
Universal Transverse Mercator (UTM) grid
vertical datum
vertical exaggeration

KEY LEARNING concepts

After completion of this lab, you should be able to:

1. *Construct* contour lines (isolines of equal elevation) and *interpret* a mapped landscape.
2. *Construct* a topographic profile and *calculate* an appropriate vertical exaggeration for a profile.
3. *Describe* and *use* the legend and marginal labels and information on a topographic quadrangle map.

Materials/Sources Needed

pencils
colored pencils
calculator
compass

protractor
ruler
stereoscope topographic map of local area

Lab Exercise and Activities

SECTION 1

Contour Lines and Topographic Maps

The USGS depicts information on *quadrangle maps*, rectangular maps bounded by parallels and meridians rather than by political boundaries. A conic map projection, the **Alber's projection** (equal-area), is used as a base for these quadrangle maps (see outside of fold-out flap on back cover of this manual). Two standard parallels (where the cone intersects the globe's surface) are used to improve the accuracy in conformality and scale for the conterminous United States: 29.5°N and 45.5°N latitudes (Figure 23.1). For the U.S. topographic mapping program the standard parallels are shifted for conic projection base maps of Alaska (55°N and 65°N) and Hawai'i (8°N and 18°N).

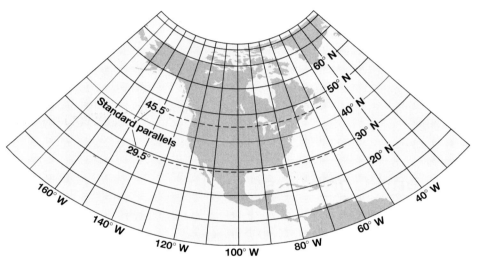

▲ Figure 23.1 Alber's equal-area conic projection with standard parallels of 29.5°N and 45.5°N, a standard for U.S. topographic maps

In mapping, a basic **planimetric map** is first prepared, showing the horizontal position of boundaries; land-use aspects; and political, economic, and social features. A highway map, such as one that you may have in your car, is a common example of a planimetric map. A vertical scale of physical features is then added to portray the terrain. The most popular and widely used of these maps are the detailed topographic maps prepared by the USGS.

Topographic maps portray physical **relief**, the change in elevation between the highest point and lowest point on the map. Local relief refers to the change in elevation between two specified points on the map. Relief is indicated through the use of elevation contour lines. A contour line connects all points at the same elevation above or below a stated reference level. This reference level—usually mean sea level—is called the **vertical datum**. The **contour interval** is the difference in elevation between two adjacent contour lines. (In Figure 23.2, the contour interval is 20 ft or 6.1 m; in Figure 23.4, the contour interval is 40 ft or 12.2 m.)

More specifics on contour lines:

a) Contour lines are typically printed in *brown color* on topographic maps.
b) All points along a contour line are at the same elevation—a contour line is an *isoline* depicting equal elevation.
c) Contour lines separate areas of higher (upslope) elevation from areas of lower (downslope) elevation.
d) Given a large enough map, contour lines always form a closed polygon, although on a smaller map contour lines may run off the margin.
e) The *contour interval* is the difference in elevation between two adjacent contour lines.
f) Contour lines only touch where there is a steep cliff and never cross unless the cliff face has an overhanging ledge. (Hidden contours are then depicted as dashed lines.)
g) Contour lines spaced closer together depict a steeper slope, whereas a wider spacing of contour lines depicts a gentler slope.
h) Concentric, closed contours denote a hill or summit, whereas a similar pattern with *hachure marks* (small tick marks) on the downslope side depicts a closed depression.
i) When a map is showing a depression, the first hachured contour line is the same elevation as the adjacent lower contour line.
j) Contour lines form an upstream-pointing, V-shaped pattern wherever contour lines cross a stream.
k) Contour lines occur in pairs from one side of a valley to the other side.
l) Index contours are thicker and usually have their elevation labeled.

The topographic map in Figure 23.2 shows a hypothetical landscape that demonstrates how contour lines and intervals depict slope and relief. Slope is indicated by the pattern of lines and the space between them. The steeper a slope or cliff, the closer together the contour lines appear in Figure 23.2b; note the narrowly spaced contours representing the cliff. A more gradual slope is portrayed by a wider spacing of these contour lines, as you can see from the widely spaced lines on the beach.

Use the hypothetical landscape and topographic map in Figure 23.2 to answer the following questions:

1. What is the contour interval?

   How can you tell?

2. In terms of local relief (the difference between the highest and the lowest elevation): what is the relief on the west (left) side of the highway? _____ on the east (right) side of the highway?

3. What is the highest point on the map, and what is its elevation?

4. Imagine that you are planning a *low-exertion* walk from the church across the river to the high point on the west side of this landscape. First on Figure 23.2a and then on Figure 23.2b, use a red colored pencil to draw the route you would take; give your reasons for choosing this easier route in terms of elevation change per distance traveled.

(a)

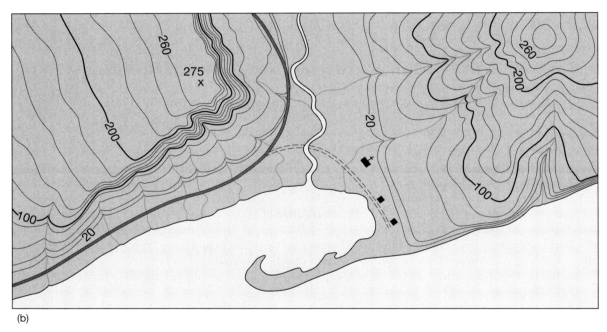

(b)

▲ Figure 23.2 (a) Perspective view of a hypothetical landscape; (b) topographic map of that landscape. (After the U.S. Geological Survey)

5. On each figure use a blue pencil to draw in streams, using arrows to show the direction of stream flow. Note that there is a main stream in the valley and there are at least four tributary streams or creeks. How do the contour lines indicate the direction of flow?

Contour lines and contour maps are two-dimensional images portraying relief, which is a three-dimensional concept. A **stereoscopic contour map** (also called a *stereogram*) helps to show these contour lines in three dimensions. Figure 23.3 shows two views of a hill and valley area: (a) is a contour map, and (b) is a stereoscopic contour map.

Use stereolenses to view Figure 23.3b in three dimensions and compare this view with features marked on the contour map Figure 23.3a.

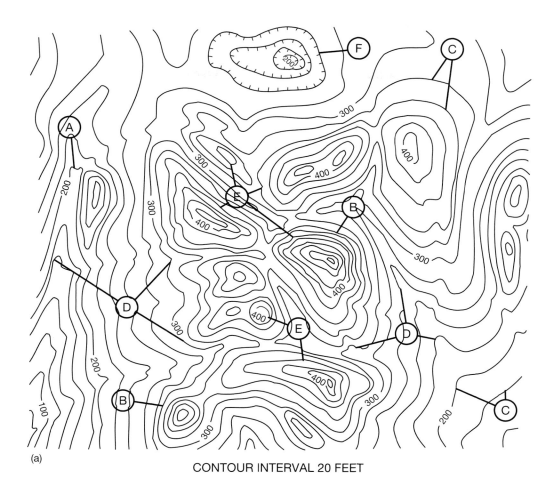

(a) CONTOUR INTERVAL 20 FEET

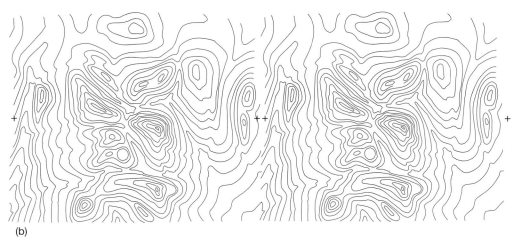

(b)

▲ Figure 23.3
Contour map (a) and stereoscopic contour map (b) of the same hill and valley landscape. (From Horace MacMahan, Jr., *Stereogram Book of Contours.* Copyright © 1995 by Hubbard Scientific Company, pp. 8–9. Reprinted by permission of American Educational Products, Hubbard Scientific.)

A. Each contour line connects points of equal elevation.
B. Varying degrees of steepness are obvious: the steepest areas are where the lines are closest together.
C. Widely spaced contours depict where the slope is gentle.
D. Contour lines crossing stream valleys form a V pointing upstream.
E. Hills are indicated by a series of closed, concentric contours.
F. Depression contours have hachure marks on the downslope side. The hachures are not marked on the stereo pairs in Figure 23.3b as they are on the contour map in Figure 23.3a, but the depression is clearly evident as you view the stereo pair in 3-D through the stereoscope.

Once you become adept at reading contour lines, you will be able to "see" the relief, even though the maps are only in two dimensions. Stereograms will be used again in later lab activities when we examine landforms.

SECTION 2

Constructing Contour Lines

Figure 23.4 presents elevation values for a selected landscape. The contour interval is 40 feet, and **index contours** (thicker lines—on this map every fifth contour, marked with the line's elevation) at 200 feet, 400 feet, and 600 feet are drawn for you in bold isolines. A river channel is noted on the map flowing along a valley separated from the ocean by a ridge. The elevation of the channel is marked at three locations.

Virtual Tour
Stewarts Point

http://goo.gl/o48qGe

Use Figure 23.4 to do the following work and answer the questions.

1. Using a pencil, sketch 40-foot-interval contour lines using the specific site elevations given. Note that we have drawn the 40-ft contour line for you. You must *interpolate* (estimate) elevations between known values. Draw your contour lines lightly at first as you determine the best portrayal, then darken in your work and erase stray pencil marks. (Do not worry if your lines go through the elevation labels.) Remember the contour line basics given in Section 1.
    The coastline is the datum—mean sea level. *Begin your work at the coast and work inland.* (*Hint:* Make this task easier by color-coding the elevations in 40-foot intervals. Circle all elevations from 0–40 feet in green, 41–80 feet in yellow, etc. Then, when drawing the contour lines, you will be "grouping" the colors, with the contour lines as separators between the color groups.)

2. Three vertical control bench marks, indicated by the letters "BM" and marked with an "x," are noted on the map. What are their elevations?

3. Three other specific spot elevations are noted with an "x." What are their elevations?

4. What effect do the intermittent (periodically dry) stream channels have on the topography of the region?

5. Given the trend and location of the 40-foot elevation contour, if you were walking along the water's edge, how would you characterize the topography and relief of the coastline? Steep? Gentle relief?

6. In what compass direction does the river flow? How can you tell? (See Section 1, question 5.)

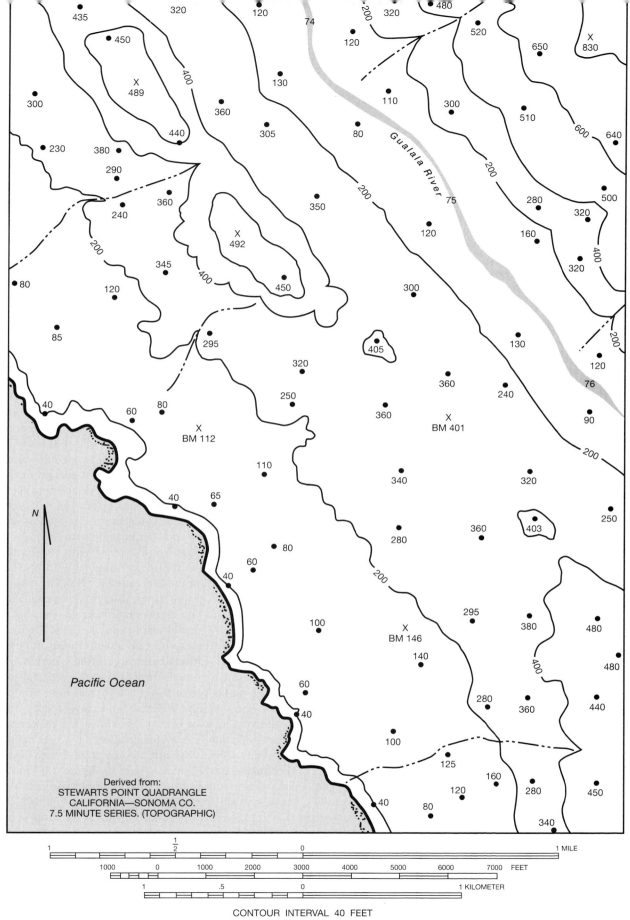

▲ Figure 23.4
A portion of the Stewarts Point quadrangle, enlarged and adapted with 200-, 400-, and 600-foot index contours highlighted.

7. Which portion (NW, NE, SE, or SW) of the map has the greatest relief ?

8. **Challenge question**: If you were assigned the task of building a single general aviation runway 3700 feet in length, where would you place it? Draw your runway plan on the map to the correct scale. (Make it 100 feet wide.) Give your reasons for this site selection.

9. Google Earth™ activity, Stewarts Point quadrangle. For the KMZ file and questions go to mygeoscienceplace.com. Then click on the cover of *Applied Physical Geography: Geosystems in the Laboratory*.

SECTION 3

Topographic Profile

Relief refers to vertical elevation differences in a local landscape. In Figure 23.4 of this exercise, you constructed contour lines and completed an analysis of a coastal landscape. The local topography included a coastal marine platform (terrace), low ridge, stream valley, and hill. The maximum relief on the map was approximately 700 feet, averaging about 400 feet.

An important method of topographic map analysis is construction of a **topographic profile**, the graphic representation of graduated elevations along a line segment drawn on a map. A **map view**, or plan view, is the normal way we view a map as if we were looking straight down from above ("bird's-eye" view). A cross section of a landscape—as if you sliced through the strata and obtained a side view— is a **profile view** and is demonstrated in Figure 23.5. The profile view shows you the "shape" of the land and demonstrates a line-of-sight perspective. As you work with topographic profiles, allow yourself to develop methods with which you are comfortable.

One method of preparing a profile is to draw a line connecting two points (a *transect*) along which you want to obtain the profile (Figure 23.5, Step 1). Take a piece of paper, fold the paper over and crease it for a stiff straight edge. The paper can be plain or graph paper. (Graph paper divided 10 squares to the inch is included at the back of this manual.) Lay the folded edge along the drawn line. Make tick marks at each point where a contour line makes contact with the edge of the paper and note the elevation at that point. These will represent elevation points on the profile graph (Step 2). Carry the tick marks down to the graph at the appropriate elevation, plotting each point; connect the points with a smooth curved line to complete the profile (Step 3).

Construct your first topographic profile using Figure 23.6. Portrayed is an enlarged portion of the Palmyra, New York, quadrangle. The western New York region was blanketed during several advances of continental glaciers. The geomorphic feature you are profiling was deposited by a glacier.

A prepared graph is supplied with the figure; however, let's briefly discuss how it is set up. Note in Figure 23.6 that the horizontal scale of the map is 1:12,000, or 1″ = 888′. If we used this same scale for the vertical axis on the topographic profile graph, the local relief of 192 feet would only cover 0.192 inch on the graph. To construct a readable and useful profile a technique of **vertical exaggeration** is employed. In this figure a vertical scale of 1:1200, or 1′ = 100′, is used and represents a 10 * (times) exaggeration of the horizontal scale. Other landscapes require different vertical exaggerations; the lower the maximum relief of a landscape, the greater the exaggeration should be to conveniently fit on a graph for analysis. For instance, the greater relief in Figure 23.7 made an exaggeration of 3.3 * (times) more appropriate as compared to the 10 * exaggeration in Figure 23.6. The horizontal scale should be left at the same scale as the map.

Follow the procedure illustrated in Figure 23.6 to construct a profile along the line drawn in Figure 23.6, making tick marks at each point where the drawn line and contour lines intersect. Carry the elevations denoted by the tick marks down to the graph, plotting each point. The steeper the slope, the closer the points are placed; a gentle slope places them farther apart. The last step is to connect the points with a smooth curved line to visualize the relief and topography of this landscape. You may want to shade the area below the line to better display the profile.

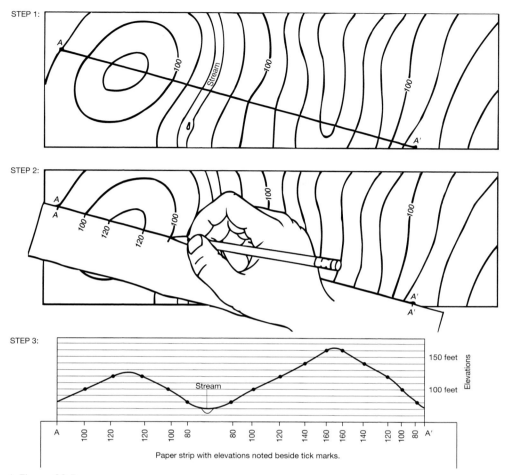

▲ Figure 23.5
Constructing a topographic profile using the edge of a piece of paper. (From Busch, Richard M., editor, *Laboratory Manual in Physical Geology*, 3rd ed., Macmillan Publishing Company © 1993.)

Use the profile constructed in Figure 23.6 to answer the following questions and completion items.

1. What distance does this topographic profile cover?

2. What is the maximum relief along this profile?

3. The landform feature you have profiled is a drumlin. Drumlins are formed by glacial deposits and are streamlined in the direction of the glacier's movement. They have a blunt end upstream, a tapered end downstream, and a rounded summit. Using your protractor, what direction was the glacier flowing?

4. Using the 500-foot contour, what is the width of this feature?

5. If we used a vertical scale identical to the horizontal scale, how many squares on the graph would accommodate the maximum relief along the profile?

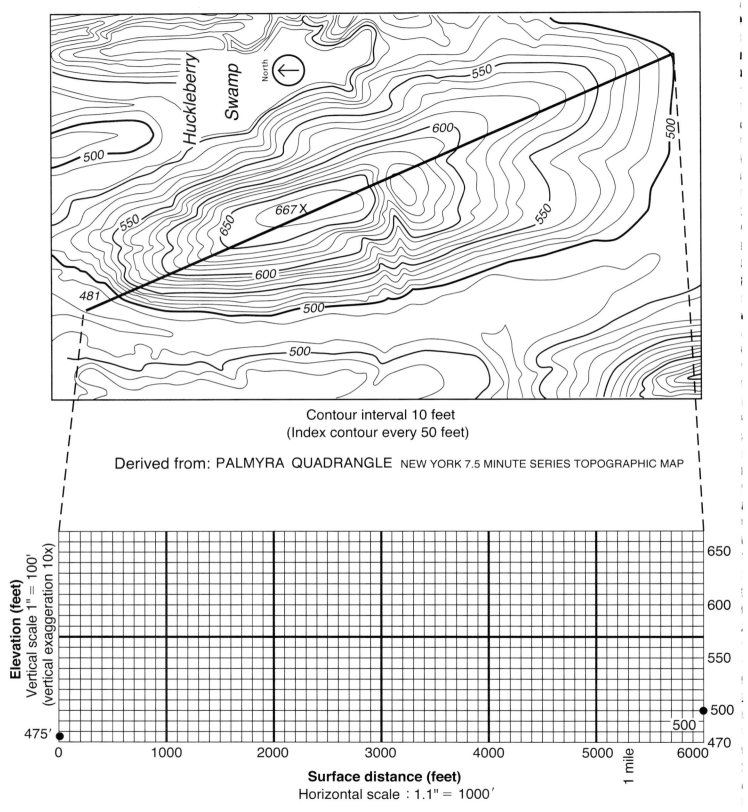

▲ Figure 23.6
Topographic profile from a portion of the Palmyra, New York, quadrangle (enlarged).

230 Lab Exercise 23

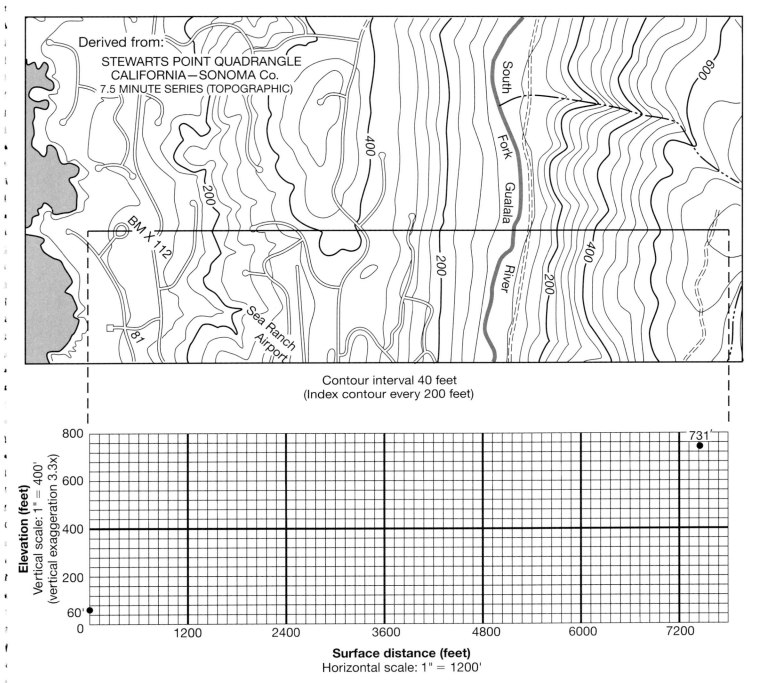

▲ Figure 23.7
Topographic profile from a portion of the Stewarts Point, California, quadrangle map (enlarged). From west to east along the profile, this landscape features a coastal marine terrace, a ridge, a river valley, and a hill rising to 731 feet above sea level. The coastline forms a contour line of equal elevation.

6. Google Earth™ activity, Palmyra quadrangle drumlin. For the KMZ file and questions go to mygeoscienceplace.com. Then click on the cover of *Applied Physical Geography: Geosystems in the Laboratory*.

A second example of a vertical profile is in Figure 23.7. Using the same procedure as you did on the previous assignment, construct a topographic profile. Label the South Fork of the Gualala River (see Figure 23.5, Step 3), the bench mark at 112 feet, and various road crossings on your profile. The coastline forms a contour line of equal elevation—sea level. Note the change in contour interval and scale from that of Figure 23.6.

Use Figure 23.7 to answer further questions about preparing a topographic profile.

7. What is the length of this topographic profile in miles? in kilometers?

8. How wide is the marine terrace (a relatively flat platform or "shelf" along the coast) in miles? in kilometers?

9. If we used a vertical scale identical to the horizontal scale, how many squares on the graph would accommodate the maximum relief along the profile?

10. What is the relief (elevation difference between the highest and lowest points, in a local landscape) in the river valley along southwest slopes?

    along the northeast slopes?

11. Use your ruler to draw a line on the elevation profile in Figure 23.7 from the 731′ dot to the crest of the hill on the transect. This line represents your line of sight from the 731′ elevation. Lightly shade the locations that you cannot see from the 731′ elevation.

12. Can you physically see the shoreline when standing at the summit at 731 feet elevation? Explain.

    Can you see the marine terrace from the summit?

    Can you see the northwest end of the runway at Sea Ranch Airport (a point just south of the profile) from the summit? Explain.

SECTION 4

Interpreting Topographic Map Information

Topographic maps in the series of quadrangle maps prepared by the USGS, or Centre for Topographic Information in Canada, provide a wealth of information. Have your lab instructor show you an index map for your state or province that portrays all the quadrangle map coverage available. Your lab instructor will provide you with a topographic map sheet to examine as you follow along with this description of its elements.

The name of the topographic map quadrangle is given in the upper-right corner, with the state name and 15-minute or 7.5-minute series designation. If it is a 7.5-minute quad, it also signifies which portion of the published 15-minute quad it represents ("SW/4, NE/4," etc.). Other grids, in addition to latitude/longitude, are in use, so geographic coordinates of the **Universal Transverse Mercator (UTM)** grid system are listed near the corner margins of the quad map. The Public Lands Survey township and range system used in maps of the western United States is also noted.

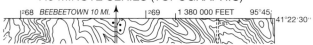

The lower-right corner carries the name of the quad; portion of 15-minute quad, if applicable; **geographic index number**; and the year of the map. The geographic index number is derived from the latitude and longitude of the lower-right corner of the map. Each corner also features complete geographic coordinates.

COUNCIL BLUFFS NORTH, IOWA—NEBR.
N4115–W9545/7.5
1956
PHOTOREVISED 1969 AND 1975
AMS 6866 IV SE—SERIES V876

The upper-left corner always carries the same credit line for the primary government agent for U.S. mapping.

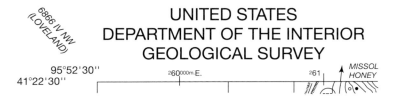

A more complete credit line and map preparation history is in the lower-left margin. If cooperative assistance from another entity was only minor, this may be listed in this place. A variety of such information might be featured in this label.

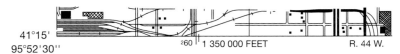

Mapped, edited, and published by the Geological Survey

Control by USGS, USC&GS, and USCE

Topography from aerial photographs by Kelsh plotter
Aerial photographs taken 1952. Field check 1956

Polyconic projection. 1927 North American datum
10,000-foot grid based on Iowa coordinate system, south zone
1000-metre Universal Transverse Mercator grid ticks,
zone 15, shown in blue

Red tint indicates area in which only landmark buildings are shown

Revisions shown in purple compiled from aerial photographs
taken 1969 and 1975. This information not field checked

Purple tint indicates extension of urban areas

Where applicable, the names of adjoining topographic quadrangles are given in parentheses at each corner and along the sides. If this reference appears, you can assume that the map name refers to a quad at the same scale and the same series.

Earth's geographic North Pole and magnetic North Pole locations do not coincide. In addition, the magnetic pole slowly migrates from year to year. The lower left-center margin features a magnetic declination diagram for the year of the map field survey or date of map revision. This declination is taken from an **isogonic map** (Figure 3.3) that maps magnetic declination. This diagram allows you to correct compass bearing readings when using the topographic map. A state map graphically shows the location of the quad in the lower right-center margin. Here are some samples of this designation:

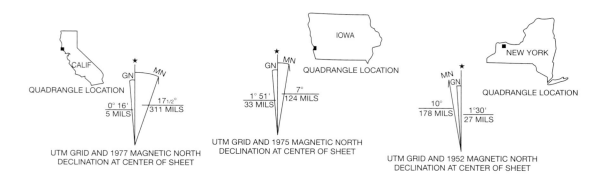

The map scale (representative fraction and graphic or bar scale), contour intervals, any supplementary contour intervals, vertical datum ("National Geodetic Vertical Datum of 1929" appears on post-1975 maps), depth or sounding information in bodies of water, and sales information are shown in the lower-center margin of the map sheet.

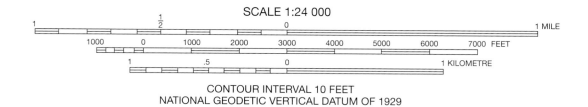

A complete map symbol key should be used for interpreting the map (inside front cover of this lab manual), although road symbols used for the quad are shown in the lower-right corner.

INTERIOR—GEOLOGICAL SURVEY, RESTON, VIRGINIA–1977
269 000m. E.

4570000m. N.
41°15'
95°45'

(MINEOLA)
6866 II NW

ROAD CLASSIFICATION

Heavy-duty ____ _____   Light-duty ____ _____

Unimproved dirt = = = = = = =

U.S. Route   State Route

Interstate Route

Use a topographic map *provided by your lab instructor* to answer the following questions and completion items.

1. Name, state, and series of the map:

2. If a 7.5-minute map, give its location and position relative to the 15-minute series quad for this region:

3. Did any other agencies or entities participate in the preparation of this map?

4. List the geographic index number for this quad:

5. Denote the latitude and longitude of each corner of the quad map:

   NE corner:

   SE corner:

   SW corner:

   NW corner:

6. List the nearest UTM (Universal Transverse Mercator) coordinate for the upper-right corner of the map:

7. Describe the complete map history and credit line:

8. If you needed to obtain the maps adjoining this quadrangle, what are their names? Label these names along the appropriate sides and corners on the quadrangle.

Copyright © 2015 Pearson Education, Inc.

9. What is the representative fraction scale for this map?

   Convert this to a written (verbal) scale. (Show your work.)

10. Using the graphic or bar scale on the map, determine the distance between two points selected by your instructor. The points are: _____ and _____
    Distance in kilometers _____ ; and miles _____

11. Calculate the total area of the map in square kilometers. (Show your work.)

    Now calculate the area in square miles. (Show your work.)

12. What is the magnetic declination for your quadrangle map?

    Which direction (and how many degrees) would you have to turn in order to adjust for this declination?

    To what azimuth would your compass needle point?

    Hikers and backpackers frequently include topographic maps and compasses among their equipment. Magnetic declination must be taken into account when using a compass, and adjustments must be made. To do this, simply superimpose the magnetic compass points (based on the declination arrow) over the true north compass points (N, E, W, S). It is easy to see that, if the magnetic declination is 10°E, you must turn 10° west, so that your compass north arrow is pointing to 350° (360°−10° = 350°).

13. List any additional information provided in the margins of this particular map quadrangle:

    Your instructor may want to provide you with additional questions pertaining to the topographic map that has been selected for this activity. Questions may include features specific to the chosen map and use of symbols for which you might need to consult the topographic map legend on the inside front cover of this manual.

MasteringGeography™ | Looking for additional review and lab prep materials? Go to www.masteringgeography.com for pre lab videos and pre and post lab quizzes.

Name: _____ Laboratory Section: _____

Date: _____ Score/Grade: _____

Video
Exercise 24
Pre-Lab Video

LAB EXERCISE

Scan to view the Pre-Lab video

http://goo.gl/JZKYs3

24 Topographic Analysis: Fluvial Geomorphology

One of the benefits you receive from studying physical geography is an enhanced appreciation of the scenery. As we visit Earth's varied landscapes, we witness the active physical processes that sculpt the land. *Endogenic processes* produce internal forces within our planet that build landforms. However, as the landscape is formed, a variety of *exogenic processes* produce external forces that simultaneously wear these features down. The landscape we see is the product of this continuous struggle of building and destruction.

Geomorphology is a science that analyzes and describes landforms—their origin, evolution, form, and spatial distribution. Geomorphology is an important discipline within physical geography.

Denudation is a general term that refers to all processes that cause reduction or rearrangement of landforms. The principal denudation processes affecting surface materials include *weathering, mass movement, erosion, transportation*, and *deposition*, as produced by the agents of moving water, air, waves, and ice, and the pull of gravity.

Fluvial (moving water; overland flow and streams) erosion processes and their resulting landforms dominate land surfaces throughout the world. Specific geomorphic landforms and processes resulting from fluvial actions are analyzed in this exercise using topographic maps and illustrations. Lab Exercise 24 features three sections and three Google Earth™ activities.

Key Terms and Concepts

alluvium
anticline
backswamp
denudation
drainage basins
drainage patterns
floodplain
fluvial

geomorphology
natural levees
oxbow lakes
syncline
watershed
water gap
wind gap
yazoo tributary

After completion of this lab, you should be able to:

1. *Analyze* and *describe* fluvial geomorphic processes using topographic maps, photo stereopairs, and Google Earth™ imagery.

2. *Interpret* several fluvial processes and the characteristic landscapes that are produced, including drainage basins and floodplains.

3. *Analyze* fluvial processes in a folded landscape, Cumberland Gap, Maryland.

Materials/Sources Needed

pencil
ruler

calculator
stereolenses

Copyright © 2015 Pearson Education, Inc.   Lab Exercise 24   237

Lab Exercise and Activities

SECTION 1

Drainage Patterns and Floodplains

Stream-related processes are termed **fluvial** (from the Latin *fluvius*, meaning "river"). Fluvial systems exhibit characteristic processes and produce predictable landforms. The erosive action of flowing water and the deposition of stream-transported materials produce landforms. A stream system behaves with randomness, unpredictability, and disorder as described by the phenomena of chaos. Yet, enough regularity exists to allow some predictability.

A stream is a mixture of water and solids—carried in solution, suspension, and by mechanical transport. **Alluvium** is the general term for the clay, silt, and sand transported and then deposited by running water.

Streams may drain large regions. Consider the travels of rainfall in north-central Pennsylvania. This water feeds hundreds of small streams that flow into the Allegheny River. At the same time, rainfall in southern Pennsylvania feeds hundreds of streams that flow into the Monongahela River. The two rivers then join at Pittsburgh to form the Ohio River. The Ohio flows southwestward and at Cairo, Illinois, connects with the Mississippi River, which eventually flows on past New Orleans into the Gulf of Mexico. Each contributing tributary, large or small, adds its discharge and sediment load to the larger river.

Streams are organized into areas or regions called **drainage basins**. A drainage basin is the spatial geomorphic unit occupied by a river system, defined by ridges that form *drainage divides*, i.e., the ridges are the dividing lines that control into which basin precipitation drains (Figure 24.1).

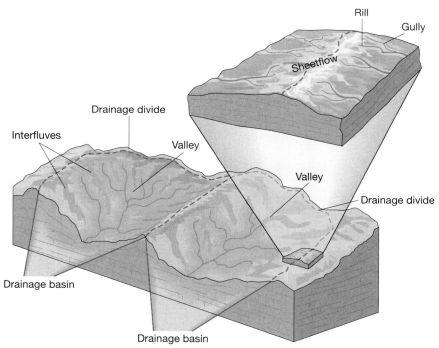

▲ Figure 24.1
The drainage basin and its watershed, separated from other basins by a drainage divide.

Drainage divides define a **watershed**, the catchment area of the drainage basin. A major drainage basin system, such as the one created by the Mississippi-Missouri-Ohio river system, is made up of many smaller drainage basins, which in turn comprise even smaller basins, each divided by specific watersheds. Surface runoff concentrates in *rills*, or small-scale indentations, which may develop further into *gullies* and *stream courses* with distinct drainage patterns. The resultant **drainage pattern** is an arrangement of channels determined by slope, differing rock resistance, climatic and hydrologic variability, and structural controls imposed by the landscape.

Figure 24.2 shows seven common drainage patterns. A most familiar pattern is *dendritic drainage* (a). This treelike pattern (from the Greek word dendron, meaning "tree") is similar to that of many natural systems, such as capillaries in the human circulatory system, or the vein patterns in leaves, or tree roots. Energy expended by this drainage system is efficient because the overall length of the branches is minimized. The *trellis* drainage pattern (b) is characteristic of dipping or folded topography, which exists in the nearly parallel Ridge and Valley Province of the eastern United States, where drainage patterns are influenced by rock structures of variable resistance and folded strata. We see this later in this lab, in the Cumberland topographic map and photo stereopair in Section 3.

The sketch in Figure 24.2b suggests that the headward-eroding part of one stream could break through a drainage divide and *capture* the headwaters of another stream in the next valley, and indeed this does happen. The sharp bends in two of the streams in the illustration are called *elbows of capture* and are evidence that the stream has breached a drainage divide. This type of capture, or *stream piracy*, occurs in other drainage patterns.

The remaining drainage patterns in Figure 24.2 are caused by other specific structural conditions. A radial drainage pattern (c) results from streams flowing off a central peak or dome, such as occurs on a volcanic mountain. We will see this later in the Mt. Rainier topographic map in Lab Exercise 25, Section 2. *Parallel* drainage (d) is associated with steep slopes. A *rectangular* pattern is formed by a faulted and jointed landscape, directing stream courses in patterns of right-angle turns (e). Structural domes, with concentric patterns of rock strata guiding stream courses, produce *annular* patterns (f). In areas having disrupted surface patterns, such as the glaciated shield regions of Canada and northern

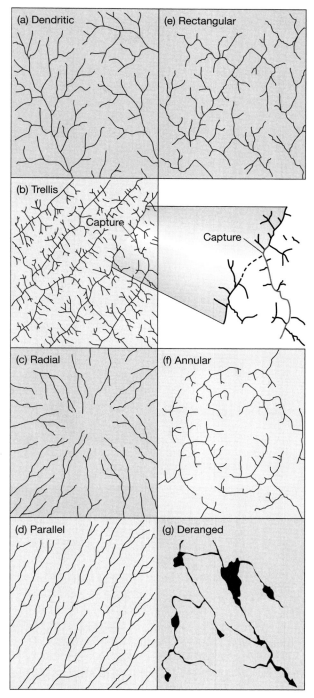

▲ Figure 24.2
Seven most common drainage patterns [After A. D. Howard, "Drainage analysis in geological interpretation: A summation," *Bulletin of American Association of Petroleum Geologists* 51 (1967), p. 2248. Adapted by permission.]

Europe, a *deranged* pattern (g) is in evidence, with no clear geometry. This will be seen in the Jackson, Michigan, topographic map in Lab Exercise 25, Section 3.

The low-lying area near a stream channel that is subjected to recurrent flooding is a **floodplain**, formed when the river leaves its channel during times of high flow. Thus, when the river channel changes course or when floods occur, the floodplain is inundated with water. When the water recedes, alluvial deposits generally mask the underlying rock. Episodic flood events are often devastating, as we witnessed during the Midwest floods of summer 1993; Ohio River floods of late winter 1996–1997; and Red River floods that drowned Fargo and Grand Forks, North Dakota, and the regions surrounding Winnipeg, Manitoba, in spring 1997; or North Carolina in 1999, following the landfall of three hurricanes. Figure 24.3 illustrates a characteristic floodplain, with the present river channel embedded in the plain's alluvial deposits.

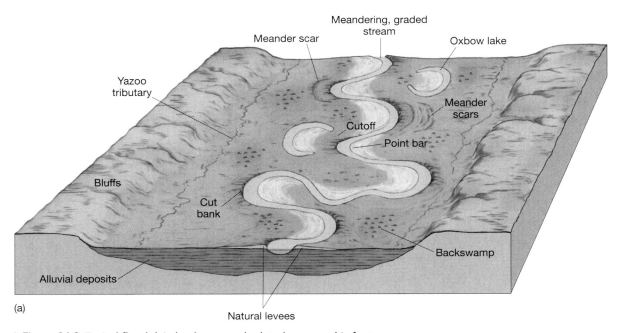

▲ Figure 24.3 Typical floodplain landscape and related geomorphic features

On either bank of most streams, **natural levees** develop as by-products of flooding. When flood waters arrive, the river overflows its banks, loses velocity as it spreads out, and drops a portion of its sediment load to form the levees. Notice on Figure 24.3 an area labeled **backswamp** and a stream called a **yazoo tributary**. The natural levees and elevated channel of the river prevent this yazoo tributary from joining the main channel, so it flows parallel to the river and through the backswamp area. When a meandering stream erodes its outside bank as the curve migrates downstream, the neck of land created by the looping meander eventually erodes through and forms a *cutoff*. When the former meander becomes isolated from the rest of the river, the resulting **oxbow lake** may gradually fill in with silt or may again become part of the river when it floods.

Drainage Patterns and Drainage Basins

Figures 24.4 through 24.7 show some of the different types of drainage basins. Use Figure 24.2 to help determine which type of pattern is shown in each figure.

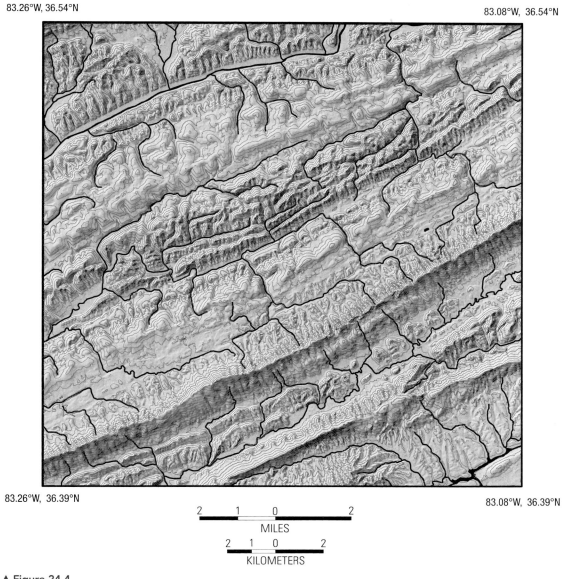

▲ Figure 24.4

1. Which type of drainage pattern is shown in Figure 24.4? Briefly describe the structural conditions that have created this pattern.

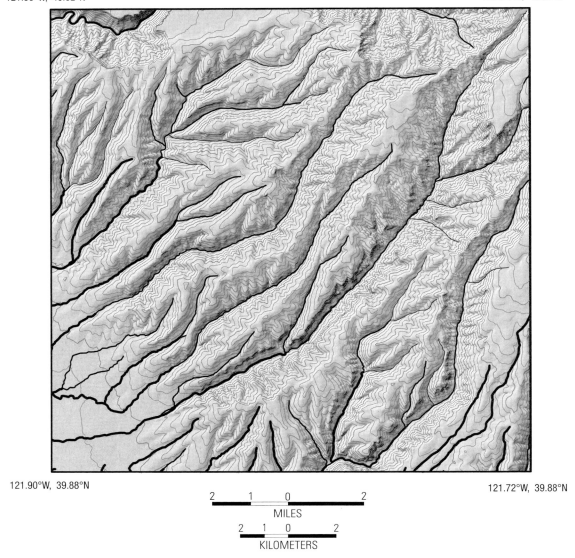

▲ Figure 24.5

2. Which type of drainage pattern is shown in Figure 24.5? Briefly describe the structural conditions that have created this pattern.

122.26°W, 41.46°N   122.13°W, 41.46°N

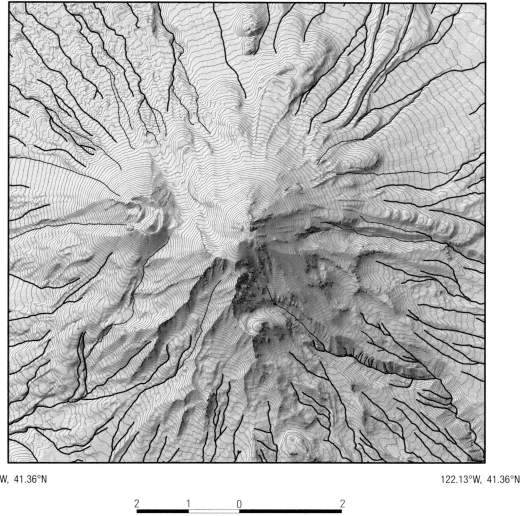

122.26°W, 41.36°N   122.13°W, 41.36°N

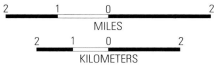

▲ Figure 24.6

3. Which type of drainage pattern is shown in Figure 24.6? Briefly describe the structural conditions that have created this pattern.

119.55°"W, 37.85°N   119.26°W, 37.85°N

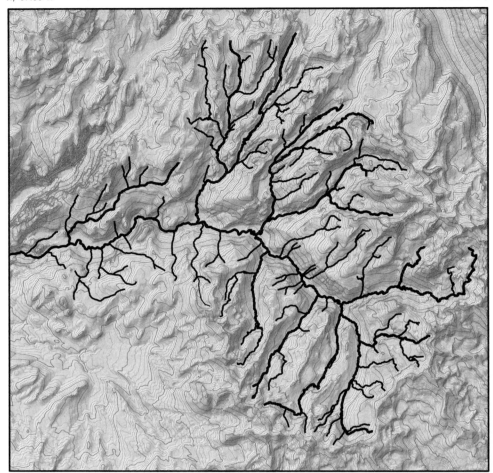

119.55°W, 37.62°N   119.26°W, 37.62°N

▲ Figure 24.7

4. Which type of drainage pattern is shown in Figure 24.7? Briefly describe the structural conditions that have created this pattern.

Philipp, Mississippi, Quadrangle

Topographic Map #2 is a portion of the Philipp, Mississippi, 7.5-minute quadrangle in northwestern Mississippi. The Tallahatchie River flows through the landscape from west to east. (Remember to use the topographic map symbol legend inside the front cover of this manual.)

Analyze **Topographic Map #2** closely and answer the following questions and completion items (Figure 24.4).

Virtual Tour
Philipp, MS

http://goo.gl/o48qGe

5. What is the contour interval on this map segment?

   Given this value, determine the maximum relief shown:

   What is the highest spot elevation ("X") on the map and in what PLSS (Public Land Survey Section) section is it located?

6. How are the levees along the river portrayed by the elevation contours? What is the elevation at the top of the levees?

7. What is the elevation of the water surface of the lake north of the Jones Chapel Cemetery?

8. What type of geographic feature is the water body called Fish Lake, and how did it form? Use the illustration in Figure 24.3 to identify this water feature.

9. With a dark blue or black colored pencil or pen, outline all the water features that were formed by the same process as Fish Lake. What is the name of the widest of these water features on the map?

10. There is a stream connecting Possum Bayou (shown on the map as just *Possum*) with the smaller lake called Fish Lake. With a dark blue or black colored pencil or pen, draw an arrow indicating which direction the water is flowing.

11. Do the fluvial features have any effect on road patterns? Explain and give some examples. Highlight your examples with a colored pencil or pen on the topographic map.

12. Would you go fishing in the area labeled Shegogue Lake in LeFlore County? What kind of feature is the area labeled Shegogue Lake in LeFlore County? Circle or highlight at least two other areas on the map that are the same type of feature.

13. In the space below, sketch the topographic map symbol for a marsh or swamp. With a blue colored pencil or pen, lightly shade the backswamp areas on the topographic map.

14. Optional Google Earth™ activity, Drainage patterns. For the KMZ file and questions go to mygeoscienceplace.com. Then click on the cover of *Applied Physical Geography: Geosystems in the Laboratory*.

SECTION 2

Sacramento River, California

Virtual Tour
Sac River

http://goo.gl/o48qGe

Streams often are used as natural political boundaries, but it is easy to see how disagreements might arise when boundaries are based on river channels that shift around. For example, the Ohio, Missouri, and Mississippi Rivers can shift their positions quite rapidly during times of flood, thus creating a mismatch between the boundary and river.

The boundary separating Glenn and Butte Counties in California provides us with an example, as shown in Topographic Map #3. The Glenn–Butte border was originally placed midchannel in the meandering Sacramento River. The river has shifted course, but the state boundary line still follows the former channel meander, placing parts of Glenn County on the same side of the river as Butte County and vice versa!

Topographic Map #3 is a portion of the Ord Bend 7.5-minute quadrangle, in northern California. The Sacramento River flows through the landscape from north to south.

Analyze **Topographic Map #3** and answer the following completion items.

1. What is the contour interval on this map segment?

    Given this value, determine the relief from Dunning Slough in the north to Perch Slough in the south.

    What are the elevations of several bench marks ("BM") on the map?

2. Use a piece of string and the scale bar on the map to measure the river from the north end of the map to the south end. How many feet did you measure? What is the straight-line distance from the river at the north end of the map to the south end in feet?

The ratio of river distance to straight-line distance is the sinuousity of the river. For meandering streams, it is greater than 1.5. What value would you expect to find, based on the type of river this is? What value did you find?

3. Shade in areas on Topographic Map #3 north of 39° 43' on the Glenn County (west) side of the river but legally in Butte County and vice versa.

4. What kind of features are labeled Murphy Slough and Jenny Lind Bend? Use a green colored pencil or pen and color in the other features of that type on the map.

5. Use a red colored pencil or pen and outline at least two meander scars on the map.

6. Which set of features was formed first, and which set is older: those named in question #4 or question #5? Briefly explain how those features are formed by the river. Begin with the river.

7. Why do you think Phelan Island and Golden State Island have the names they do? Describe how they were formed.

8. Use a blue colored pencil or pen and shade in the backswamp areas on the map.

9. At the meander bend downstream from Golden State Island in Section 5, draw arrows on the cut bank side of the river to indicate the direction that the river will migrate initially. Use a dark blue colored pencil or pen and shade in the point bar region where deposition is occurring now.

10. In the space below, make a sketch showing how the meander bend will become a cutoff, an oxbow lake, and a meander scar.

11. Optional Google Earth™ activity, Sacramento River and political boundaries. For the KMZ file and questions go to mygeoscienceplace.com. Then click on the cover of *Applied Physical Geography: Geosystems in the Laboratory.*

SECTION 3

Cumberland Quadrangle and Photo Stereopair

Virtual Tour Cumberland, MD

http://goo.gl/o48qGe

In folded landscapes, ridges and valleys are the prominent *topographic features*, reflecting the **anticlines** and **synclines** that are the *structural landscape controls*. Anticlines are the upfolds, and synclines are the downfolds. In the Ridge and Valley geomorphic province of the Appalachians, **water gaps** occur where a river cuts through an *anticlinal ridge*. Such a geomorphic feature is thought to form when a preexisting stream course flows across a landscape with buried geologic structures. As the overburden cover is removed, the downcutting stream cuts into the underlying structure. The stream valley is described as *superposed* where Wills Creek cuts through an anticlinal ridge (Haystack Mountain and Wills Mountain) in this example from Cumberland, Maryland. The importance of these water gaps to early transportation and migration is obvious. Note the abandoned canal in the southeast portion of the map.

Wind gaps are passes that were formerly occupied by a stream that subsequently was captured and incorporated into another stream drainage. These abandoned stream valleys are generally dry. Two remarkable examples of these features are on the portion of the Cumberland 7.5-minute quadrangle presented as **Topographic Map #4**.

Figure 24.8 shows two views (a contour map and a stereoscopic contour map) of a hypothetical anticline and syncline landscape. As you did in Lab Exercise 23, use stereolenses to view Figure 24.8b in three dimensions and note selected features marked on the contour map. The anticlines (A) are separated by synclines (B), and the heavy lines (C) and (D) mark the axes along which the folding occurred forming the ridges and troughs. The arrows at (E) point downhill, away from the crest of the anticline, and the arrows at (F) also point downhill, towards the trough of the syncline.

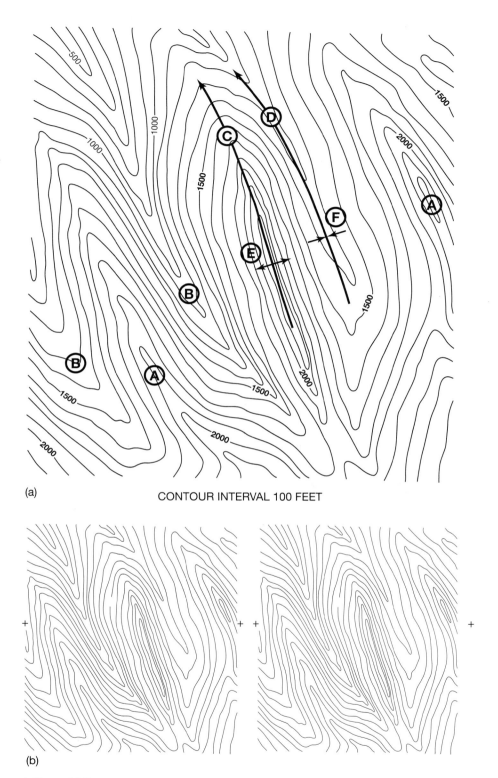

(a) CONTOUR INTERVAL 100 FEET

(b)

▲ Figure 24.8
Contour map (a) and stereoscopic contour map (b) of the same anticline and syncline landscape. (From Horace MacMahan, Jr., *Stereogram Book of Contours*. Copyright © 1995, by Hubbard Scientific Company, pp. 24–25. Reprinted by permission of American Educational Products, Hubbard Scientific.)

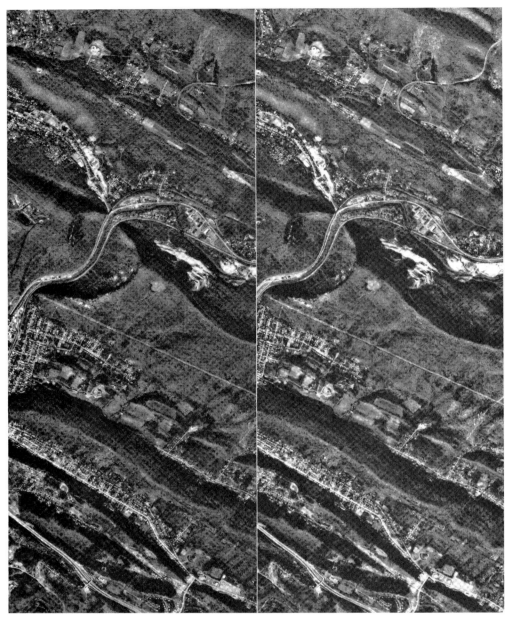

▲ Figure 24.9
Cumberland, Maryland, photo stereopair. North is to the right, west to the top. (NAPP, USGS)

Refer to Section 1 and the review of drainage basins and drainage patterns. Viewing the stereoscopic contour map, it is easy to imagine the direction in which water would flow down the anticlines into streams that will occupy the synclines, establishing the *trellis* drainage pattern in these ridge and valley regions. Keep this stereoscopic image and the trellis drainage pattern in mind when analyzing Topographic Map #4.

Analyze **Topographic Map #4** and the set of stereo photos of the same area in Figure 24.9 to answer the following questions and completion items.

1. What is the contour interval on this map segment?　　　　What is the scale of this map?

2. Wills Creek at The Narrows is a *superposed stream* forming a *water gap*. How wide is the water gap gorge (distance between the 1560-foot contour to the north and the 1480-foot contour to the south)?

   How deep is the water gap at The Narrows?

3. What is the relief between the floor of the wind gap that Braddock Road runs through and the floor of the valley to the north in which the stream Braddock Run flows?

   A wind gap was formed when the lower course of Braddock Run was diverted (captured) by a tributary of Wills Creek (stream in the water gap). How wide is the wind gap, measured from the 1400' contour line on the north rim to the 1500' contour line on the south rim?

   How deep is the gap, measured from the water tank on the north rim to Braddock Road in the gap?

4. What is the elevation of the floor of the wind gap? What is the elevation of the valley to the north of the wind gap (the same valley that Braddock Run and Wills Creek run through)?

   What is the elevation of the valley to the south of the wind gap (the valley with the Allegheny County/Mineral County lines)? What is the vertical relief between the wind gap and the valley to the north? What is the vertical relief between the floor of the wind gap and the valley to the south?

5. Imagine a topographic profile from west of Narrows Park, along Braddock Run at the 700-foot contour line, to the 760-foot contour line south of Allegany High School at the southeast corner of the cemetery. How long is this profile line in straight distance?

   What is the highest point along the profile?

   What is the relief from Allegany High School to the highest point along the profile (south of the water gap)?

6. What topographic features appear to be controlling the stream course of the north branch of the Potomac River (joins Wills Creek along the southeast portion of the map), and what are the names of the structural landscape controls?

7. Optional Google Earth™ activity, Cumberland Gap. For the KMZ file and questions go to mygeoscienceplace.com. Then click on the cover of *Applied Physical Geography: Geosystems in the Laboratory*.

MasteringGeography™ | Looking for additional review and lab prep materials? Go to www.masteringgeography.com for pre lab videos and pre and post lab quizzes.

Name: _____ Laboratory Section: _____

Date: _____ Score/Grade: _____

Video
Exercise 25
Pre-Lab Video

http://goo.gl/46Pujm

LAB EXERCISE 25

Scan to view the Pre-Lab video

Topographic Analysis: Glacial Geomorphology

About 77% of Earth's freshwater is frozen, with the bulk of that ice in just two places—Greenland and Antarctica. The remaining ice covers various mountains and fills some alpine valleys. More than 32.7 million cubic kilometers (7.8 million cubic miles) of water is tied up as ice.

A **glacier** is a large mass of ice, resting on land or floating as an ice shelf in the sea adjacent to land. Glaciers are not frozen lakes or groundwater ice. Instead, they form through the continual accumulation of snow that recrystallizes under its own weight into an ice mass.

Glacial erosion and deposition produce distinctive landforms that differ greatly from the way the land looked before the ice came and went. Worldwide, glacial ice is in retreat, melting at rates exceeding anything in the recent ice record. In the European Alps alone, some 75% of the glaciers have receded in the past 50 years, losing more than 50% of their ice mass since 1850. At this rate, the European Alps will have only 20% of their preindustrial glacial ice left by 2050.

In this lab exercise, topographic map interpretation continues. Specific exogenic geomorphic processes and the resulting landforms from glacial (moving ice) action are analyzed using topographic and stereoscopic contour maps, a photo stereopair, and illustrations. A complete legend to topographic map symbols is inside the front cover of this lab manual. Lab Exercise 25 features three sections and one Google Earth™ activity.

Key Terms and Concepts

alpine glacier
arêtes
cirque
col
continental glacier
drumlin
end moraine
esker
glacier
horn

kame
kettle
medial moraine
moraine
outwash plain
paternoster lakes
tarn
terminal moraine
till plains
valley glacier

After completion of this lab, you should be able to:

1. *Analyze* and *describe* geomorphic processes by using topographic maps and Google Earth™ imagery.
2. *Use* data to *graph, calculate,* and *analyze* glacial mass balance.
3. *Interpret* and *describe* several glacial processes and the characteristic landscapes that result, including erosional and depositional features of alpine and continental glaciers.

Materials/Sources Needed

pencil
ruler
stereolenses
string
calculator

Lab Exercise and Activities

SECTION 1

Glaciated Landscapes

With few exceptions, a glacier in a mountain range is called an **alpine glacier**, or *mountain glacier*. The name comes from the Alps, where such glaciers abound. On a larger scale than individual alpine glaciers, a continuous mass of ice is known as a **continental glacier** and in its most extensive form is called an *ice sheet*. Two additional types of continuous ice cover associated with mountain locations are designated as *ice caps* and *ice fields*.

Alpine glaciers form in several subtypes. One prominent type is a **valley glacier**, an ice mass constricted within a valley that was originally formed by stream action. As a valley glacier flows slowly downhill, the mountains, canyons, and river valleys beneath its mass are profoundly altered by its passage. Most alpine glaciers originate in a mountain snowfield that is confined in a bowl-shaped recess. This scooped-out erosional land form at the head of a valley is called a **cirque**. In the cirques where the valley glaciers originated, small mountain lakes, called **tarns,** may form after the valley glacier fully melts. Some cirques may contain small, circular, stair-stepped lakes, called **paternoster** ("our father") lakes, for their resemblance to rosary (religious) beads. As the cirque walls erode away, sharp ridges form, dividing adjacent cirque basins. These **arêtes** ("knife edge" in French) become the sawtooth, serrated ridges in glaciated mountains. Two eroding cirques may reduce an arête to a saddle-like depression or pass, called a **col**. A **horn** (pyramidal peak) results when several cirque glaciers gouge an individual mountain summit from all sides. The deposition of glacial sediment produces a specific landform, a **moraine**. Several types of moraines are associated with alpine glaciation. A lateral moraine forms along each side of a glacier. If two glaciers with lateral moraines join, a **medial moraine** may form. Other moraines are associated with both alpine and continental glaciation. Eroded debris that is dropped at the glacier's farthest extent is a **terminal moraine**. **End moraines** may also be present, formed at other points where a glacier paused after reaching a new equilibrium between accumulation and ablation. Figure 25.1 shows a cross section of a typical retreating alpine and Figure 25.2 shows a typical post-glaciated landscape.

Virtual Tour
Mt. Darwin

http://goo.gl/o48qGe

Mount Darwin, California, Quadrangle

Analyze **Topographic Map #5**. The topographic map was made in 1994. The NAPP photos were taken in 1999.

1. A cirque basin may serve as an accumulation zone for alpine glaciers. Using contour lines on the mountains and on the glaciers, how many ice-filled cirques (scooped out amphitheater-like basins) do you identify on the map? _____ What is the elevation of the highest glacial ice surface on the map?

2. What type of glacial feature is the unnamed peak, with an elevation of 13,253 ft, north of Mount Darwin?

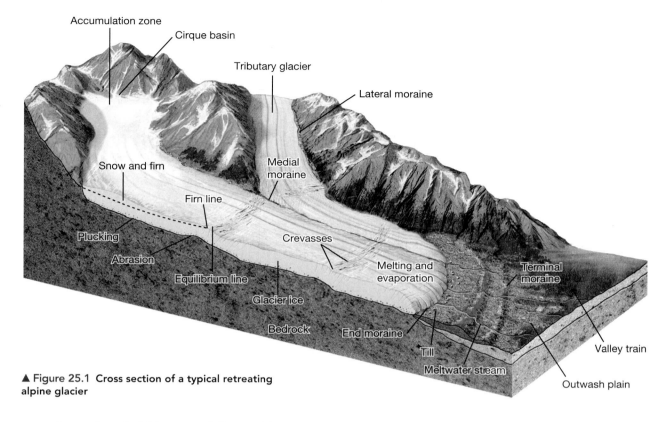

▲ Figure 25.1 Cross section of a typical retreating alpine glacier

3. What is the relief (elevation difference) between Goethe Peak and Goethe Lake?

4. a) What is the linear distance between Goethe Peak and the south shoreline of Goethe Lake?

___

___

b) Take the relief in feet that you determined in question #3 and divide it by the distance between the two locations in miles (divide the distance you found in 4 a) by 5280 to find the distance in miles) In terms of slope, state this distance in feet per mile: (Show your work.)

___

___

c) Take the relief in feet that you determined in question #3 and divide it by the distance between the two locations in feet from 4 a) to determine the percent grade (relief divided by distance). What percent grade did you determine? (Show your work.)

___

___

5. Locate Sky High Lake in Section 34 on the topo map. What type of feature is Sky High Lake? How was this lake formed?

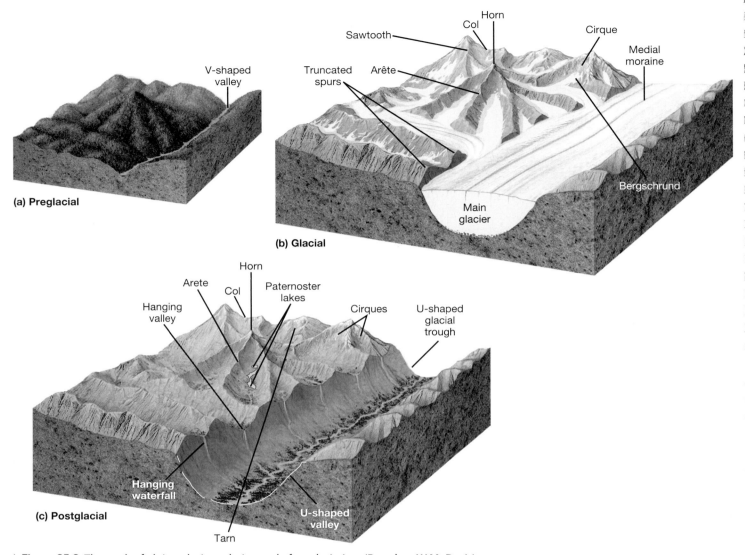

▲ Figure 25.2 The work of alpine glaciers, during and after glaciation. (Based on W.M. Davis)

6. What is the term for the string of lakes in Darwin Canyon?

7. South of Goethe Lake find Mount Goethe on (Topographic Map #5) and photo stereopairs Figure 25.3b.

   a) How many cirque glaciers do you count that surround Mount Goethe? _____

   b) How many are on the north side? _____ South side? _____

   c) Describe the orientation of the slopes with snow on them on the topographic map and the photo stereopair. Is there more snow on north facing slopes or on south facing slopes? Why would there be more snow on one slope than another? How would this affect the amount of glaciation that would occur on each slope?

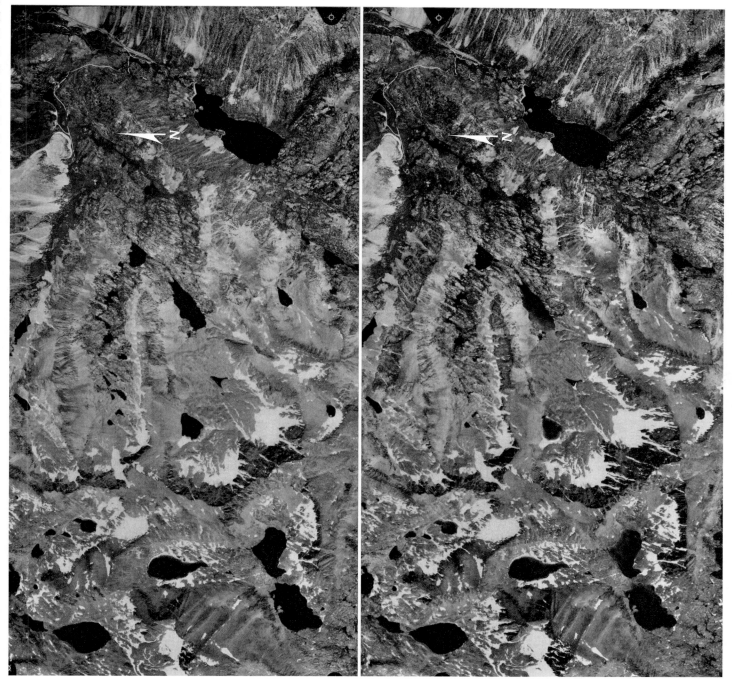

▲ Figure 25.3
Photo stereopair of the Mount Darwin area of California. In this arrangement north is at the left, west is at the bottom.

    d) After looking at the photo stereopair, describe the steepness of the north- and south-facing slopes of Mount Goethe and Mount Lamark. Which side has steeper slopes? Why do you suppose this is?

_____

_____

_____

8. a) What term would you use to describe the ridge west of Mount Goethe? _____

_____

b) What is the term for the feature found on the ridge between Mount Goethe and Muriel Peak?

_____

9. Optional Google Earth™ activity, Mount Darwin, California. For the KMZ file and questions go to mygeoscienceplace.com. Then click on the cover of *Applied Physical Geography: Geosystems in the Laboratory*.

SECTION 2

Alpine Glacier and Mount Rainier Quadrangle

Virtual Tour
Mt. Rainer

http://goo.gl/o48qGe

An alpine (valley) glacier has profound effects on the landscape. Figure 25.2 illustrates views during and after a glacier's passage. A typical river valley has a characteristic **V**-shape and stream-cut tributary valleys. During active glaciation, glacial erosion and transport actively remove much of the regolith (weathered bedrock) and the soils that covered the stream valley landscape. When climates warm and the ice has retreated, the valleys are exposed as **U**-shaped glacially carved valleys, greatly changed from their previous stream-cut form. You can see the oversteepened sides and the straightened course of the valley.

Figure 25.4a shows the contour lines for an alpine glacier scene with key portions labeled, whereas Figure 25.4b is a stereocontour map of an alpine glacier area to view with stereolenses. Please compare this to the alpine glaciation illustration in Figure 25.2. Mount Rainier is featured in **Topographic Map #6**—a dormant composite volcano in the Cascade Range that is covered by several glaciers and glacial features.

Activities and completion items related to alpine glacier geomorphology:

1. Figure 25.4 shows a contour map and a stereoscopic map of a hypothetical landscape that has been shaped by alpine glaciation. As glaciers carved their way down preexisting V-shaped stream valleys, they widened the valleys into U-shaped valleys separated by sharp ridges. Hanging valleys/troughs left stranded above the main valley floor will be occupied by streams that form picturesque waterfalls as they plunge over the steep edge.

   Glaciers may scour out the bedrock of the valley floor, leaving depressions that, as the glacier retreats and a stream reoccupies the valley, subsequently fill with water, forming glacial lakes called tarns. Steep-walled cirques mark the origin of the glaciers high on the mountain slope. Keep this stereoscopic contours map, along with the stereopairs and Mount Darwin topo map, in mind when analyzing the topographic map of Mount Rainier.

   After reviewing Figures 25.1 and 25.2, answer the following questions about Figure 25.4a:

   If you wanted to go swimming in a tarn, which letters would you look for?

   If you were a rock climber and wanted to scale the steep sides of the valley, which letters would you look for?

   Which letters indicate the location of a pass on the ridge between two valleys?

   If you wanted to climb the horn created by three glaciers, which letter would you look for?

   What is the elevation of the peak?

   Find and label the following features on Figure 25.4a: a hanging valley; the U-shaped valley formed by the main glacier; a cirque; a horn; an arête (the narrow ridge that separates two valleys); a col (a pass on an arête); and a chain of paternoster lakes (if they were filled with water).

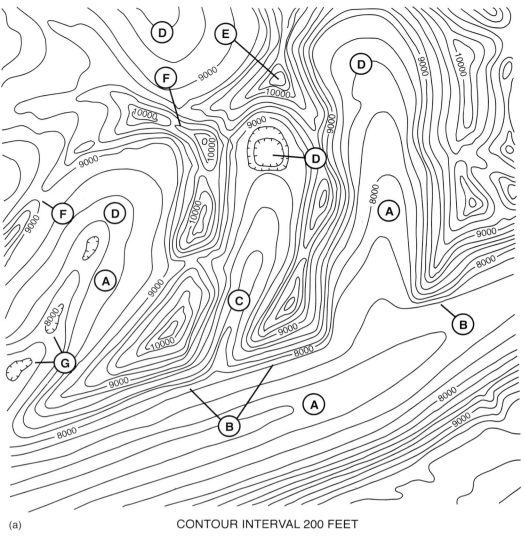

(a) CONTOUR INTERVAL 200 FEET

(b)

▲ Figure 25.4 Alpine glacier landscape
Contour map (a) and stereoscopic contour map (b) of the same alpine glacial landscape. (Adapted from Horace MacMahan, Jr., *Stereogram Book of Contours*. Copyright © 1995 by Hubbard Scientific Company, pp. 12–13. Reprinted by permission of American Educational Products—Hubbard Scientific.)

2. Refer to the Mount Rainier National Park topographic quadrangle—**Topographic Map #6**. What is the scale of this map as a representative fraction (1:???)? as a written scale (one inch equals ??? miles)?

   What is the contour interval of the topo map?

3. Name several glaciers on which you identify medial and lateral moraines. Use a brown colored pencil or dark pen and color down the middles of at least three medial moraines.

4. What type of feature is Sunset Amphitheater?

5. What type of feature is labeled Nisqually Cleaver? List several other features that are the same type, including at least two that are not named Cleaver.

6. The summit of Mount Rainier is at 14,410 ft (4392 m). What is the lowest elevation shown on this map?

   Where is this location? Give its name.

   Therefore, the relief on this map segment is

   What is the linear distance between these high and low points?

   In terms of slope, state this distance and relief in *feet per mile*:

   State this in terms of *percent grade* along an ideal slope between these two points (relief in feet ÷ distance in feet): _____ %.

   Show your work:

SECTION 3

Continental Glaciation and the Jackson, Michigan, Topographic Quadrangle

Continental glaciers advanced and retreated over North America and Europe, producing many erosional and depositional features. Figure 25.5 illustrates some of the most common depositional features associated with the passage of a continental glacier. The unsorted and unstratified deposits of gravel, sand, and clay form moraines, including ground and terminal moraines. Many relatively flat plains of *unsorted* coarse till are formed behind terminal moraines. These **till plains** typically have low, rolling relief, deranged drainage patterns, and the following depositional features. **Drumlins**, smooth hills made of till shaped by the ice, are oriented in the direction of the glacier's movement. **Eskers** are curving, narrow deposits of coarse gravel left by meltwater stream deposits in tunnels beneath the ice. Often eskers end in a delta. **Kames** are small hills of poorly sorted sand and gravel that collected in depressions in the surface of a glacier.

Beyond the morainal deposits, glacio-fluvial **outwash plains** of *stratified drift* feature stream channels that are meltwater-fed, braided, and overloaded with debris deposited across the landscape. **Kettles** are depressions left by melting blocks of ice that were buried in the drift and are found in both ground moraines and in outwash plains.

Virtual Tour
Jackson, MI

http://goo.gl/o48qGe

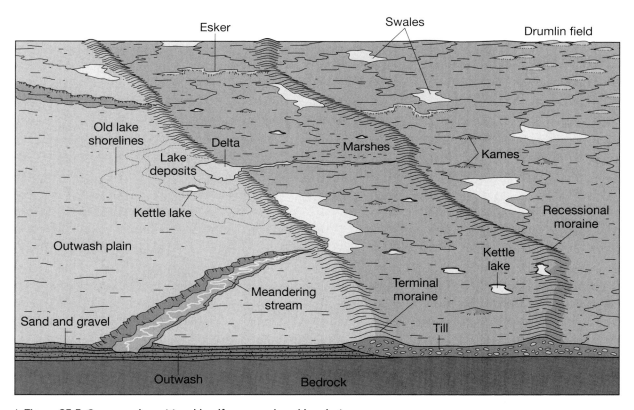

▲ Figure 25.5 Common depositional landforms produced by glaciers

Activities and completion items related to continental glaciation:

1. Figure 25.6 shows two views (a contour map and a stereoscopic contour map) of a hypothetical landscape that has been shaped by continental glaciers. In Figure 25.6a, note the unsorted and unstratified deposits of gravel, sand, and clay that form moraines, including terminal moraines (A) and interlobate moraines (B). Use stereolenses to view the typical continental features in Figure 25.6b in three dimensions.

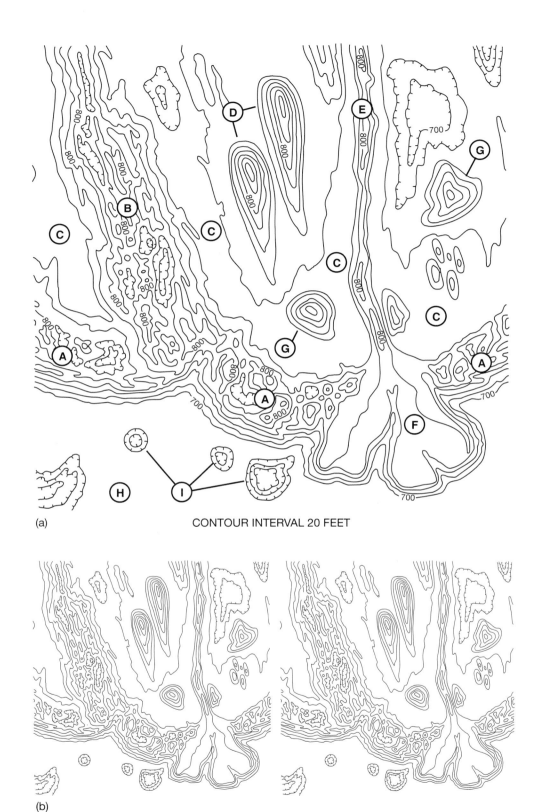

▲ Figure 25.6 Continental glaciated landscape. Contour map (a) and stereoscopic contour map (b) of the same continental glaciated landscape
(From Horace MacMahan, Jr., *Stereogram Book of Contours*. Copyright © 1995 by Hubbard Scientific Company, pp. 14–15. Reprinted by permission of American Educational Products, Hubbard Scientific.)

After reviewing Figures 25.5 and 25.6, answer the following questions about Figure 25.6a:

If you were looking for drumlins, which letters would you look for?

Which letter shows kames?

If you wanted to walk along the crest of an esker, which letter would you look for?

If you were looking for kettles, which letters would you look for?

Which letter is located on the ground moraine?

Which letter is located on the outwash plain?

Once again refer to Figure 24.2g in Lab Exercise 24 and the review of drainage basins and drainage patterns. Viewing the stereoscopic contour map, imagine the deranged drainage patterns that will result from the disruption of former surface patterns by the erosional and depositional work of continental glaciation.

Keep Figures 25.5 and 25.6 in mind when analyzing **Topographic Map #7** of the formerly glaciated area around Jackson, Michigan.

2. Refer to the Jackson, Michigan, topographic quadrangle—**Topographic Map #7**. What is the scale of this map?

    Convert this to a verbal scale. Show your work.

    What is the contour interval?

3. A sinuously curving, narrow ridge of coarse sand and gravel is called an esker. Eskers form along the channel of a meltwater stream that flows beneath a glacier, in an ice tunnel, or between ice walls beneath the glacier. As a glacier retreats, the steep-sided esker is left behind in a pattern roughly parallel to the path of the glacier.

    Locate and identify by name a prominent esker on the map:

    What is the average elevation of this esker?

    How long is this esker, in miles? (Use a string placed along the esker, then pull it straight and compare it to the scale.)

4. Identify the drainage pattern for the portion of the topo map that is generally covered by the Kalamazoo moraine (southwest part of map). Explain what evidence you used to determine this.

5. Use a dark blue colored pencil or pen to trace the drainage of the Grand River. Add arrows to indicate the direction it is flowing.

6. Is this area well or poorly drained? Explain your answer.

7. Sometimes an isolated block of ice, perhaps more than a kilometer across, persists in a *ground moraine*, an outwash plain, or valley floor after a glacier retreats. Perhaps 20 to 30 years are required for it to melt. In the interim, material continues to accumulate around the melting ice block. When the block finally melts, it leaves behind a steep-sided hole. Such a feature then frequently fills with water. This is called a kettle. Locate several named kettles that are ponds or lakes on this map and identify their general location:

8. What type of feature is Hatt Hill? Briefly describe how it was formed. Use a red colored pencil or pen and circle several other of these features on the map.

9. If you were walking from Hammer Lake to Blue Ridge, would you be walking over a till plain or an outwash plain? Explain.

10. What is the highest elevation on the map segment? Location and name:

    What is the lowest elevation on the map segment? Location and name:

Name: _____ Laboratory Section: _____ Video
Date: _____ Score/Grade: _____ Exercise 26
Pre-Lab Video

http://goo.gl/Q200B1

LAB EXERCISE 26

Scan to view the Pre-Lab video

Topographic Analysis: Coastal and Arid Geomorphology

In addition to streams of water and ice—rivers and glaciers—carving their way across Earth's surface, other exogenic processes are also at work. Waves are relentless in their efforts to reshape the contours of the continents, whether sandy **beaches** or rocky cliffs, along the narrow contact zone between the landmasses and oceans. Anthropogenic (human-caused) processes are also very important, as we will examine in the development of the New Jersey coast. And landforms that are characteristic of arid climates are limited to global areas where adequate precipitation is always lacking, allowing wind and intermittent streams to remodel the surface.

Topographic map interpretation continues in this lab exercise. Specific geomorphic processes and the resulting landforms from coastal (wave) action and the work of streams in arid climates are analyzed using topographic maps, photo stereopairs, and illustrations. Once again, refer to the complete legend of topographic map symbols located on the inside front cover of this lab manual. Lab Exercise 26 features four sections and four Google Earth™ questions.

Key Terms and Concepts

alluvial fan
barrier spit
bajada
basin and range
bay barrier
beach
horst
graben

interior drainage
lagoon
littoral zone
playa
sea cliff
tombolo
wave-cut platform

After completion of this lab, you should be able to:

1. *Analyze* and *describe* geomorphic processes using topographic maps and Google Earth™ imagery.

2. *Interpret* and *describe* several coastal processes and the characteristic landscapes that result, including erosional and depositional features.

3. *Interpret* and *analyze* patterns of development in sensitive coastal areas and *suggest* appropriate development strategies.

4. *Interpret* and *describe* several arid climate processes and the characteristic landscapes that are produced, including alluvial fans, buttes, and mesas.

5. *Analyze* and *evaluate* the water budget of an exotic stream and *predict* future water issues.

Materials/Sources Needed

pencil, colored pencils, or pens
ruler

calculator
stereolenses

Lab Exercise and Activities

SECTION 1

Coastal Features and the Point Reyes Quadrangle and Stereophotos

Most of Earth's coastlines are relatively new and are settings for continuous change. The land, ocean, atmosphere, Sun, and Moon interact to produce tides, currents, and waves that create erosional and depositional features along the continental margins. A dynamic equilibrium among the energy of waves, wind, and currents; the supply of materials; the slope of the coastal terrain; and the fluctuation of relative sea level produces coastline features of infinite variety. The coastal environment is called the **littoral zone**. (Littoral comes from the Latin word *litus*, meaning "seashore.") The littoral zone spans both land and water. Landward, it extends to the highest water line that occurs on shore during a storm. Seaward, it extends to the point at which storm waves can no longer move sediments on the seafloor (usually at depths of approximately 60 m, or 200 ft). The specific contact line between the sea and the land is the *shoreline*, and adjacent land is considered the *coast*.

The active margins of the Pacific along the North and South American continents are characteristic coastlines affected by erosional landform processes. Erosional coastlines tend to be rugged, of high relief, and tectonically active, as expected from their association with the leading edge of drifting lithospheric plates (see the plate tectonics discussion in Lab Exercise 19).

Figure 26.1 presents features commonly observed along an erosional coast. **Sea cliffs** are formed by the undercutting action of the sea. As indentations are produced at water level, such a cliff becomes notched, leading to subsequent collapse and retreat of the cliff. A **wave-cut platform**, or *wave-cut terrace*, is formed by wave action as a horizontal bench in the tidal zone.

If the relationship between the land and sea level has changed over time, multiple platforms or terraces may arise like stairsteps back from the coast. Other erosional forms evolve along cliff-dominated coastlines, including *sea caves, sea arches*, and *sea stacks*. As erosion continues, arches may collapse, leaving isolated stacks in the water.

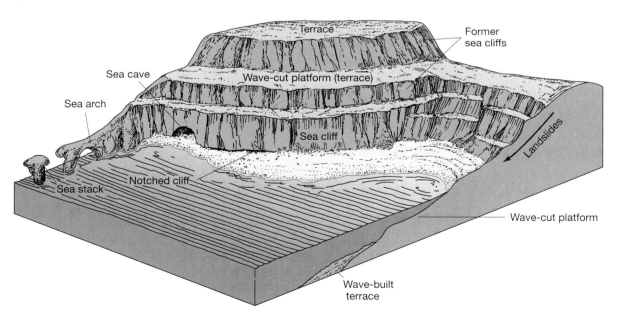

▲ Figure 26.1 Characteristic coastal erosional landforms

Depositional coasts generally are located near onshore plains of gentle relief, where sediments are available from many sources, although such features can occur along all coasts. Characteristics of wave- and current-deposited landforms are illustrated in Figure 26.2 and may involve sediments of varying sizes. A **barrier spit** consists of material deposited in a long ridge, attached at one end, extending out from a coast; it partially blocks the mouth of a bay, except for a *tidal inlet*. A spit becomes a **bay barrier**, sometimes referred to as a *baymouth bar*, if it completely cuts off the bay from the ocean and forms an inland **lagoon**. Tidal flats and salt marshes are characteristic low relief features wherever tidal influence is greater than wave action.

A **tombolo** occurs when sediment deposits connect the shoreline with an offshore island or sea stack. A tombolo forms when sediments accumulate on a wave-built terrace that extends below the water.

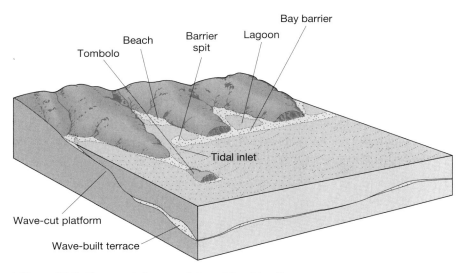

▲ Figure 26.2 Characteristic coastal depositional landforms

Activities and completion items related to coastal geomorphology:

1. Refer to the Point Reyes topographic quadrangle in **Topographic Map #8**.

   What is the scale of this map?

   What is the contour interval?

2. Give the name and location of a barrier spit on the map segment.

3. List the names of one lagoon and two estuaries shown on the map. Note that "estero" means estuary in Spanish.

4. Using the topographic map symbol key on the inside front cover of your lab manual, what is the composition of Limantour Spit?

Virtual Tour
Pt. Reyes

http://goo.gl/o48qGe

5. Marine terraces are remarkable indicators of an emerging coastline. What evidence do you find of such a leveled terrace on the map? Describe.

6. Looking at the contour lines, can you determine the prevailing, effective wind direction along Point Reyes Beach? Explain. Using a black colored pencil or pen, draw arrows on the topographic map to indicate the prevailing wind direction.

7. From Drakes Beach around to Chimney Rock are examples of wave-cut cliffs and hanging valleys. Determine the height of several of these features above sea level.

8. If you were in a small boat near Point Reyes and wanted to come ashore, would you consider docking near Chimney Rock? Explain why or why not.

9. In the photo stereopair in Figure 26.3, locate the lagoons and bays behind Limantour Spit (the bay barrier and barrier spit) that you saw on the topographic map. The topo map was prepared in 1954 and revised in 1971, whereas the photo pair was made 39 years later, in 1993.

   a) Describe some specific differences you see in comparing the map and the photos. _____

   _____

   _____

   _____

   b) Are there any cultural features that appear on the photos and not on the map? Describe. _____

   _____

   _____

   _____

10. Optional Google Earth™ activity, Pt. Reyes, California. For the KMZ file and questions, go to mygeoscienceplace.com. Then click on the cover of *Applied Physical Geography: Geosystems in the Laboratory*.

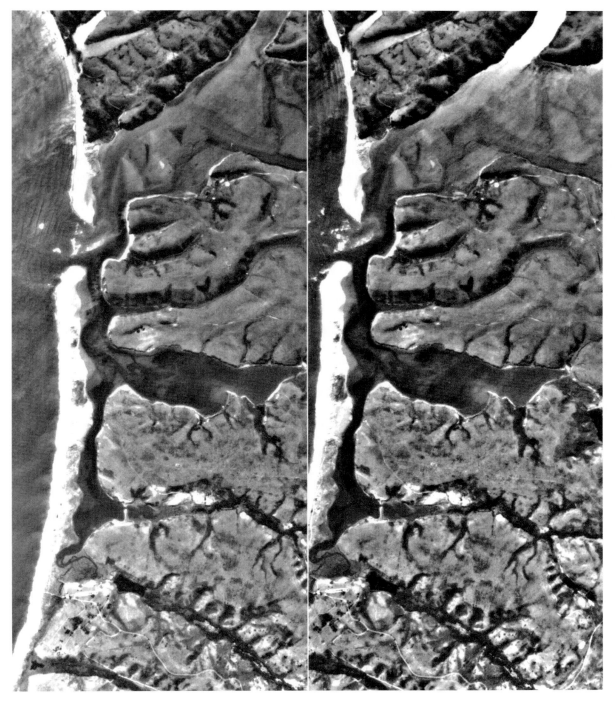

▲ Figure 26.3
Photo stereopair of a portion of Point Reyes National Seashore, California. In this arrangement, north is to the right, west to the top. (Photos by NAPP, USGS, 1993)

SECTION 2

Coastal Geomorphology and Land-Use Planning

The coastal environment on the east coast of the United States has been heavily developed and extensively studied. Ian McHarg, in his book *Design with Nature*, proposed ecological planning principles to guide development of coastal resources. In this Google Earth™ exercise, you will use historic and current topographic maps, as well as Google Earth™ imagery, to analyze physical changes along the New Jersey coast and evaluate past planning decisions and the potential effects of global climate change. For the KMZ file and questions, go to mygeoscienceplace.com. Then click on the cover of *Applied Physical Geography: Geosystems in the Laboratory*.

SECTION 3

Death Valley, California

Dry climates occupy about 26% of Earth's land surface and, if all semiarid climates are considered, perhaps as much as 35% of all land, constituting the largest single climatic region on Earth. Despite the general dryness, water remains the major erosional force in arid and semiarid regions.

In arid climates, with their intermittent water flow, a particularly noticeable fluvial landform is the **alluvial fan**, or *alluvial cone*, which occurs at the mouth of a canyon where it exits into a valley (Figure 26.4). The fan is produced by flowing water that loses velocity as it leaves the constricted channel of the canyon and, therefore, drops layer upon layer of sediment along the base of the mountain block. Water then flows over the surface of the fan and produces a braided drainage pattern, shifting from channel to channel with each moisture event.

An interesting aspect of an alluvial fan is the natural sorting of materials by size. Near the mouth of the canyon, boulders and gravels are deposited, grading slowly to pebbles and finer gravels with distance out from the mouth. Then sands and silts are deposited, with the finest clays and salts carried in suspension and solution all the way to the valley floor. In the wet season, a shallow lake may form that will evaporate and form a dry lake bed, or **playa**. Over time, as more sediment is deposited by the shifting stream channels, the alluvial fans will begin to coalesce into a continuous surface called a **bajada**. **Interior drainage** occurs when streams do not flow to the ocean but instead end in inland basins or seas. Playas are an indication of interior drainage.

Earth's crust is being stretched by tectonic activity, forming the **basin and range** topography of the American west. The stress of tension forms normal faults, which creates a landscape of mountain ridges called **horsts** and down-dropped fault blocks called **grabens**. Death Valley, California, is one such

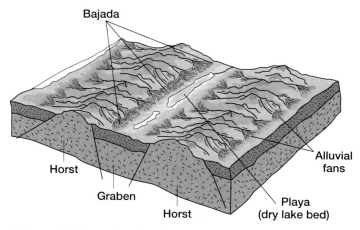

▲ Figure 26.4 Typical arid landscape features.

graben formed by the stretching of Earth's crust. Tectonic forces have created a juxtaposition of the lowest spot in North America, -86 m (-282 ft.) at Badwater in Death Valley, only 240 km (150 mi.) from the highest point in the contiguous lower 48 United States, Mt. Whitney, at 4421m (14,421 ft.).

Examples of these features are found in Death Valley, California, presented as **Topographic Map #9**.

Analysis and completion items for **Topographic Map #9**:

Virtual Tour
Death Valley

http://goo.gl/o48qGe

1. What is the contour interval on this map segment?

   What is the scale of this map?

2. Although Death Valley receives less than 5 cm (2 in.) of rain each year, water is an important agent in modifying the landscape. List by name at least three water features formed by running or standing water (not springs) and trace or outline them on the topographic map with a blue colored pencil or pen.

3. Using a black colored pencil or pen, label the following features on **Topographic Map #9**: bajada, alluvial fan (may be part of a bajada), playa, and graben.

4. Alluvial fans are important sources of water in arid climates. Rain from storms percolates through the alluvium of the fan before reaching an impermeable layer. Some of the signs of this are springs and vegetation at the bottom of alluvial fans. Identify by name at least four locations on **Topographic Map #9** with springs or vegetation and circle them on the topographic map with a blue colored pencil or pen.

5. What is the linear distance from the 2000-ft elevation contour in Trail Canyon to the 0-ft contour? What is the relief between these two points?

   In terms of slope, state this distance and relief in feet per mile:

   State this in terms of percent grade:

6. Use a sheet of graph paper from the back of your lab manual to plot the topographic profile of the line on **Topographic Map #9** that runs between Township 25½ N and Township 25 N. It is just south of the Devils Speedway in The Devils Golf Course. Begin in the west at the 500-m contour line and end in the east at the 500-m contour line.

7. Using a black colored pencil or pen, trace the contour lines of the alluvial fan at the mouth of Trail Canyon. What is the lowest-value elevation contour line on the fan? Using a blue colored pencil or pen, trace the streams on the Trail Canyon alluvial fan.

8. Briefly describe how alluvial fans are formed. Why do you think there is more than one stream on the Trail Canyon alluvial fan? Relate your answer to the process by which alluvial fans are formed.

9. Explain why you think the alluvial fans on the west side of Death Valley are larger than the alluvial fans on the east side of Death Valley.

10. Optional Google Earth™ activity, Cedar Creek Alluvial Fan in Ennis, Montana. For the KMZ file and questions, go to mygeoscienceplace.com. Then click on the cover of *Applied Physical Geography: Geosystems in the Laboratory*.

SECTION 4

Mitten Buttes, Arizona, Quadrangle and Photo Stereopair

Virtual Tour
Mitten Buttes, AZ

http://goo.gl/o48qGe

Monument Valley straddles the Utah–Arizona border west of the Four Corners area. It is almost entirely within the Navajo Indian Reservation and is accessed through the Navajo Tribal Park off U.S. 163. Part of this region is shown on **Topographic Map #10** and in the photo stereopair in Figure 26.4.

The mesas, buttes, and pinnacles (listed in decreasing size and increasing age) of arid landscapes are resistant horizontal rock strata that have eroded differentially. Removal of the less-resistant sandstone strata produces unusual desert sculptures—arches, windows, pedestals, and delicately balanced rocks. Specifically, the upper layers of sandstone along the top of an arch or butte are more resistant to weathering and protect the sandstone rock beneath.

The removal of all surrounding rock through differential weathering leaves enormous buttes as residuals on the landscape. If you imagine a line intersecting the tops of the Mitten Buttes shown in Figure 26.5, you can gain some idea of the quantity of material that has been removed. These buttes exceed 300 m (1000 ft) in height, similar to the Chrysler Building in New York City or First Canadian Place in Toronto.

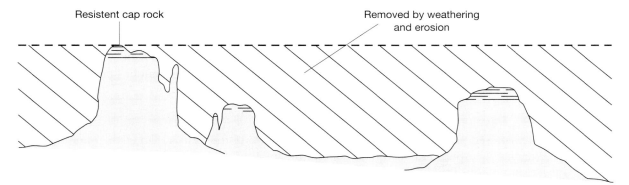

▲ Figure 26.5
Rock removal by weathering, erosion, and transport in Monument Valley

Figure 26.5 is a photo stereopair of the buttes and part of a mesa in Monument Valley in **Topographic Map #10**. Examine the photos with stereolenses and the topographic map to answer the following.

1. What is the contour interval on this map segment?

   What is the scale of this map?

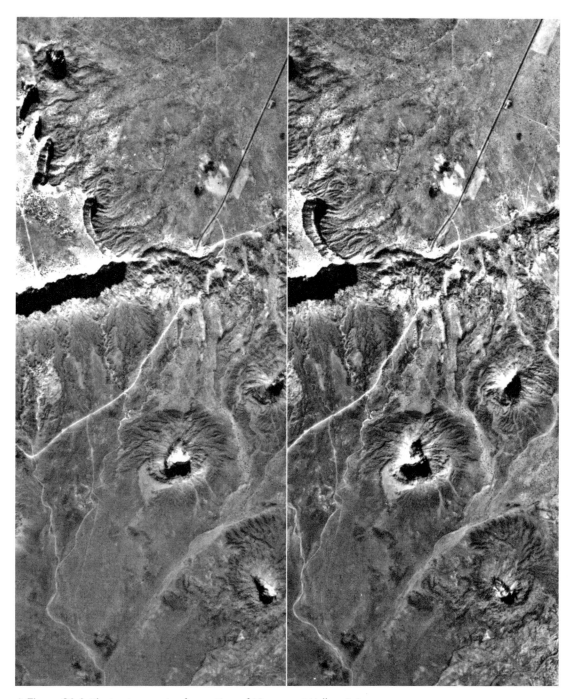

▲ Figure 26.6 Photo stereopair of a portion of Monument Valley, Arizona.
In this arrangement, north is to the right, south to the left, and west at the top. (Photos by NAPP, USGS)

2. What is the relief between the top of the mesa on the west side of the map (part of Mitchell Mesa) and the lower elevations along the east side of the map?

3. Of the two Mitten Buttes and Merrick Butte, which is the tallest, and what is its elevation?

4. Examining these features in the photo stereopair and on the topo map, can you describe any features that indicate differing rock resistances to weathering (slopes = less resistance; cliffs = more resistance)? Describe and explain.

5. The road to the lookout at the campground is paved. Can you find the dirt road on the photo that leaves this paved road that goes out across Monument Valley?

   The campground loop road?

   At what elevation is the visitor center?

6. Earlier we stated that water is the major erosional force in the desert; although occurring infrequently, when it rains, the downpour is usually intense, and floods flash across the sparsely vegetated surfaces swiftly. Given this reality, and using the map and photo stereopair, describe several examples that demonstrate to you that water is the major erosional force and is occurring infrequently. List at least four examples (look for dry washes, sharp features, and alluvial deposits).

7. Optional Google Earth™ activity, Mitten Buttes, Arizona. For the KMZ file and questions, go to mygeoscienceplace.com. Then click on the cover of *Applied Physical Geography: Geosystems in the Laboratory*.

MasteringGeography™ | Looking for additional review and lab prep materials? Go to www.masteringgeography.com for pre lab videos and pre and post lab quizzes.

Name: _____ Laboratory Section: _____

Date: _____ Score/Grade: _____

Video
Exercise 27
Pre-Lab Video

Scan to view the Pre-Lab video

http://goo.gl/6tVVe1

LAB EXERCISE

27 Topographic Analysis: Karst Landscapes

Limestone is abundant on Earth, and many landscapes are composed of it. These areas are quite susceptible to chemical weathering. Such weathering creates a specific landscape of pitted- and bumpy-surfaced topography, limited surface streams due to infiltration and well-developed solution channels (dissolved openings and conduits) underground. Remarkable labyrinths of underworld caverns also may develop due to weathering and erosion caused by groundwater.

These are the hallmarks of **karst topography**, named for the Krš Plateau in Slovenia, where karst processes were first studied. Approximately 15% of Earth's land area has some karst features, with outstanding examples found in southern China, Japan, Puerto Rico, Cuba, the Yucatán of Mexico, Kentucky, Indiana, New Mexico, and Florida. As an example, approximately 38% of Kentucky has **sinkholes** and related karst features noted on topographic maps.

For a limestone landscape to develop into karst topography, there are several necessary conditions:

- The limestone formation must contain 80% or more calcium carbonate for solution processes to proceed effectively.

- Complex patterns of joints in the limestone are needed for water to form routes to subsurface drainage channels.

- There must be an air-filled zone between the ground surface and the water table.

- Vegetation cover supplies varying amounts of organic acids and carbon dioxide that enhance the solution process.

The role of climate in providing optimum conditions for karst processes remains in debate, although the amount and distribution of rainfall appears important. Karst occurs in arid regions, but here it was primarily formed in prior times with greater humidity and precipitation.

Areas of karst may exhibit diversion of surface water to underground flows. A stream may appear and disappear along its channel, called a **disappearing stream**. Groundwater and surface waters dynamically interact through the jointed rock structures and permeable rock. Lab Exercise 27 features two sections and two Google Earth™ activities.

Key Terms and Concepts

karst topography
sinkhole

After completion of this lab, you should be able to:

1. *Analyze* and *describe* karst landscapes by using topographic maps and a photo stereopair.
2. *Relate* these surface features to groundwater and the underground world of solution caverns.

Copyright © 2015 Pearson Education, Inc. Lab Exercise 27 275

Materials/Sources Needed

pencil
calculator
ruler
stereolenses

Lab Exercise and Activities

SECTION 1

An idealized karst landscape with karst topography

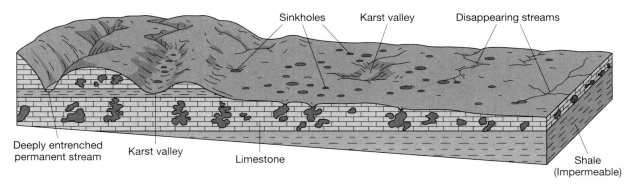

▲ Figure 27.1 Idealized features of a karst landscape

Oolitic, Indiana, Quadrangle and Stereophotos

Virtual Tour
Oolitic, IN

http://goo.gl/o48qGe

The weathering of limestone landscapes creates many sinkholes, which form in circular depressions. Traditional studies may call a sinkhole a *doline*. Water from the surface can flow through joints or fractures and slowly dissolve the rock around entry, forming a pit called a solution sinkhole. Another type of sinkhole is formed if an underground cavern's roof collapses. Everything above the cavern (rocks, soil, trees, buildings) will drop into fill cavern and form a collapse sinkhole. A gently rolling limestone plain might be pockmarked by sinkholes with depths of 2 to 100 m (7 to 330 ft) and diameters of 10 to 1000 m (33 to 3300 ft). Using **Topographic Map #11** and the photo stereopair (Figure 27.2) of part of the mapped area, answer the following.

1. Refer to the Oolitic, Indiana, topographic quadrangle—**Topographic Map #11**. What is the scale of this map?

   What is the contour interval?

2. In the space provided draw the topographic map symbol that indicates a sinkhole; sketch a contour line and illustrate this symbol. (Check out the chart of topographic map symbols inside the front cover of this manual.)

Find Section 31 on the map (west of the highway, west center of map). Section 31 is marked on the north and east by a red section line, on the south by a black-dashed township line, and on the west by 500 West Road.

Answer the following about Section 31:

3. What are the highest and lowest elevations in Section 31?

    Highest:

    Lowest:

4. How many sinkholes are there in the 1 mi² section?

    Do any of these sinkholes have water in them (implying that they must have impermeable soil or rock along their bases)?

5. Describe any economic activity specific to this karst environment that you see on the topo map.

Examine with stereolenses the photo stereopair on the next page in Figure 27.2 for this same area around Oolitic, Indiana.

6. Describe how the sinkholes appear in the photos.

7. Can you distinguish between older and newer mining areas by using the stereophotos? How were you able to determine this observation? Discuss.

8. This limestone region is deeply dissected by Salt Creek, flowing along the east side of the map and photos. The entrenched meanders east of Oolitic are remarkable in their geometric (rock-structure controlled) bends. Such an entrenched stream is important in the development of a karst region because it permits a continuous movement of water through the jointed limestone landscape. What is the local relief between the town and the creek? Examine the contour lines carefully—they are tightly spaced along the cliffs.

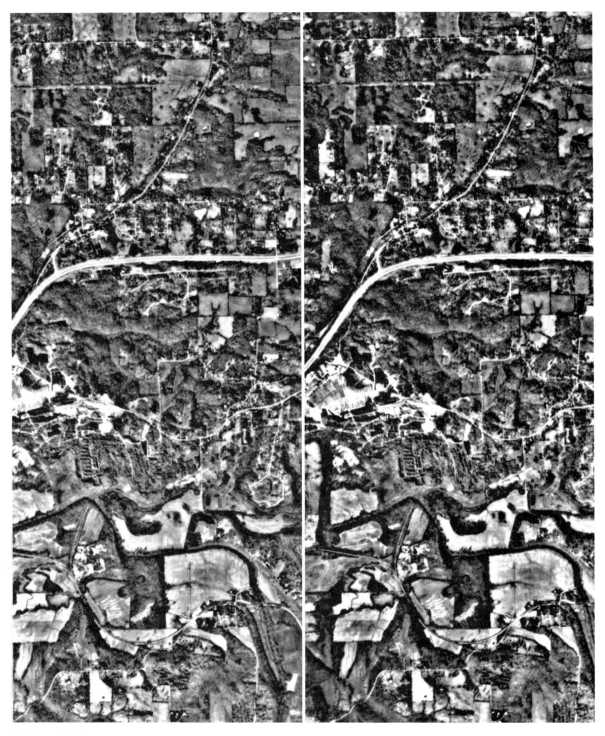

▲ Figure 27.2
Photo stereopair of an area west of Oolitic, Indiana. In this arrangement, north is to the right, west to the top. (Photos by NAPP, USGS)

9. Optional Google Earth™ activity, Oolitic, Indiana. For the KMZ file and questions, go to mygeoscienceplace.com. Then click on the cover of *Applied Physical Geography: Geosystems in the Laboratory.*

SECTION 2

Lake Wales, Florida, Topographic Quadrangle Map

In Florida, several sinkholes have made news because water tables have been lowered by groundwater pumping have caused their collapse into underground solution caves, taking with them homes, businesses, and even new cars from an auto dealership. One such sinkhole collapsed in a suburban area in 1981 and others in 1993 and 1998. The area around Lake Wales, Florida (south of Orlando and north of the Everglades), is characteristic of limestone with karst topography, high water tables, some dry depressions, limited surface streams, marsh lands, and groves of citrus trees—see **Topographic Map #12**. Complete the following items relative to this map.

Virtual Tour
Lake Wales, FL

http://goo.gl/o48qGe

1. Refer to the Lake Wales, Florida, topographic quadrangle—**Topographic Map #12**. What is the scale of this map?

    What is the contour interval?

2. How many sinkhole depressions contain lakes in the map segment?

3. Do any of these lakes sit in a depression (give a count) 10 ft deep?

    15 ft deep?

    20 ft deep?

    More than 25 ft deep?

4. Water will flow downslope between sinkholes. Using lake- and pond-surface elevation as an indicator, in which direction do you think groundwater flows through this area? Explain your thinking. Using a dark blue colored pen, draw arrows to indicate the direction of flow of groundwater.

5. A hospital is shown south (SSE) of Lake Wales. Determine the elevation of the hospital as well as the elevation of the lake surface of Lake Wales.

    Hospital: Lake Wales:

    If you were standing at the hospital, could you see the lake? Explain.

6. Evaluate the status of the elevation of the water table over time on this map. Is the water table rising or falling? Using a dark green colored pencil or pen, circle at least three features on the map that support your statement. Describe why these three features support your assertion.

7. Do you see any evidence that there is a retirement community in Lake Wales (mobile home parks or small subdivisions of houses in a compact arrangement)? Locate and describe.

8. What is the primary activity on the landscape? Take a look at Sections 25 and 36. Use the topographic map symbol key inside the front cover of this manual to help identify the activity.

9. Do you find any surface streams on the topographic map?   Explain your observation. What would produce this hydrologic situation?

10. Optional Google Earth™ activity, Lake Wales, Florida. For the KMZ file and questions, go to mygeoscienceplace.com. Then click on the cover of *Applied Physical Geography: Geosystems in the Laboratory*.

MasteringGeography™ | Looking for additional review and lab prep materials? Go to www.masteringgeography.com for pre lab videos and pre and post lab quizzes.

Name: _____ Laboratory Section: _____ Video
Date: _____   Score/Grade: _____ Exercise 28 Pre-Lab Video

LAB EXERCISE 28

Soils

Scan to view the Pre-Lab video

http://goo.gl/vB9fe8

Earth's landscape generally is covered with soil. **Soil** is a dynamic natural material composed of fine particles in which plants grow, and it contains both mineral fragments and organic matter. The soil system includes human interactions and supports all human, other animal, and plant life. If you have ever planted a garden, tended a house plant, or been concerned about famine and soil loss, this lab exercise will interest you. You may discover that you have made some casual observations that can be applied in this exercise.

Soil science is interdisciplinary, involving physics, chemistry, biology, mineralogy, hydrology, taxonomy, climatology, and cartography. Physical geographers are interested in the spatial patterns formed by soil types and the environmental factors that interact to produce them. **Pedology** concerns the origin, classification, distribution, and description of soil (from the Greek *pedon*, meaning "soil" or "earth"). Pedology is at the center of learning about soil as a natural body, but it does not dwell on its practical uses. *Edaphology* (from the Greek *edaphos*, meaning "soil" or "ground") focuses on soil as a medium for sustaining higher plants. Edaphology emphasizes plant growth, fertility, and the differences in productivity among soils. Pedology gives us a general understanding of soils and their classification, whereas edaphology reflects society's concern for food and fiber production and the management of soils to increase fertility and reduce soil losses.

Observing soils firsthand in the field is the best way to learn about their properties: Construction sites, excavations on your campus or nearby, and roadcuts along highways all provide opportunities for seeing soils in their natural surroundings. In many locales, an *agricultural extension service* can provide specific information and perform a detailed analysis of local soils. Soil surveys and local soil maps are available for most counties in the United States and for the Canadian provinces. Your local phone book may list the U.S. Department of Agriculture, Natural Resources Conservation Service (http://www.nrcs.usda.gov/), or Agriculture Canada's Soil Information System (http://sis.agr.gc.ca/cansis/intro.html).

This exercise gives you an opportunity for some hands-on experience with soils and for using some of the tools and methods that soil scientists use in their work. Lab Exercise 28 features six sections. (*Note: The results for many activities in this exercise will vary, depending on the samples, local sites, and other materials available to your lab.*)

Key Terms and Concepts

clay
humus
loam
pedology
pedon
peds
permeability
polypedon
porosity
regolith
sand
silt

soil
soil classification
soil color
soil consistence
soil horizon
soil pH (acidity–alkalinity)
soil profile
soil properties
soil science
soil structure
soil texture
solum

KEY LEARNING concepts

After completion of this lab, you should be able to:

1. *Identify* basic components of soil and soil properties.
2. *Determine* main components of soil sample by color.
3. *Identify* major soil texture categories and *classify* soils by texture.
4. *Use* the "feel method" and *determine* texture of soil samples.
5. *Demonstrate* and *observe* how soil texture affects porosity and movement of water through soil.
6. *Discern* the characteristics and properties of horizons in soil profiles and *identify* horizons.
7. *Measure* pH level in soil samples and *determine* the soil pH (acidity or alkalinity).
8. *Identify* specific soil series that are featured as state soils.

Materials/Sources Needed

pencils
color pencils
ruler
soil samples
Munsell Color Chart (optional, 175 colors)
glass jars or beakers

water
soup can with both ends removed (optional)
pH test equipment (handheld meter or color test kit)
soil samples, or soil sample photographs, or Internet site of soil profiles
Internet access (optional)

Lab Exercise and Activities

SECTION 1

Soil Color

Soil properties are the characteristics or traits of soil, some of which include **soil color**, **texture**, **structure**, **consistence**, **porosity**, moisture, and chemistry. We examine a few of these properties, beginning with color.

Soil color is one of the most obvious traits, suggesting composition and chemical makeup in mineral soils. If you look at exposed soil, color may be the most obvious trait. Among the many possible hues are the reds and yellows found in soils of the southeastern United States (high in iron oxides); the blacks of prairie soils in portions of the U.S. grain-growing regions and Ukraine (richly organic); and white-to-pale hues found in soils containing silicates and aluminum oxides. Reduced iron imparts gray and greenish colors. In dry areas, calcium carbonate or other water-soluble salts give the soil a white color.

However, color can be deceptive: Soils of high humus content, organic materials from decomposed plant and animal litter, are often dark, yet clays of warm-temperate and tropical regions with less than 3% organic content are some of the world's blackest soils.

To standardize color descriptions, soil scientists describe a soil's color by comparing it with a *Munsell Color Chart* (developed by artist and teacher Albert Munsell in 1913). These charts display 175 colors arranged by *hue* (H, the dominant spectral color, such as red), *value* (V, degree of darkness or lightness), and *chroma* (C, purity and saturation of the color, which increase with decreasing grayness).

Let's examine some essentials relative to soil color in four soil samples.

1. Using soil samples (observed/collected in the field or provided by your instructor, or soil photographs), note the predominant soil color and indicate the likely soil component responsible for the color. (Answers will vary, depending on the samples/photographs provided.) Be sure and note whether the sample is wet, moist, or dry.

   a) Soil sample A _____

   _____

   b) Soil sample B _____

   _____

   c) Soil sample C _____

   _____

   d) Soil sample D _____

   _____

Using These Same Four Samples

In the Munsell color system, every color has three qualities: *hue* (the dominant spectral color, such as red), *value* (degree of darkness or lightness), and *chroma* (purity and saturation of the color, which increase with decreasing grayness). The complete Munsell notation for a chromatic color is written symbolically, like this: H V/C. For example, for a strong red having a *hue* of 5R, a *value* of 6, and a *chroma* of 14, the complete Munsell notation is 5R 6/14 (5R is hue, 6 is value, and 14 is chroma). As another example, a pale brown is 10YR 6/3. A dark brown is noted as 10YR 2/2. More refined divisions of any of the attributes use decimals.

The light you use when you view a sample is important and can affect your assessment of the color notation. It is best to view the chart and the sample with the Sun over your shoulder, shining on the sample, with you facing away from the Sun. If you are in a lab or classroom under artificial light, try to have a light source as close to white light (all spectrum) as possible. Don't be too concerned at this stage; more practice will improve your assessment under varying light conditions. You will find low values and low chromas the most difficult to match against the color chips.

Note: When doing actual field work with a soil **pedon** (the complete soil profile and basic sampling unit in soil surveys), you will find different colors in each horizon, and maybe more than one color in a single horizon.

These details should be noted in your assessment.

2. Using a *Munsell Color Chart*, note the *color description* and the *Munsell notation* for each of the four samples you assessed in question #1.

   a) Soil sample A _____

   _____

   b) Soil sample B _____

   _____

   c) Soil sample C _____

   _____

   d) Soil sample D _____

   _____

SECTION 2

Soil Texture and Soil Structure

Soil texture refers to the mixture of sizes of its individual particles and the proportion of different sizes of soil separates (individual particles of soil). Figure 28.1 illustrates particle size comparisons, and Table 28.1 lists the standards for soil particle grades (sizes). Particles smaller than gravel are considered part of the soil, while larger particles such as gravel, pebbles, or cobbles are not. As you can see from the table, sand is further graded—ranging from very coarse to very fine.

If you have been to a beach, you have felt the texture of **sand**: It has a "gritty" feel. **Silt**, on the other hand, feels smooth—somewhat soft and silky, like flour used in baking bread. When wet, **clay** has a sticky feel and requires quite a bit of pressure to squeeze it, like the clay used in making pottery.

Soils nearly always consist of more than one particle size. By determining the relative amounts of sand, silt, and clay in a particular soil sample, it can be placed into one of twelve classes, as shown on the soil texture triangle in Figure 28.2. Each side presents percentages of a particle grade. See the line from each side of the triangle (following the direction indicated by the orientation of the numbers on each axis). You see that a soil consisting of 36% sand, 43% silt, and 21% clay is classified as **loam**, a term for soils consisting of mostly sand and silt with a relatively smaller amount of clay. This determination is a mechanical analysis or particle size analysis of the soil.

Soil structure refers to the arrangement of these soil separates. The smallest natural lump or cluster of particles is a **ped**. Soil structure is described as crumb or granular, platy, blocky, or prismatic or columnar.

Keep this distinction in mind when assessing soil: Soil separates of individual mineral particles are the *soil texture*, and peds, which are the arrangement of soil particles, comprise the *soil structure*.

Soil Particle Grade	Diameter (mm)	Diameter (in.)
Gravel	>2.0	>0.08
Very coarse sand	1.0–2.0	0.04–0.08
Coarse sand	0.5–1.0	0.02–0.04
Medium sand	0.25–0.5	0.01–0.02
Fine sand	0.10–0.25	0.004–0.01
Very fine sand	0.05–0.10	0.002–0.004
Silt	0.002–0.05	0.00008–0.002
Clay	<0.002	<0.00008

▲Table 28.1 Soil texture grades
(U.S. Department of Agriculture, Natural Resources Conservation Service)

You can see in the table that soil grades from sand to silt to clay, from most-coarse to most-fine mineral particle sizes. Soils that represent the best particle size mix for plant growth are those that balance the three sizes. This is the loam described earlier in this section. Remember, pedology gives us a general understanding of soils and their classification, whereas edaphology reflects society's concern for food and fiber production and the management of soils to increase fertility and reduce soil losses.

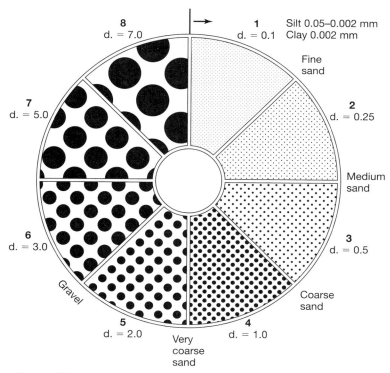

▲ Figure 28.1
Sizes of soil particles larger than silt in approximate diameters (d) of millimeters (mm). (Adapted from *Soil Survey Manual*, USDA, Handbook No. 18, Natural Resources Conservation Service, October 1993, p. 137)

1. Use the soil texture triangle to name the following by its correct texture class (named on Figure 28.2):

   a) 17% sand, 28% silt, 55% clay: _____

   b) 31% sand, 55% silt, 14% clay: _____

2. Determine the percentage of each particle size for A and B examples plotted on the triangle (Figure 28.2):

   a) A: _____ % sand, _____ silt, _____ clay

   b) B: _____ % sand, _____ silt, _____ clay

Please read the appropriate section on soil horizons in *Geosystems* or *Elemental Geosystems*. Soil textures vary through the horizons, so the soil texture triangle can be used for samples in each horizon. For instance, assume a silt loam in Ohio is sampled in the A, B, and C horizons. The mix of mineral particles occurring at each horizon gives you a complete sense of the soil. The textural analysis is as follows:

Sample points	% Sand	% Silt	% Clay
1 = A horizon	22	63	15
2 = B horizon	31	25	44
3 = C horizon	42	34	24

▲Table 28.2 Textural analysis of a silt loam soil

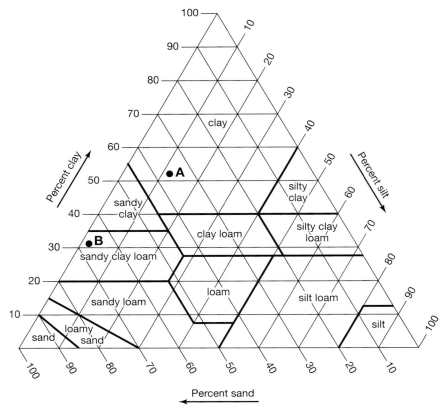

▲ Figure 28.2 Soil texture triangle
(USDA, Natural Resources Conservation Service, same source as Figure 28.1, p. 138)

3. Using the numbers **1, 2,** and **3** to designate each of the three soil horizons, place these numbers in the appropriate locations on the soil texture triangle in Figure 28.2. In the textural analysis of mineral particle sizes, the A horizon is principally

   the B horizon is in the majority

   and the C horizon is in the majority

   In a soil profile, is this what you would expect the particle size distribution to be? Explain.

4. In a less quantitative way than using sieves to physically sort soil particles, soil texture can be determined in the field by feeling the soil and estimating the percentages of sand, silt, and clay. Try this method using the following procedure with available soil samples, recording your observations and results through each of these steps. Your instructor may direct you to use the four original soil samples from previous sections of this exercise.

   Step 1: Fill your palm with dry soil, moistening it with enough water so that it sticks together sufficiently to be worked with your fingers. Add the water gradually: if it becomes too runny or if it sticks to your fingers, add more dry soil. You want a "plastic" mass that you can mold, somewhat like putty. (*Note:* If the soil is gritty and loose and won't form a ball but falls apart when rubbed between your fingers, it is a "sandy" soil.)

   Step 2: Knead the soil between your thumb and fingers, removing any pebbles that may be present and breaking up any peds (aggregates or clumps).

Step 3: The soil should form a ball when you squeeze it. Position the ball between your thumb and forefinger and gently push the ball with your thumb, squeezing it into a ribbon of uniform thickness and width. The ribbon will extend over your finger, eventually breaking from its own weight. Continue until several ribbons have formed and broken off. (If you have difficulty with this, roll the soil into a cylinder that is uniform in diameter and then squeeze the cylinder of soil into a ribbon.) The *clay content* will determine the length of the ribbon before it breaks:

- Ribbon length less than 1.5 in. (3.8 cm): clay content is <27%
- Ribbon length 1.5 to 3 in. (3.8 to 7.6 cm): clay content is 27 to 40%
- Ribbon length more than 3 in. (7.6 cm): clay content is >40%

Step 4: Now put a small pinch of the soil sample into your palm and add enough water to excessively wet it. Rub it with your forefinger, *estimating* the sand content by the amount of *grittiness* you feel (silt has a smoothness, clay a stickiness in feel):

- Sand is the dominant texture you feel: sand content is >50%
- Sand is noticeable, but not dominant: sand content is 20–50%
- Predominant feel is smooth, indicating high silt content: sand content is <20%

Step 5: Combine the estimates of sand and clay content and use Table 28.3 to place your sample in one of the classes. (*Note:* If neither a gritty/sand nor smooth/silt feeling is predominant in Step 4, it will fall into the center column.)

Step 6: Record your observations in the spaces provided after Table 28.3. Your instructor will determine how many samples to use in your work. Space is provided here for three sample reports.

		Estimated Percentage of Sand (predominant feeling)		
		>50 (gritty)	20–50 (neither)	<20 (smooth)
Estimated Percentage of Clay	>40	Sandy clay	Clay	Clay
				Silty clay
	27–40	Sandy clay loam	Clay loam	Silty clay loam
	<27	Sandy loam	Loam	Silt loam
		Loamy sand		
		Sand		

▲Table 28.3 Simplified soil texture table

Post your results from **Step 4** and **Step 5** and Table 28.3 here:

a) Soil sample A _____

_____

b) Soil sample B _____

_____

c) Soil sample C _____

_____

Note that in soil science, the term consistence is used to describe the consistency of a soil or cohesion of its particles. Consistence is a product of texture (particle size) and structure (ped shape). You are working with soil consistence in the exercise above. Consistence reflects a soil's resistance to breaking and manipulation under varying moisture conditions:

- A wet soil is sticky between the thumb and forefinger, ranging from a little adherence to either finger, to sticking to both fingers, to stretching when the fingers are moved apart. *Plasticity*, the quality of being moldable, is roughly measured by rolling a piece of soil between your fingers and thumb to see whether it rolls into a thin strand.

- A moist soil is filled to about half of field capacity (the usable water capacity of soil), and its consistence grades from loose (noncoherent), to friable (easily pulverized), to firm (not crushable between thumb and forefinger).

- A dry soil is typically brittle and rigid, with consistence ranging from loose, to soft, to hard, to extremely hard.

SECTION 3

Soil Texture and Porosity

Soil texture affects the movement of water through the soil and, therefore, the amount of water available to plants. Sandy soils have large air spaces between the particles, making them very porous, allowing the water to percolate through the soil quickly. Thus, little water is retained, causing the soil to dry out quickly. You may have observed this at a beach as water quickly drained through the sand, leaving the surface dry and loose.

The small particles in clay soils fit closely together, leaving few air spaces between them so that water percolates very slowly, collecting on the soil surface or running off if there is any slope. Once the water is absorbed into the clay soil, it is held so tightly that little is available to the plants.

Therefore, the texture and the structure of the soil dictate available pore spaces, or porosity. The property of the soil that determines the rate of soil-moisture recharge is its **permeability**. Permeability depends on particle sizes and the shape and packing of soil grains.

Loam soils have a loose, porous texture due to their mixture of particle sizes. As a result, loams retain water well without becoming waterlogged, leaving the water readily available to plants. This is one reason why farmers consider loamy soils ideal. Table 28.4 presents water absorption rates for various soil textures.

Soil Texture	Absorption Rate per Hour
Sandy	>5 cm (+2 in.)
Loam	0.6–5 cm (0.25–2 in.)
Clay	<0.6 cm (<0.25 in.)

▲Table 28.4 Water absorption rates in soil

1. You can perform a simple demonstration to observe the effect of soil texture and structure on porosity. Your instructor may have you use the same soil samples from earlier sections in this exercise.

    Step 1: Obtain samples of several soils of differing textures and put each in a glass jar or beaker. Fill the jars to the same level, leaving several inches at the top, taking care not to compact the soil in the container.

    Step 2: Measure 5 cm (2 in.) of water in another container of the same size. Measure the same amount for each sample. Pour the water over the soil in each container and observe how quickly the water percolates through the soil to the bottom of the jar. Roughly time the percolation rates. (*Note:* The main focus of this demonstration is observation, not exact measurement and quantification of the percolation rates.)

Step 3: Use the spaces below to make a notation of your observations. Room for three soil sample observations is given. Assess the texture of each sample, based on these observations.

a) Soil sample A _____

_____

b) Soil sample B _____

_____

c) Soil sample C _____

_____

2. Optional: A similar demonstration can be done "in the field." Selecting a site where the soil is not compacted, trim any vegetation completely to the ground and carefully remove any loose organic materials. Pound a soup can (both ends removed) into the soil several inches. If necessary, place a board over the top of the can to prevent the can from crumpling while you do this. Pour water into the can, filling it to the top. After a period of time, observe how much water has been absorbed. You can do this demonstration at several sites and compare the results. (Again, while you could take careful measurements, the purpose of this investigation is primarily observation and relative absorption rates.)

a) Site A _____

b) Site B _____

c) Site C _____

SECTION 4

Soil Profiles of Soil Horizons

Just as a book cannot be judged by its cover, so soils cannot be evaluated at the surface only. Instead, a soil profile should be studied from the surface to the deepest extent of plant roots, or to where regolith or bedrock is encountered. Such a profile, called a pedon, is a hexagonal column measuring 1–10 m² in top surface area. At the sides of the pedon, the various layers of the soil profile are visible in cross section and are labeled with letters. *A pedon is the basic sampling unit used in soil surveys.*

Many pedons together in one area make up a **polypedon**, which has distinctive characteristics differentiating it from surrounding polypedons. A polypedon is comprised of an identifiable series of soils in an area. It can have a minimum dimension of about 1 m² and no specified maximum size. *The polypedon is the basic mapping unit used in preparing local soil maps.*

Each distinct layer exposed in a pedon is a **soil horizon**. Each layer or horizon is distinct from the one directly above or below, with visible boundaries distinguished by differences in soil properties, some of which are color, texture, porosity, moisture, and the presence or absence of certain moisture, and the presence or absence of certain minerals or other materials. The group of horizons from the O horizon down through the B horizon together are the **solum**, considered the true soil profile.

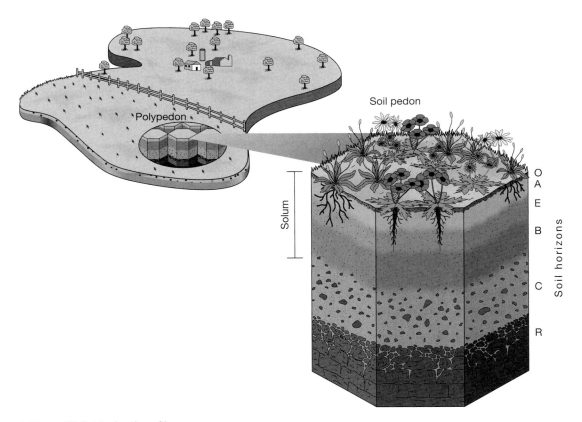

▲ Figure 28.3 Ideal soil profile

Figure 28.3 illustrates the horizons in a model or ideal profile. From the surface down the soil horizons are as follows:

- **O horizon**—Named for the organic materials from plant and animal litter. This layer is often subdivided into two layers: the topmost is O_1, consisting of leaves and other largely undecomposed organic material; just beneath is O_2, which is the decomposed organic debris called **humus**.
- **A horizon**—The uppermost mineral horizon. This dark-colored layer is rich in fine clay-sized particles and organic material derived from the humus above as well as plant roots.
- **E horizon**—Layer named for the process of *eluviation*, in which clays, aluminum and iron oxides, and organic matter are leached (washed out) to lower layers. This horizon is lighter in color and comprised largely of sand and coarse silt particles.
- **B horizon**—Characterized by the accumulation of materials leached from above. This deposition process is called *illuviation*. Presence of oxides may give this layer reddish or yellowish hues, and the accumulated fine clay and organic particles make this layer quite dense.
- **C horizon**—Not considered part of true soil. Consisting of **regolith**, or weathered parent material, this layer lacks biological activity.
- **R horizon**—The bottom layer of the profile, and like the C horizon above, not considered to be part of true soil. This horizon is either unconsolidated (loose) material or solid bedrock.

The Simplified Guide to Soil Taxonomy, available for download at http://www.nrcs.usda.gov/wps/portal/nrcs/site/soils/home/ has many images of soil horizons.

1. Using photographs of three soil profiles (either in the soils chapter of a textbook or provided by your instructor) and working in groups with your lab partners, identify the various soil horizons in each profile. Observe horizon characteristics such as thickness, color, texture, and structure, and note variations from one profile to another. Speculate on the factors that combined to produce the appearance of each.

   a) Soil profile A _____

   _____

   b) Soil profile B _____

   _____

   c) Soil profile C _____

   _____

2. You may be able to locate one or more sites on your campus or nearby where you can see soil profiles in their natural setting. As you did in the first activity, observe the profile's characteristics; if more than one profile is available, note variations between them and see if you can determine the major factors influencing the formation of each.

   a) Profile site A _____

   _____

   b) Profile site B _____

   _____

   c) Profile site C _____

   _____

SECTION 5

Soil Acidity and Alkalinity

Soil pH, the **acidity** or **alkalinity** of the soil as expressed on the pH scale, strongly affects soil fertility (Figure 28.4). Nutrient availability is low in soils that are either very acidic or very alkaline. A soil rich in hydrogen ions (H+) is an acid soil. On the other hand, a soil high in base cations (calcium, magnesium, potassium, sodium) is a basic, or alkaline, soil.

Pure water is nearly neutral, with a pH of 7.0. Readings below 7.0 represent increasing acidity. Readings above 7.0 indicate increasing alkalinity. Acidity usually is regarded as strong at 5.0 or lower, whereas 10.0 or above is considered strongly alkaline.

Several factors influence soil acidity. The chemistry of soil parent materials, as well as crop fertilization and harvesting, can increase soil acidity. However, the major contributor to soil acidity in this modern era is acid precipitation (rain, snow, fog, or dry deposition). Acid rain actually has been measured below pH 2.0—an incredibly low value for natural precipitation, as acid as lemon juice. Increased acidity in the soil solution accelerates the chemical weathering and depletion rates of some mineral nutrients, and it can also decrease the availability of other nutrients. Because most crops are sensitive to

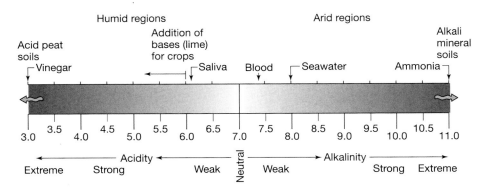

▲ Figure 28.4 pH scale

specific pH levels, acid soils below pH 6.0 require treatment to raise the pH. This soil treatment is accomplished by the addition of bases in the form of minerals that are rich in base cations, usually lime (calcium carbonate, $CaCO_3$).

Soil pH is regularly tested at many locations using an *electrometric method* (a probe into a soil-water mixture) or using a *dye method* (organic compounds in drops or test paper that react and indicate pH through color changes). More accurate work can be done in a laboratory. Care must be taken in the field when reading the pH of a soil sample because values may vary over a short distance, perhaps in response to previous fertilizer applications, or reflect topographic effects. Despite these variations, soil pH is related to so many biological and chemical factors that influence both soil processes and plant response that it is an important parameter to know.

1. Using pH test equipment provided by your instructor, evaluate two soil samples, perhaps one from your home and one from campus, or from two different areas on your campus. Ideally, you would use samples from sites with different vegetation. Record their pH readings on Figure 28.4 above. How far apart on the pH scale are the two readings?

2. Are either of the samples strongly acidic? What remedial actions could be taken to make them more pH neutral, and under what circumstances might you want to do this?

3. Review Focus Study 3.2 in *Geosystems* or Focus Study 2.2 in *Elemental Geosystems*. How has acid deposition affected soils, forests, and lakes in the United States and Europe? Do you think that the remedial actions you outlined in item #2 are practical for large-scale problems affecting entire regions? What do you think would be the best solution at these larger spatial scales? (*Hint:* Consider what the sources of acidifying materials are.)

SECTION 6

In the **soil classification** system developed by the Natural Resources Conservation Service, there is a hierarchy of soil categories. Table 28.5 presents the categories of the U.S. Soil Taxonomy.

Figure
Soil Taxonomy

http://goo.gl/YBCCXa

Soil Category	Number of Soils Included
Orders	12
Suborders	47
Great groups	230
Subgroups	1200
Families	6000
Series	15,000

▲Table 28.5   U.S. Soil Taxonomy

1. Using the map of the world distribution of the Soil Taxonomy systems in *Geosystems*, or in *Elemental Geosystems*, find your present approximate location. Although this is only a general assessment because of the scale of the world map, use the color map legend and identify the one or two soil orders and their main characteristics that seem to represent your region. Record your findings here:

Web page
Soil Surveys

http://goo.gl/a8ChLn

Optional—Internet access required

2. You can see from Table 28.5 that there are some 15,000 soil series identified in the soil taxonomy for the United States and Canada. The USDA-Natural Resources Conservation Service has a website about its soil surveys: http://websoilsurvey.sc.egov.usda.gov/App/HomePage.htm.

To find out about the horizons, color, texture, structure, consistence, mineral, and chemical composition of your soil—or any other soil—go to this website.

The website allows you to set an area of interest and explore it in amazing detail. Zoom in to your school and use the Area of Interest tool to select your campus. The Soil Map tab displays all the individual soils found within the Area of Interest you set as an overlay on top of the satellite image of your campus, as well as a clickable list. The list of map units provides a wealth of data, including elevation, mean annual precipitation and temperature, frost-free period, farmland classification, landforms associated with that map unit, slope, drainage class, depth to water table, and water capacity. The Soil Data Explorer tab allows you to research the suitablity of soils for various uses, such as building site development, construction materials, vegetative productivity, and water management.

Zoom in to your campus and use the tools to set the Area of Interest to your campus.

Answer the following questions for your campus: how many map units are on your campus? Which covers the largest area? How many acres and what percent of your campus does it cover? How well drained is it? What is its farmland classification?

Note: When doing actual field work with a soil pedon (the complete soil profile and the basic sampling unit in soil surveys), you will find different colors in each horizon and perhaps a couple colors in one horizon. If you identify more than one color in the soil samples provided for this exercise, be sure to note it in your assessment.

MasteringGeography™ | Looking for additional review and lab prep materials? Go to www.masteringgeography.com for pre lab videos and pre and post lab quizzes.

Name: _____ Laboratory Section: _____

Date: _____ Score/Grade: _____

Video
Exercise 29
Pre-Lab Video

LAB EXERCISE 29

Biomes: Analyzing Global Terrestrial Ecosystems

Scan to view the Pre-Lab video

The Earth systems science approach embodied in physical geography synthesizes content from across the disciplines to create a holistic perspective. Exciting progress toward an integrated understanding of Earth's physical and biological systems is under way. Physical geography plays an important role in mapping and analyzing Earth's **terrestrial ecosystems**.

The diversity of organisms is a response to the interaction of the atmosphere, hydrosphere, and lithosphere, producing diverse conditions within which the biosphere exists. A first-ever-international attempt to protect biodiversity (species richness) is now ratified through the auspices of the United Nations.

The biosphere includes myriad ecosystems from simple to complex, each operating within general spatial boundaries. An **ecosystem** is a self-regulating association of living plants and animals and their non-living physical environment. In an ecosystem, a change in one component causes changes in others, as systems adjust to new operating conditions. Interacting populations of plants and animals in an area form a **community**. Each plant and animal occupies an area in which it is biologically suited to live—its habitat—and within that **habitat** it performs a basic operational function—its **niche**.

Earth itself is the largest ecosystem within the natural boundary of the atmosphere. Natural ecosystems are open systems for both energy and matter, with almost all ecosystem boundaries functioning as transition zones rather than as sharp demarcations.

Plants are the most visible part of the biotic landscape, a key aspect of Earth's terrestrial ecosystems. In their growth, form, and distribution, plants reflect Earth's physical systems: its energy patterns; atmospheric composition; temperature and winds; air masses; water quantity, quality, and seasonal timing; soils; regional climates; geomorphic processes; and ecosystem dynamics. The net photosynthesis for an entire plant community (photosynthesis minus respiration) is its *net primary productivity*. This is the amount of stored chemical energy (biomass) that the community generates for the ecosystem. **Biomass** is the net dry weight of organic material and varies among ecosystems.

A large, stable terrestrial ecosystem is known as a **biome**. Specific plant and animal communities and their interrelationship with the physical environment characterize a biome. Each biome is usually named for its *dominant vegetation*. We can generalize Earth's wide-ranging plant species into six broad biomes: *forest, savanna, grassland, shrubland, desert,* and *tundra*. Because plant distributions are responsive to environmental conditions and reflect variation in climatic and other abiotic factors, the world climate map and the global terrestrial biome map are presented together inside the cover of this lab manual.

Earth's diversity is expressed in 300,000 plant species. Despite this complexity of diverse plant and animal communities and their interrelationships, we can generalize Earth's ecosystems into 10 global terrestrial biome regions on the global terrestrial ecosystem map (inside back cover of this lab manual) and in the comprehensive Table 29.1. The table synthesizes many aspects of physical geography, integrating them under the 10 biomes. These biomes are included in the glossary and are presented on the biome map and integrative table. Lab Exercise 29 features four sections and one optional Google Earth™ activity.

Key Terms and Concepts

arctic tundra
biodiversity
biogeography
biomass
biome
community
ecosystem
equatorial and tropical rainforest
formation class
habitat
island biogeography

life zone
Mediterranean shrubland
midlatitude broadleaf and mixed forest
midlatitude grasslands
montane forest
niche
needleleaf forest
temperate rainforest
terrestrial ecosystem
tropical savanna
tropical seasonal forest and scrub

After completion of this lab, you should be able to:

1. *Identify* and *differentiate* various terrestrial ecosystem formation classes as they relate to patterns of precipitation and temperature.
2. *Compare* Köppen climate classifications from Lab Exercise 17 with the pattern of terrestrial ecosystems (see Appendix B, *Geosystems*, 9/e, or Appendix C in *Elemental Geosystems*, 8/e).
3. *Relate* vertical life zones to the latitudinal distribution of biomes.
4. *Utilize* a world biome map and an integrative table of major terrestrial biomes and their characteristics to *analyze* environmental conditions.

Materials/Sources Needed

pencil
world atlas or physical geography text

Lab Exercise and Activities

SECTION 1

Climate Controls of Ecosystem Structure and Form

Figure 29.1 illustrates the general relationship among temperature, precipitation, and vegetation. **Formation classes** are units that refer to the structure and appearance of dominant plants in a terrestrial ecosystem, for example, forest, shrubland, or grassland. Each formation includes numerous plant communities, and each community includes innumerable plant habitats. The illustration relates temperature and precipitation over hot, temperate, cool, and polar and alpine environments. As you examine the illustration, mentally review what you learned about the characteristics and distribution of climates in Lab Exercise 17.

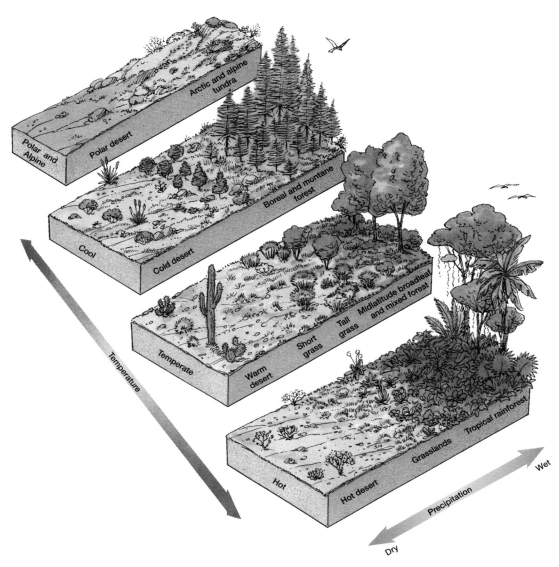

▲ Figure 29.1 Abiotic (nonliving) climate controls of ecosystem types: the generalized relationship among rainfall, temperature, and vegetation

Questions using Figure 29.1:

1. Describe in general terms the characteristic vegetation type and related temperature and moisture relationship that fit the area of your present town or school.

   Now, describe the same for the region where you were born (if substantially different).

2. Using Figure 29.1, and referring to the world biome map (inside the back cover of this lab manual), write these city names in the appropriate environmental location—locate and label each one on the figure. Discuss these with other lab members to make your determinations.

   a) Your present region
   b) Where you were born (if a different region from your present location)
   c) New York City
   d) Key West, Florida
   e) Montreal, Québec
   f) Dawson, Yukon
   g) Yuma, Arizona
   h) Omaha, Nebraska
   i) Elko, Nevada
   j) Everglades, Florida
   k) Coastal Oregon

   Recall that Köppen's climatic classification was based on two climate control regimes:

1. **Temperature**—reflecting either
   - Latitude—as in
   tropical
   subtropical
   midlatitude
   subarctic/subpolar
   polar
   icecap

   or

   - Altitude—as in *highland*

2. **Precipitation**—as in
   rainy
   wet and dry
   monsoon
   arid
   semi-arid
   humid
   dry

   In addition, global pressure and wind belts influence precipitation regimes:

- **Rainy/humid conditions**—result from one of the following:
  - instability and convectional uplift in warm *equatorial* and *subtropical* regions
  - frontal uplift in the *midlatitudes*
  - intensification of rainfall on the *windward side* of mountains

- **Dry conditions**—caused by one of the following:
  - subsiding air of the *subtropical high pressure belts*
  - isolation within *midlatitude* continental interiors
  - cold stable air of the *high latitudes*
  - drying of air masses descending the *leeward side* of mountain ranges

These are broad generalizations but can help you to see the causes of resulting climatic patterns. You will refer to these relationships in Section 3.

The descriptive names in Köppen's climate classification system (see Appendix B in *Geosystems*, or Appendix C in *Elemental Geosystems*) also often included the dominant vegetation types that those regimes supported: *Tropical Rainy* or *Rainforest*, *Tropical Wet and Dry* or *Savanna*, *Tropical/Midlatitude Arid* or *Desert*, *Tropical/Midlatitude Semi-arid* or *Steppe*, *Subtropical Dry* or *Mediterranean*, *Subarctic/Subpolar* or *Taiga*, *Polar* or *Tundra*.

SECTION 2

Earth's Major Terrestrial Biomes

The global distribution of Earth's major terrestrial biomes is portrayed on the map inside the back cover of this manual. Table 29.1 describes each biome on the map and summarizes other pertinent environmental information—a compilation from many aspects of physical geography, for Earth's biomes are a synthesis of the environment and biosphere.

Questions and completion items about this sample of Earth's terrestrial biomes.

1. From Table 29.1, the biome map, and the climate map, determine the terrestrial biome that best characterizes each of the following descriptions and write its name *on the first line* provided. *On the second line*, relate it to temperature and precipitation regimes and/or global pressure and wind belts outlined at the end of Section 1. The first one is completed for you as an example.

   a) PRECIP less than 1/2 POTET: _____[Cold desert and semidesert]_____
   _____[midlatitude continental interior and on leeward side of mountain range]_____

   b) Southern and eastern U.S. evergreen pines: _____

   c) Mollisols and aridisols: _____

   d) Characteristic of central Australia: _____

   e) Selva: _____

   f) Characteristic of the majority of central Canada: _____

   g) Transitional between rain forest and tropical steppes: _____

   h) Tallest trees on Earth: _____

   i) Sedges, mosses, and lichens: _____

   j) Characteristic of Zambia (south-central Africa): _____

   k) Four biome types that occur in Chile: _____

   l) Precipitation of 150–500 cm/year, outside the tropics: _____

   m) Characteristic of central Greenland: _____

Biome and Ecosystems (map symbol)	Vegetation Characteristics	Soil Orders (Soil Taxonomy)	Köppen Climate Designation	Annual Precipitation Range	Temperature Patterns	Water Balance
Equatorial and Tropical Rain Forest (ETR) Evergreen broadleaf forest Selva	Leaf canopy thick and continuous; broadleaf evergreen trees, vines (lianas), epiphytes, tree ferns, palms	Oxisols Ultisols (on well-drained uplands)	Af Am (limited dry season)	180–400 cm (>6 cm/mo)	Always warm (avg 25°C)	Surpluses all year
Tropical Seasonal Forest and Scrub (TrSF) Tropical monsoon forest Tropical deciduous forest Scrub woodland and thorn forest	Transitional between rain forest and grasslands; broad-leaf, some deciduous trees; open parkland to dense undergrowth; acacias and other thorn trees in open growth	Oxisols Ultisols Vertisols (in India) Some alfisols	Am Aw Borders BS	130–200 cm (>40 rainy days during 4 driest months)	Variable, always warm (>18°C)	Seasonal surpluses and deficits
Tropical Savanna (TrS) Tropical grassland Thorn tree scrub Thorn woodland	Transitional between seasonal forests, rain forests, and semiarid tropical steppes and desert; trees with flattened crowns, clumped grasses, and bush thickets; fire association	Alfisols (dry: Ultalfs) Ultisols Oxisols	Aw BS	90–150 cm, seasonal	No cold-weather limitations	Tends toward deficits, therefore fire- and drought-susceptible
Midlatitude Broadleaf and Mixed Forest (MBME) Temperate broadleaf Midlatitude deciduous Temperate needleleaf	Mixed broadleaf and needleleaf trees; deciduous broadleaf, losing leaves in winter; southern and eastern evergreen pines demonstrate fire association	Ultisols Some Alfisols	Cfa Cwa Dfa	75–150 cm	Temperate, with cold season	Seasonal pattern with summer maximum PRECIP and POTET (PET); no irrigation needed
Needleleaf Forest and Montane Forest (NF/MF) Taiga Boreal forest Other montane forests and highlands	Needleleaf conifers, mostly evergreen pine, spruce, fir; Russian larch, a deciduous needleleaf	Spodosols Histosols Inceptisols Alfisols (boralfs: cold)	Subarctic Dfb Dfc Dfd	30–100 cm	Short summer, cold winter	Low POTET (PET), moderate PRECIP, moist soils, some water-logged and frozen in winter; no deficits
Temperate Rain Forest (TeR) West Coast forest Coast redwoods (U.S.)	Narrow margin of lush evergreen and deciduous trees on windward slopes; redwoods, tallest trees on Earth	Spodosols Inceptisols (mountainous environs)	Cfb Cfc	150–500 cm	Mild summer and mild winter for latitude	Large surpluses and runoff
Mediterranean Shrubland (MSh) Sclerophyllous shrubs Australian eucalyptus forest	Short shrubs, drought adapted, tending to grassy woodlands; chaparral	Alfisols (Xeralfs) Mollisols	Csa Csb	25–65 cm	Hot, dry summers, cool winters	Summer deficits, winter surpluses
Midlatitude Grasslands (MGr) Temperate grassland Sclerophyllous shrub	Tallgrass prairies and short-grass steppes, highly modified by human activity; major areas of commercial grain farming; plains, pampas, and veld	Mollisols Aridisols	Cfa Dfa	25–75 cm	Temperate continental regimes	Soil moisture utilization and recharge balanced; irrigation and dry farming in drier areas
Warm Desert and Semidesert (DBW) Subtropical desert and scrubland	Bare ground graduating into xerophytic plants including succulents, cacti, and dry shrubs	Aridisols Entisols (sand dunes)	BWh BWk	<2 cm	Average annual temperature around 18°C, highest temperatures on Earth	Chronic deficits, irregular precipitation events, PRECIP <½ POTET (PET)
Cold Desert and Semidesert (DBC) Midlatitude desert, scrubland, and steppe	Cold desert vegetation includes short grass and dry shrubs	Aridisols Entisols	BSh BSk	2–25 cm	Average annual temperature around 18°C	PRECIP >½ POTET (PET)
Arctic and Alpine Tundra (AAT)	Treeless; dwarf shrubs, stunted sedges, mosses, lichens, and short grasses; alpine, grass meadows	Gelisols Histosols Entisols (permafrost)	ET Dwd	15–80 cm	Warmest months <10°C, only 2 or 3 months above freezing	Not applicable most of the year, poor drainage in summer
Ice			EF			

▲Table 29.1 Major terrestrial biomes and their characteristics

n) Characteristic of Iran (northeast of the Persian Gulf): _____

o) Major area of commercial grain farming: _____

p) Seasonal precipitation of 90 to 150 cm/year: _____

q) Cfa and Dfa climate types: _____

r) Spodosols and permafrost, short summers: _____

s) Southern Spain, Italy, and Greece, central California: _____

t) Characteristic of Ireland and Wales: _____

u) Just west of the 98th meridian in the United States: _____

v) Just east of the 98th meridian in the United States: _____

w) Bare ground and xerophytic plants: _____

x) East coast of Madagascar: _____

y) West coast of Madagascar: _____

z) Characteristic of northern Mexico: _____

Finally, from the **GEOGRAPHY I. D.** that you completed in the Preface of this lab manual, complete the following for your hometown or college campus location:

Place name: Köppen classification:

2. What are the main climatic influences for this station (air pressure, air mass sources, degree of continentality, temperature of ocean currents)?

Terrestrial biome characteristics:

SECTION 3

Life Zones—Conditions Changing with Elevation

Alexander von Humboldt (1769–1859)—an explorer, geographer, and scientist—deduced that plants and animals occur in related groupings wherever similar conditions occur in the abiotic environment. After several years of study in the Andes Mountains of Peru, he described a distinct relationship between elevation and plant communities, his **life zone** concept. As he climbed the mountains, he noticed that the experience was similar to that of traveling away from the equator toward higher latitudes.

This zonation of plants with altitude is noticeable on any trip from lower valleys to higher elevations. Each life zone possesses its own temperature, precipitation, and insolation relationships and therefore, its own biotic communities. The key is that temperatures decrease rapidly with increasing elevation at an average of 6.4 C° per kilometer (3.5 F° per 1000 ft), a rate known as the *normal lapse rate*.

The Grand Canyon in Arizona provides a good example. The inner gorge at the bottom of the canyon (600 m, or 2000 ft, in elevation) exhibits life forms characteristic of the lower Sonoran Desert of northern Mexico. However, the north rim of the canyon (2100 m, or 7000 ft, in elevation) is dominated by ecosystems similar to those of southern Canadian forests. On the summits of the nearby San Francisco Mountains (3600 m, or 12,000 ft, in elevation), the vegetation is similar to the arctic tundra of northern Canada.

1. Using a physical geography text, such as *Geosystems*, add appropriate labels to Figure 29.2 in the spaces provided at the bottom of Fig. 29.2: tropical rain forest, temperate deciduous forest, needleleaf forest, tundra, and ice and snow. Next, assuming the normal lapse rate is in effect, beginning at sea level and the rain forest label, assume the temperature to be at 27°C (80°F). Assume the labels are spaced at 600 m (2000 ft) intervals and calculate the temperature on this particular day at each label elevation. Calculate five temperatures decreasing with altitude (600, 1200, 1800, 2400, and 3000 m) and record your answers on the lines along the right side of the mountain in Fig. 29.2.

2. Following discussion among lab members, compare climate controls (temperature and precipitation) and the effects of altitude—Figures 29.1 and 29.2. Briefly describe the importance of each control on ecosystem character, and plants and animals.

   a) Temperature: _____

   _____

   _____

   b) Precipitation: _____

   _____

   _____

   c) Elevation: _____

   _____

   _____

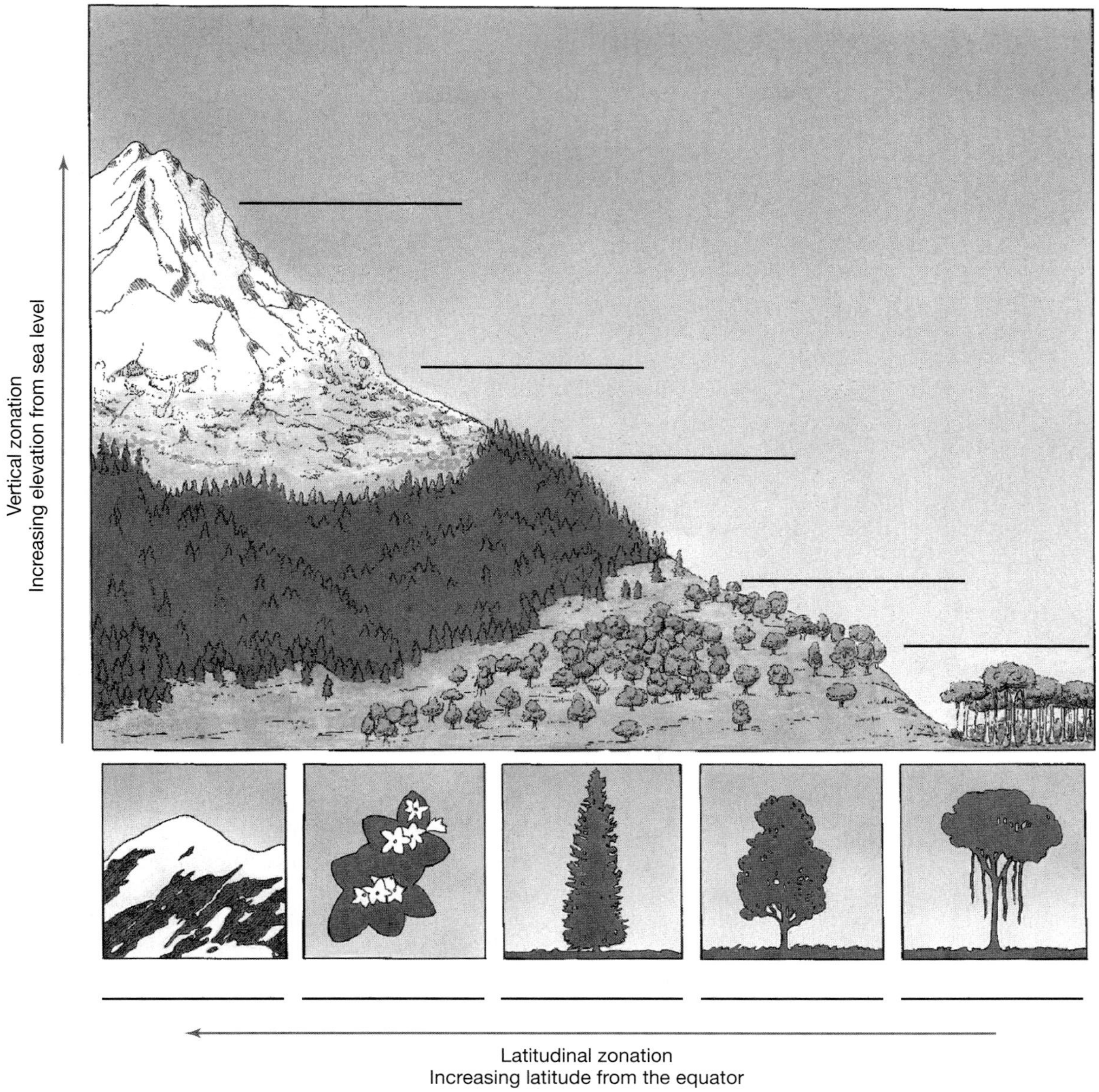

▲ Figure 29.2 Progression of generalized plant community life zones with increasing elevation or latitude.

3. Relative to the life zone concept in Bolivia, South America, speculate on these items:

   a) Given these concepts of elevation and temperature, why are there glaciers along the crest of the Andes Mountains so near the equator? _____

   _____

   _____

   b) Why are Bolivians able to grow wheat, potatoes, and barley at 4250 m (14,000 ft) at the same latitude of rain forests at low elevation where these same crops will not grow?

   _____

   _____

   _____

SECTION 4

Island Biogeography

Biogeography is the study of the temporal and spatial distribution of plants and animals. Biogeographers study aspects of a population of individuals of a species, as well as general patterns of biological communities. Biogeography can be used across a wide range of spatial and temporal scales, from a group of individuals in one particular study area to climate and geological factors operating at a global spatial scale and geologic time scale. It draws upon the fields of ecology, evolutionary biology, geology, physical geography, and GIS (Geographic information systems). Biogeographers have worked to identify factors that cause species and biological communities to vary with latitude, elevation, insolation, and habitat size. Patterns of species distributions are often understood through speciation and extinction, plate tectonics, climate change (including glaciation and changes in sea level that expose and submerge land bridges), habitat size and isolation, and insolation, which powers photosynthesis.

David Quammen described biogeography as practiced by scientists in his book *The Song of the Dodo*: "Biogeography does more than ask *Which species?* and *Where?* It also asks *Why?* and, what is sometimes even more crucial, *Why not?*"

An important part of biogeography is **island biogeography**, the study of the factors that determine the number of species in a specific community. An island is not just a location surrounded by water but can be any location surrounded by a significantly different habitat or barrier to dispersal. Island biogeography is a very powerful concept because it applies to many situations. Loss of **biodiversity**, or species richness, is an important concern in ecology since the present rate of the loss of species is equaled only by mass extinction events that in the past have been caused by global catastrophes, such as the Cretaceous–Tertiary extinction event that saw the end of 75% of all the species on Earth at the time.

Scientists have been aware of the relationship between habitat size and the number of different species, called the species–area relationship. The species–area relationship is not linear, but logarithmic, because for for the number of species to double, the habitat area needs to increase by 10 times.

In 1962 Robert MacArthur and Edward O. Wilson published *The Theory of Island Biogeography*, which refined the species–area relationship. They considered the size of an island, and also the distance of the island from the mainland, which is the source of new individuals and species. Larger islands have several advantages over smaller islands, in terms of biodiversity. Larger islands are bigger targets, so more individuals will potentially find them. Larger islands also may have more different habitats. MacArthur and Wilson hypothesized that an equilibrium biodiversity value could be calculated, based upon the rate of new species arriving and the rate of extinction for species on the island. Figure 29.3 shows four different predicted equilbrium levels based on the size of the island and its distance from the mainland. One application of island biogeography is studying the effects of habitat fragmentation on species loss in order to design natural reserves to preserve biodiversity.

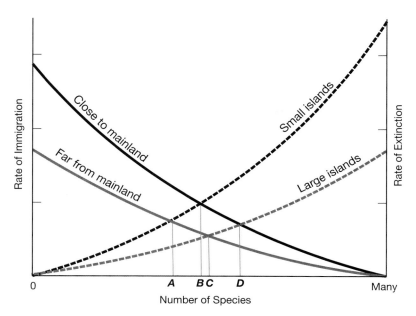

▲ Figure 29.3 Equilibrium theory of island biogeography (from MacArthur, R. H., and E. O. Wilson. 1967. *The Theory of Island Biogeography*)

1. Figure 29.3 shows four examples of species equilibrium numbers. In this hypothetical example, which combination of factors produces the highest number of species? The lowest number of species? In this example, is island size or distance from the mainland more important in producing a higher species equilibrium number?

2. What are some of the characteristics of animals that would you expect to find on an island? What are some characteristics of animals that you would expect to not find?

3. Some islands, such as the Hawaiian Islands, are far from the nearest land and were initially bare rock. Other islands, such as the Channel Islands, off the southern California coast, were part of the mainland and were isolated by rising sea levels. How might the organisms on the Hawaiian Islands differ from organisms found on the Channel Islands?

MacArthur and Wilson's theory of equilibrium species numbers took into account new species and organisms colonizing the islands from a mainland. There are, however, locations where the number of new immigrants may be zero. High mountain habitats in the western United States are one example. As climate became warmer and drier during the Pleistocene, habitats shifted up in elevation in response, creating islands of montane habitat surrounded by large expanses of desert. The desert is an impassible barrier for nonflying animals. In this example, there is no mainland source of new individuals and species. Lomolino, Brown, and Davis examined biodiversity in montane forests in the western United States. Their results are shown in Table 29.2.

	Forest Island	Species	Area (km²)
1	Mogollon	16	11134.33
2	Lasals	13	435.23
3	Chuskas	11	1507.9
4	Abajos	10	958.57
5	Sandias	10	43.85
6	Sacramentos	10	1349.8
7	Mt. Taylor	8	465.3
8	Manzanos	8	141.13
9	Blacks	8	872.96
10	Zunis	7	625.63
11	San Mateos	7	366.23
12	N. Uncompaghre	6	359.1
13	Capitans	6	84.8
14	Magdalenas	6	106.8
15	S. Kaibab	4	274.27
16	Pinalenos	4	87.09
17	Navajo	4	6.89
18	Guadalupes	3	25.71
19	Chiricahuas	2	68.13
20	Catalinas	2	29.8
21	Prescotts	2	115.8
22	Hualapais	2	18.25
23	Organs	1	8.57
24	Animas	1	10.58
25	Huachucas	1	12.7
26	Santa Ritas	1	11.9
27	N. Rincon	1	10.08

▲Table 29.2 Species numbers and habitat size for sky islands. (from Lomolino, M. V., J. H. Brown, and R. Davis. 1989. Island biogeography of montane forest mammals in the American Southwest. *Ecology* 70:180–194)

4. What are some examples of barriers to dispersal of organisms? What are examples of factors that would act as filters to only certain species? Why do you think this study looked only at nonflying mammals?

5. Plot the biodiversity and area values on Figure 29.4. The values for the first 10 forest islands and the trend line have been plotted for you.

6. What is the ratio of species/area for the largest park compared with the smallest park? How many species should be in the largest park if the species/area ratio there is the same as the value for the smallest park?

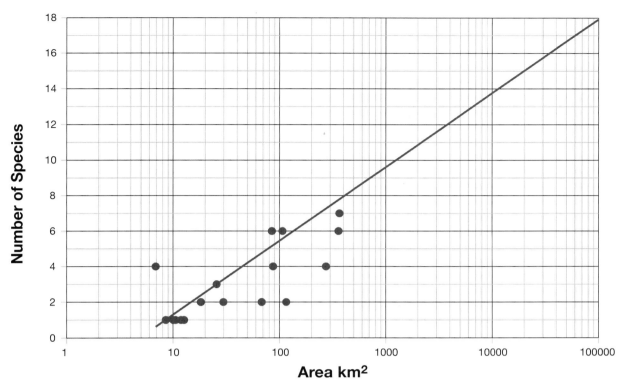

▲ Figure 29.4 Species numbers and habitat size

Use Figure 29.4 and the trend line to answer the following questions.

7. How many square kilometers of habitat would be required to support 10 species?

   How many square kilometers of habitat would be required to support 5 species?

8. How large is the largest park? How many species does it support? How many species would you expect to find in a park half that size? What percentage reduction is this over the original population?

9. How many square kilometers are required to support four species? How many species would you expect to find in a park half that size? What percent reduction is this over the original population?

10. Based on your answers to questions 7 and 8, are larger or smaller habitats more prone to species loss due to reduction in habitat size? If you were identifying habitat to protect, would you recommend protecting one 400 km² area or four 100 km² areas? What other factors might you consider? Are all species equal?

11. Optional Google Earth™ activity, Deforestation in the Amazon rainforest. For the KMZ file and questions, go to mygeoscienceplace.com. Then click on the cover of *Applied Physical Geography: Geosystems in the Laboratory*.

MasteringGeography™ | Looking for additional review and lab prep materials? Go to www.masteringgeography.com for pre lab videos and pre and post lab quizzes.

Name: _____ Laboratory Section: _____
Date: _____ Score/Grade: _____

Video
Exercise 30
Pre-Lab Video

Scan to view the Pre-Lab video

http://goo.gl/EwTuL0

LAB EXERCISE 30

An Introduction to Geographic Information Systems

Geographic information systems, known by the familiar **GIS** abbreviation, are computer-based, data-processing tools for gathering, manipulating, analyzing, and displaying geographic information. Today's sophisticated computer systems allow the integration of geographic information from direct surveys (on-the-ground mapping) and remote sensing in complex ways never before possible. Through a GIS, Earth and human phenomena are analyzed over time. The range of subjects suitable for GIS analysis is limited only by the imagination of the user. If you have used the Internet to get a map or directions, you have used GIS. Today GIS are used to track and predict patterns of disease, help public safety agencies, and plan delivery routes for shipping companies, to name just a few examples. Lab Exercise 30 features three main sections, including one Internet GIS section and one GIS analysis section.

Key Terms and Concepts

attribute data
geographic information systems (GIS)
line
point
polygon
raster
spatial data
vector

KEY LEARNING concepts

After completion of this lab, you should be able to:

1. *Describe* essential GIS concepts and *identify* the utility of construction of a GIS model.
2. *Locate* and *utilize* GIS data on the Internet.
3. Use an online GIS to analyze the risks from Mt. Rainier.

Materials/Sources Needed

Data for analysis
Internet browser access

Lab Exercise and Activities

SECTION 1

Basic GIS Concepts

Modern GIS owes a great deal to the work of Ian McHarg and his search for an ecologically based land-use planning system. He drew on clear acetate sheets to create layers of information. Each layer showed one theme—roads or land cover or existing houses, for example. He stacked multiple sheets to synthesize data and reveal complex relationships. Although he used sheets of plastic to display data (just as you will in the following exercises) rather than a computer screen, the underlying principles are the same. Figure 30.1 shows an example of how multiple layers are combined.

GIS combine location information and descriptive information for each feature in a layer. The location information tells us where each feature is. This **spatial data** is recorded in a coordinate system such as latitude–longitude. The descriptive information (*attribute data*) tells us about the qualities of each feature. **Attribute data** is stored in a spreadsheet or database and is capable of recording multiple attributes for each location.

Spatial data in a GIS are stored as **points** (Figure 30.2a), **lines** (30.2b), or **polygons** (areas, 30.2c). Features that only need location information are shown as points. Features that have length are stored as arcs, which are made by combining lines. Finally, features with length and width are stored as polygons. This point–line–polygon system is referred to as a **vector data** structure. GIS can also use data stored in another spatial data structure called **raster**. Raster, or grid, cells contain information such as elevation or remote-sensing imagery. Each cell is an intersection of rows and columns, and the attribute it contains may occur anywhere in the cell.

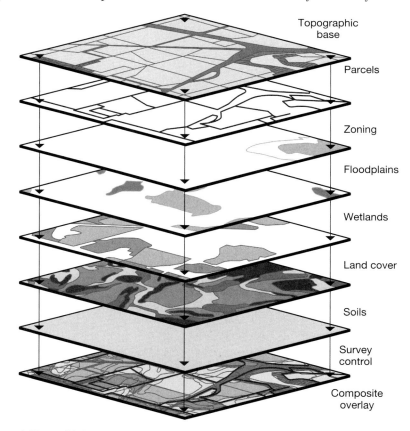

▲ Figure 30.1 Computer-processed spatial data layered in a geographic information system (GIS) produces a composite overlay for analysis. (After USGS)

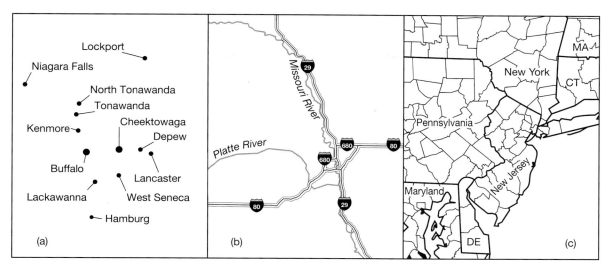

▲ Figure 30.2 Examples of point, line, and polygon features: cities (a), highways and rivers (b), and states and counties (c)

By combining layers, a GIS is capable of analyzing patterns and relationships, such as the floodplain or soil layer in Figure 30.1. GIS can create new layers of information with this overlay process as well. A research study may follow specific points or areas through the complex of overlay planes. The utility of a GIS compared with that of a fixed map is the ability to manipulate the variables for analysis and to constantly change the map—the map is alive! Before the advent of computers, an environmental impact analysis required someone to gather data and painstakingly hand-produce overlays of information to determine positive and negative impacts of a project or an event. Today, this layered information is handled by a computer-driven GIS, which assesses the complex interconnections.

1. Give some examples of GIS that you have used, perhaps without realizing it before now.

2. If you were a land planner, what questions could you answer by using the composite overlay of data layers in Figure 30.1?

3. Give examples of features in your life that would be symbolized by points, arcs, and polygons.

   Points:

   Arcs:

   Polygons:

SECTION 2

The Internet and GIS

One of the exciting GIS developments in the past few years is the spread of Internet GIS services. You have probably used many of these services without realizing that you were using a GIS. Some examples would include using the Internet to get directions to a location or to show a map. More sophisticated examples of these are Internet Map Servers (IMS) that allow you to pick which layers are displayed, as well as how they are shown. The main advantage of IMS over conventional desktop GIS is that it does not require special software to be installed on the user's computer, nor does using it require special training.

An example of a very sophisticated program than can be used with little training is Ushahidi, which uses a web interface to collect and display crowd-sourced data. It was used very successfully after the devastating earthquake in Haiti in 2010. People in Haiti who needed assistance or supplies were able to send a text message to the Ushahidi team. The text messages were received by a team of volunteers all over the world who then entered the data into a GIS that was used to coordinate rescue efforts. The rescue efforts were also assisted by OpenStreetMap (http://www.openstreetmap.org), another crowd-sourced project. OpenStreetMap is a global public street mapping project where volunteers enter and edit street data. The base data can come from remote-sensing imagery and users' GPS tracks. During the rescue efforts in Haiti, the command center's location was chosen using OpenStreetMap data. For more information about Ushahidi, visit http://www.ushahidi.com. IMS can also use real-time data, such as traffic speed and accident data, to display current road conditions on Google Maps™. Some other examples of IMS are:

> Two online galleries of user-created maps:
> https://maps.google.com/gallery/
> http://www.arcgis.com/home/gallery.html
> An excellent site showing live wildfire information:
> http://www.geomac.gov/

The Internet also has opened up access to free data that was either expensive or completely unavailable just a few years ago. For example, California has created the California Geoportal at http://portal.gis.ca.gov/geoportal, which coordinates GIS data from agencies across the state. This site allows the public to access data easily and eliminates duplication of effort among agencies. A similar site exists at the federal level at http://www.data.gov/geospatial/.

1. Using the Internet, find three websites that use online GIS. Examples might include real estate, online maps or directions, remote-sensing imagery, and wildfire tracking. Record their URLs and descriptions below.

2. Visit three websites that offer free data for downloading that could be used in a GIS. Record their URLs and descriptions below.

SECTION 3

Mt. Rainier Hazards

In this section you will use ArcGIS Online, a web-based GIS. ArcGIS Online is an example of a cloud-based GIS that allows users to perform GIS operations without needing to install programs on a computer.

Mt. Rainier is a large stratovolcano in the Pacific Northwest. It has erupted approximately nine times in the past 10,000 years. The most hazardous products of the eruptions are lahars, volcanic mud flows created by the interaction of snow or ice and lava or pyroclastic debris. For this exercise you will be analyzing lahars with 1000-year and 100-year recurrence intervals. Another hazard is pyroclastic flows of gas, ash, and rock fragments that are over 800°C (1440°F) and can travel at speeds over 160 kmph (100 mph). Figure 30.3 shows Mt. Rainier and the region of lahars and pyroclastic flow danger. In this exercise, you will produce a volcanic hazards map that will look similar to Figure 30.3.

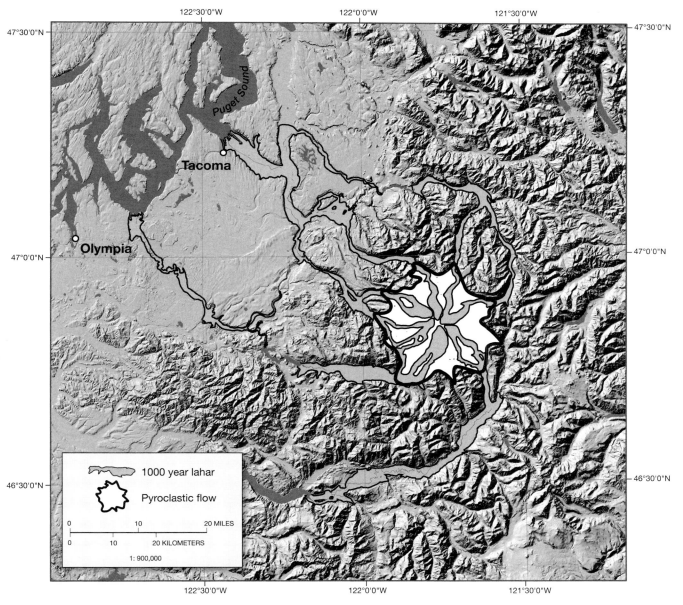

▲ Figure 30.3 Mt. Rainier

You are the newly hired emergency preparedness coordinator for the Tri-Counties Area. Your area of responsibility is the Washington counties of Thurston, Pierce, and King. Your first order of business is to identify elementary schools that are at risk from the volcanic hazards of Mt. Rainier.

Before you can begin the process, you will need to sign up for a free trial account at www.esri.com/software/arcgis/arcgisonline. The free trial will give you 30 days of access; check with your instructor to find out if your school has an agreement with ESRI that will allow you longer access. After you have created your account for ArcGIS Online, go to www.arcgis.com/home/index.html and sign in with your account information. If you have not used ArcGIS Online, you will want to watch the following tutorial videos: Creating a map; Add features from a file; and Changing symbology. After you have watched the three tutorial videos, you are ready to start making a map.

You will need to download the hazards and schools data for this exercise from http://mygeoscienceplace.com. Cloud-based GIS is a rapidly changing technology. If ArcGIS Online changes substantially, an updated exercise will be posted to http://mygeoscienceplace.com.

1. Download the files 100-year_Lahar.zip, 1000-Year_Lahar.zip, Pyroclastic_flow.zip, and Schools.zip to your computer.

2. Sign in to ArcGIS Online and try different base map layers. After you've picked a base map, use the Add tool and Search for Layers at ArcGIS Online. You won't add these layers to the map; this step is just to acquaint you with some of the data sources available. You can limit your search by using key words and specifying the region to search. What are the three most interesting layers that you found?

3. Use the Add tool and Search for Layers at ArcGIS Online and search for topographic maps. You will need a layer that has elevation contour lines, such as the USA Topo Maps layer or the USA Topographic Maps (Mature Support) layer. You can add it to your map or use it as a base map. After you have added an appropriate topographic map, use the Add tool and Add Layer from File for each of the four files you downloaded.

4. Turn off the 100-year lahar and 1000-year lahar layers. Turn each layer on and off so that you can see the extent of each. Modify the symbology of each file so that you can distinguish the pyroclastic flow layer from the 100-year and 1000-year lahar layers.

5. Which layer covers the largest area?

6. You can rearrange the order that the layers draw by dragging and dropping them in the Contents window. In order to best see both lahar layers, which layer should be on the bottom?

7. You can also adjust the transparency of each layer so you can see your data and the base map underneath. What levels of transparency did you use for the pyroclastic flow, 1000-year lahar, and 100-year lahar layers?

8. Use the topographic map and the pyroclastic flow layers to find the relief from summit to the farthest extent of the 100-year lahar flow. How far does the pyroclastic flow extend from the summit? What is the lowest elevation that is covered by potential pyroclastic flow?

9. How far does the 100-year lahar extend from the summit?

10. How far does the 1000-year lahar extend from the summit?

11. What is the name of the school that is closest to the summit and in a lahar area? What is the distance from the summit to the closest school? What is the vertical relief from the summit to the closest school?

12. If lahars can travel at 100 kmph (60 mph), how much time would it take for a flow to reach the school closest to the summit?

13. Identify three schools at risk for 1000-year lahar but not 100-year lahar.

14. Identify three schools at risk from both 1000-year and 100-year lahars.

15. Use the Add tool to search for Washington counties and add that layer to your map. How many schools are at risk from a 1000-year lahar in each county? Which county at greatest risk?

16. This exercise involved only elementary schools. What other types of structures would you analyze for hazard potential?

17. Increase the transparency of the lahar layers so you can see the terrain underneath. What terrain features appear to control the flow of the lahars?

18. When you have finished editing your map, use the Share tool to send a link to your instructor.

Data sources:
School data: National Map, http://viewer.nationalmap.gov/viewer/
Mt. Rainier hazards data: Open-File Report 98–428 Volcano Hazards from Mount Rainier, Washington, Revised 1998, http://pubs.usgs.gov/of/1998/0428/

MasteringGeography™ | Looking for additional review and lab prep materials? Go to www.masteringgeography.com for pre lab videos and pre and post lab quizzes.

Glossary

The Lab Exercise in which each term appears is **boldfaced**, followed by a specific definition related to the usage of the term in *Applied Physical Geography*.

actual evapotranspiration The actual amount of evaporation and transpiration that occurs; derived in the water balance by subtracting the deficit from potential evapotranspiration (ACTET).

adiabatic Pertaining to the heating and cooling of a descending or ascending parcel of air through compression and expansion, without any exchange of heat between the parcel and the surrounding environment.

air pressure Pressure produced by the motion, size, and number of gas molecules and exerted on surfaces in contact with the air. Normal sea level pressure, as measured by the height of a column of mercury (Hg), is expressed as 1013.2 millibars, 760 mm of Hg, or 29.92 inches of Hg. Air pressure can be measured with mercury or aneroid barometers.

albedo Reflectivity of a surface. The darker a surface, the lower the albedo (the less the surface will reflect the Sun's rays) and the higher the absorption of the Sun's energy.

Alber's projection Alber's equal-area conic projection, used as a base for U.S. quadrangle maps. Two standard parallels are used for the conterminous United States—29.5°N and 45.5° N latitudes. The standard parallels are shifted for conic projections of Alaska (55°N and 65°N) and Hawai'i (8°N and 18°N).

alluvial fan A fan-shaped fluvial landform at the mouth of a canyon, particularly noticeable in arid landscapes where streams are intermittent.

alluvial terraces Level areas that appear as topographic steps above a stream, created by the stream as it scours with renewed downcutting into its floodplain; composed of unconsolidated alluvium (see *alluvium*).

alluvium General descriptive term for clay, silt, and sand transported by running water and deposited in sorted or semi-sorted sediment on a floodplain, delta, or in a stream bed.

alpine glacier A glacier confined in a mountain valley or walled basin, consisting of three subtypes: valley glacier (within a valley), piedmont glacier (coalesced at the base of a mountain, spreading freely over nearby lowlands), and outlet glacier (flowing outward from a continental glacier).

altimetry The measurement of altitude using air pressure. A *pressure altimeter* is an instrument that measures altitude based on a strict relationship between air pressure and altitude. Air pressure is measured within the altimeter by an aneroid barometer capsule with the instrument graduated in increments of altitude.

altitude The angular distance between the horizon (a horizontal) and the Sun (or any point).

analemma A convenient device to track the passage of the Sun, the Sun's declination, and the positive and negative equation of time throughout the year.

analog Using a different method to portray some type of information. Relating to GIS, the use of manually prepared graphics (e.g., transparencies) to depict spatial information.

aneroid barometer A device to measure air pressure using a partially emptied, sealed cell (see *air pressure*).

anticline The upfolds in a folded landscape, often creating ridges between the downfolded *synclines*.

apparent temperature The temperature subjectively perceived by each individual, also known as sensible temperature.

arctic tundra A biome in the northernmost portions of North America and northern Europe and Russia, featuring low, ground-level herbaceous plants as well as some woody plants.

arêtes A sharp ridge that divides two cirque basins. Derived from "knife edge" in French, these form sawtooth and serrated ridges in glaciated mountains.

asthenosphere The region of the upper mantle known as the plastic layer; the least rigid portion of Earth's interior; shatters if struck yet flows under extreme heat and pressure.

attribute table Nonspatial information about a feature being mapped in a GIS (e.g., age, use, type, value), as contrasted to *spatial data*.

available water The portion of capillary water that is accessible to plant roots; usable water held in soil moisture storage.

azimuth A compass direction expressed as an arc area clockwise from the north through 360°.

backswamp A low-lying, swampy area of a floodplain; adjacent to a river, with the river's natural levees on one side and the sides of the valley on the other (see *yazoo tributary*).

bajada A continuous apron of coalesced alluvial fans, formed along the base of mountains in arid climates; presents a gently rolling surface from fan to fan. (See Alluvial fan.)

barometer A device to measure air pressure; either mercury or aneroid type (see *air pressure*).

barrier spit A depositional form that develops when transported sand in a barrier beach or island is deposited in long ridges that are attached at one end to the mainland and partially cross the mouth of a bay.

basin and range A region of dry climates, few permanent streams, and interior drainage patterns in the western United States; a faulted landscape composed of a sequence of horsts and grabens.

bay barrier An extensive sand spit that encloses a bay, cutting it off completely from the ocean and forming a lagoon; produced by littoral drift and wave action; sometimes referred to as a *baymouth bar*.

beach The portion of the coastline where an accumulation of sediment is in motion.

biodiversity A principle of ecology and biogeography which says that the more diverse the species population in an ecosystem (in number of species, quantity of members in each species, and genetic content), the more risk is spread over the entire community, which results in greater overall stability, greater productivity, and increased use of nutrients, as compared to a monoculture of little or no diversity.

biomass The total mass of living organisms on Earth or per unit area of a landscape; also, the weight of the living organisms in an ecosystem.

biome A large terrestrial ecosystem characterized by specific plant communities and formations; usually named after the predominant vegetation in the region.

boreal forest See *northern needleleaf forest*.

capillary water Soil moisture, most of which is accessible to plant roots; held in the soil by surface tension and cohesive forces between water and soil (see also *available water, field capacity*, and *wilting point*).

cardinal points The four principal compass points: north, south, east, and west.

cartography The making of maps and charts; a specialized science and art that blends aspects of geography, engineering, mathematics, graphics, computer science, and artistic specialties.

cinder cone A volcanic landform of pyroclastics and scoria, usually small and cone shaped and generally not more than 450 m (1500 ft) in height, with a truncated top.

circle of illumination The division between lightness and darkness on Earth; a day-night great circle.

cirque A scooped-out, amphitheater-shaped basin at the head of an alpine glacier valley; an erosional landform.

classification The process of ordering or grouping data or phenomena in related classes; results in a regular distribution of information; a taxonomy.

climate The consistent, long-term behavior of weather over time, including its variability; in contrast to *weather*, which is the condition of the atmosphere at any given place and time.

climate change A large and lasting change in the statistical distribution of weather patterns. The time scale of these changes may range from decades to millions of years. The changes include average weather conditions, or in the distribution of weather around the average conditions (i.e., more or fewer extreme weather events).

climatic regions Areas of similar climate, which contain characteristic regional weather and air mass patterns.

climatology A scientific study of climate and climatic patterns and the consistent behavior of weather and weather variability and extremes over time in one place or region and the effects climate and climate change have on human society and culture.

climograph A graph that plots daily, monthly, or annual temperature and precipitation values for a selected station; may also include additional weather information.

col Formed by two headward-eroding cirques that reduce an arête (ridge crest) to form a high pass.

community A convenient biotic subdivision within an ecosystem; formed by interacting populations of animals and plants in an area.

compass points The spots marking the 32 directional points on a compass.

composite volcano A volcano formed by a sequence of explosive volcanic eruptions; steep-sided, conical in shape; sometimes referred to as a stratovolcano, although *composite* is the preferred term (compare *shield volcano*).

conic A class of map projections based on the geometric shape of a cone.

continental drift A proposal by Wegener in 1912 stating that Earth's landmasses have migrated over the past 200 million years from a supercontinent he called Pangaea to the present configuration; a widely accepted concept today (see *plate tectonics*).

continental glacier A continuous mass of unconfined ice, covering at least 50,000 km² (19,500 mi²); most extensive as ice sheets covering Greenland and Antarctica.

continentality A qualitative designation applied to stations that lack the temperature-moderating effects of the sea and that exhibit a greater range between minimum and maximum temperatures, both daily and annually.

contour interval The vertical distance between adjacent contour lines (see *contour lines*).

contour lines Isolines on a topographic map that connect all points at the same elevation relative to a reference elevation called the vertical datum. The spaces between contour lines portray slope and are known as *contour intervals.*

convergent boundary A region where two tectonic plates converge, causing compression or subduction of one or both plates involved.

Coordinated Universal Time (UTC) The official reference time in all countries, formerly known as Greenwich Mean Time outside the United Kingdom; now measured by primary standard atomic clocks whose time calculations are collected at the Paris, France, International Bureau of Weights and Measures (BIPM). UTC assumes a prime meridian (0° longitude) passing through Greenwich, London, England.

core The deepest inner portion of Earth, representing one-third of its entire mass; differentiated into two zones—a solid iron inner core surrounded by a dense, molten, fluid metallic-iron outer core.

crater A circular surface depression formed by volcanism; built by accumulation, collapse, or explosion; usually located at a volcanic vent or pipe; can be at the summit or on the flank of a volcano.

crust Earth's outer shell of crystalline surface rock, ranging from 5 to 60 km (3 to 38 mi) in thickness from oceanic crust to mountain ranges. Average density of continental crust is crust is 2.7 g per cm^3, whereas oceanic crust is 3.0 g per cm^3.

cylindrical A class of map projections based on the geometric shape of a cylinder.

day length The duration of exposure to insolation, varying during the year depending on latitude; an important aspect of seasonality.

daylight saving time A period in the Northern Hemisphere during which time is set ahead one hour in the spring and set back one hour in the fall. Only Hawai'i, Arizona, and Saskatchewan exempt themselves. In Europe, the last Sundays in March and September generally are used.

declination The latitude that receives direct overhead (perpendicular) insolation on a particular day; migrates annually through the 47° of latitude between the tropics.

deficit In a water balance, the amount of unmet, or unsatisfied, potential evapotranspiration (DEFIC).

denudation A general term that refers to all processes that cause degradation of the landscape: weathering, mass movement, erosion, and transportation; at the heart of exogenic processes.

desert biomes Arid landscapes of uniquely adapted dry-climate plants and animals.

dew-point temperature The temperature at which a given mass of air becomes saturated, holding all the water it can hold. Any further cooling or addition of water vapor results in active condensation.

dike An intrusive igneous feature formed when magma intrudes across layers of existing rock.

discharge The measured volume of flow in a river that passes by a given cross section of a stream in a given unit of time; expressed in cubic meters per second or cubic feet per second.

divergent boundary A region where upwelling material from the mantle forms new crust and tectonic plates spread apart in a constructive process.

drainage basins The basic spatial geomorphic unit of a river system; distinguished from a neighboring basin by ridges and highlands that form divides delimiting the catchment area of the drainage basin, or its watershed.

drainage patterns A geometric arrangement of streams in a region; determined by slope, differing rock resistance to weathering and erosion, climatic and hydrologic variability, and structural controls of the landscape.

drumlin A depositional landform related to continental glaciation; an elongated hill created from till embedded in the bottom of the glacial ice that is plastered, layer upon layer, in a growing mound.

dry adiabatic rate (DAR) The rate at which a parcel of air that is less than saturated cools (if ascending) or heats (if descending); a rate of 10C° per 1000 m (5.5F° per 1000 ft) (see also *adiabatic*).

ecosystem A self-regulating association of living plants, animals, and their nonliving physical and chemical environment.

effusive eruption An eruption characterized by low-viscosity, basaltic magma, with characteristic low-gas content readily escaping. Lava pours forth onto the surface with relatively small explosions and little tephra; tends to form shield volcanoes.

elastic-rebound theory A concept describing the faulting process in Earth's crust, in which the two sides of a fault appear locked despite the motion of adjoining pieces of crust; with accumulating strain, they rupture suddenly, snapping to new positions relative to each other, generating an earthquake.

empirical classification A classification based on weather statistics or other data; used to determine general climate categories.

environmental lapse rate The actual lapse rate in the lower atmosphere at any particular time under local weather conditions; may deviate above or below the average normal lapse rate of 6.4C° per 1000 m (3.5F° per 1000 ft).

equal area A trait of a map projection; indicates the equivalence of all areas on the surface of the map, although shape is distorted.

equatorial and tropical rain forest A lush biome of tall broadleaf evergreen trees and diverse plants and animals. The dense canopy of leaves is usually arranged in three levels.

equilibrium line The area of a glacier where accumulation (gain) and ablation (loss) are balanced.

equinox March 20–21 and September 22–23, when the circle of illumination divides Earth through the poles, and all places have days and nights of equal length.

esker A sinuously curving, narrow deposit of coarse gravel that forms along a meltwater stream channel, developing in a tunnel beneath the glacier. A retreating glacier leaves such ridges behind in a pattern that usually parallels the path of the glacier; may appear branching.

evaporation The movement of free water molecules away from a wet surface into air that is less than saturated; the phase change of water to water vapor; vaporization below the boiling point of water.

evapotranspiration The merging of evaporation and transpiration water loss into one term (see *potential evapotranspiration* and *actual evapotranspiration*).

explosive eruption A violent and unpredictable volcanic eruption, the result of magma that is thicker (more viscous), stickier, and higher in gas and silica content than that of an effusive eruption; tends to form blockages within a volcano; produces composite volcanic landforms (see *composite volcano*; compare *effusive eruption*).

extratropical cyclone See *midlatitude cyclone*.

extrusive igneous rock A rock, such as basalt, that solidifies and crystallizes from a molten state as it extrudes onto the surface.

felsic A type of igneous rock, such as granite, that is high in silicate minerals, such as feldspar and quartz (pure silica), and has a low melting point. Their name is derived from *fel*dspar and *si*lica. Rocks formed from felsic minerals generally are lighter in color and less dense than those from mafic minerals.

field capacity Water held in the soil by hydrogen bonding against the pull of gravity, remaining after water drains from the larger pore spaces, or storage capacity; the available water for plants. Field capacity is specific to each soil type and is an amount that can be determined through soil surveys.

flood A high water level that overflows the natural riverbank along any portion of a stream.

flood basalts An accumulation of horizontal flows formed when lava spreads out from elongated fissures onto the surface in extensive sheets; associated with effusive eruptions.

floodplain A low-lying area near a stream channel that is subject to recurrent flooding; alluvial deposits generally mask underlying rock.

fluvial Stream-related processes; from the Latin *fluvius*, for "river" or "running water."

formation class The dominant vegetation type extending across a region.

gage height The height of the water surface above the gage datum (zero point).

genetic classification A type of classification that uses causative factors to determine climatic regions; for example, an analysis of the effect of interacting air masses on climate.

geographic grid A network of intersecting lines running north–south (through the poles) and east–west (parallel to the equator) that provides a system of global location (see *latitude* and *longitude*).

geographic index number The identification number on a topographic quadrangle, derived from the latitude and longitude of the lower-right corner of the map.

Geographic Information Systems (GIS) A computer-based methodology that merges computer cartography with data management and allows the collection, manipulation, and analysis of geographic/spatial information.

geographic North (or South) Pole The geographic North (or South) Pole—the point at which all meridians converge—giving a reference point for true north (or south); also known as *true north* (or *true south*).

geography (Preface) The science that studies the interdependence among geographic areas, natural systems, processes, society, and cultural activities over space—a spatial science. The five themes of geographic education are location, place, movement, regions, and human–Earth relationships.

geomorphology The science that analyzes and describes the origin, evolution, form, classification, and spatial distribution of landforms.

glacier A large mass of perennial ice resting on land or floating shelflike in the sea adjacent to land; formed from the accumulation and recrystallization of snow, which then flows slowly under the pressure of its own weight and the pull of gravity.

gnomonic projection A planar map projection on which all straight lines are great circle routes.

great circle Any circle of circumference drawn on a globe with its center coinciding with the center of the globe. An infinite number of great circles can be drawn, but only one parallel is a great circle—the equator.

Greenwich Mean Time (GMT) Former world standard time, now known as Coordinated Universal Time (UTC) (see *Coordinated Universal Time*).

habitat The physical location in which an organism is biologically suited to live. Most species have specific habitat parameters, or limits.

hair hygrometer An instrument for the measurement of relative humidity; based on the principle that human hair will change as much as 4% in length between 0% and 100% relative humidity.

heat index An index that indicates the human body's reaction to air temperature and water vapor. The level of humidity in the air affects our natural ability to cool through evaporation from skin.

horn A pyramidal, sharp-pointed peak that results when several cirque glaciers gouge an individual mountain summit from all sides.

horst and graben Landscape features in the Basin and Range Province. The term *horst* applies to upward-faulted blocks; *graben* refers to downward-faulted blocks.

hot spots Individual points of upwelling material originating in the asthenosphere, not necessarily associated with spreading centers, although Iceland is astride a mid-ocean ridge. They tend to remain fixed relative to migrating plates; some 50 to 100 are identified worldwide, exemplified by Yellowstone National Park and Hawai'i (see *plumes*).

humidity Water vapor content of the air. The capacity of the air to hold water vapor is mostly a function of temperature.

humus A mixture of organic debris in the soil, worked by consumers and decomposers in the humification process; characteristically formed from plant and animal litter deposited at the surface.

hurricane A tropical cyclone that is fully organized and intensified in inward-spiraling rainbands; ranges from 160 to 960 km (100 to 600 mi) in diameter, with wind speeds in excess of 119 kmph (65 knots, or 74 mph); a name used specifically in the Atlantic and eastern Pacific (compare *typhoon*).

hydrology The science that studies the flow of water, ice, and water vapor from place to place, as water flows through the atmosphere, across the land, where it is also stored as ice, and within groundwater; includes interactions with human activity.

ice core A column of glacial ice used to study past climate.

igneous rock One of the three basic rock types; it has solidified and crystallized from a hot molten state (either magma or lava) (compare *metamorphic rock* and *sedimentary rock*).

insolation Solar radiation that is intercepted by Earth.

interior drainage In regions where rivers do not flow into the ocean, the outflow is through evaporation or subsurface gravitational flow. Portions of Africa, Asia, Australia, and the western United States have such drainage.

internal drainage Drainage in regions where rivers do not flow into the ocean, and the outflow is through evaporation or subsurface gravitational flow. Portions of Africa, Asia, Australia, and the western United States have such drainage.

International Date Line The 180° meridian; an important corollary to the prime meridian on the opposite side of the planet; established by the treaty of 1884 to mark the place where each day officially begins.

intrusive igneous rock A rock, such as granite, that solidifies and crystallizes from a molten state as it intrudes into crustal rocks, cooling and hardening below the surface.

island biogeography The study of island communities, which are special places for study because of their spatial isolation and the relatively small number of species present. Islands resemble natural experiments because the impact of individual factors, such as civilization, can be more easily assessed on islands than over larger continental areas.

isobar An isoline connecting all points of equal atmospheric pressure.

isogon An isoline connecting points of equal magnetic declination (see *magnetic declination*).

isogonal map A map showing isogons (see *isogon*).

isoline A line on a map that connects all points of equal value.

isotherm An isoline that connects all points of equal air temperature.

isotope analysis A technique for long-term climatic reconstruction that uses the atomic structure of chemical elements—specifically the relative amounts of their isotopes—to identify the chemical composition of past oceans and ice masses.

kame A small hill of poorly sorted sand and gravel that accumulates in crevasses or in ice-caused indentations in the surface; may also be deposited by water in ice-contact deposits.

karst topography Distinctive topography formed in a region of chemically weathered limestone with poorly developed surface drainage and solution features that appear pitted and bumpy; originally named after the Krš Plateau of Yugoslavia.

kettle A geographic feature that forms when an isolated block of ice persists in a ground moraine, an outwash plain, or valley floor after a glacier retreats; as the block finally melts, it leaves behind a steep-sided hole that may frequently fill with water.

kinetic energy The energy of motion in a body; derived from the vibration of the body's own movement and stated as temperature.

lagoon A portion of coastal seawater that is virtually cut off from the ocean by a bay barrier or barrier beach; also, the water surrounded and enclosed by an atoll.

landfall The location along a coast where a storm moves onshore.

land-water heating differences The differences in the way land and water heat, as a result of contrasts in transmission, evaporation, mixing, and specific heat capacities. Land surfaces heat and cool faster than water and are characterized as having aspects of continentality, whereas water is regarded as producing a moderating marine influence.

latent heat The heat energy absorbed or released in the phase change of water from one state to another. The energy is absorbed or released in each phase change from one state to another. Heat energy is absorbed as the latent heat of melting, vaporization, or evaporation. Heat energy is released as the latent heat of condensation and freezing (or fusion).

latitude The angular distance measured north or south of the equator from a point at the center of Earth. A line connecting all points of the same latitudinal angle is called a *parallel*.

lava Magma that issues from volcanic activity onto the surface; the extrusive rock that results when magma solidifies (see *magma*).

life zone An altitudinal zonation of plants and animals that form distinctive communities. Each life zone possesses its own temperature and precipitation relationships.

lifting condensation level The height at which an air parcel will reach 100% relative humidity (RH) when it is cooled by dry adiabatic lifting.

line A GIS vector feature with one dimension-length.

littoral zone A specific coastal environment that is the region between the high-water line during storms and the depth at which storm waves cannot move sea-floor sediments.

loam A soil that is a mixture of sand, silt, and clay in almost equal proportions, with no one texture dominant; an ideal agricultural soil.

local Sun time Time based on actual longitudinal distance and the prime meridian; also referred to as Sun time or solar time.

longitude The angular distance measured east or west of a prime meridian from a point at the center of Earth. A line connecting all points of the same longitude is called a meridian.

mafic A type of igneous rock, such as basalt, that is derived from *m*agnesium and *f*errum (the Latin word for iron). Mafic rocks are lower in silica, higher in magnesium and iron, have higher melting points, are darker in color, and have greater density than those from felsic minerals.

magnetic declination The horizontal angle between geographic (true) north and magnetic north.

magnetic North (or South) Pole The point on Earth's surface indicated by the north-seeking point of a magnetic compass needle.

mantle An area within the planet representing about 80% of Earth's total volume, with densities increasing with depth and averaging 4.5 g per cm^3; occurs above the core and below the crust; is rich in iron and magnesium oxides and silicates.

map, map projection The reduction of a spherical globe onto a flat surface in some orderly and systematic realignment of the latitude and longitude grid.

march of the seasons The annual passage of Earth as it revolves around the Sun, producing seasonal changes in daylength and Sun altitude.

marine Descriptive of stations that are dominated by the moderating effects of the ocean and that exhibit a smaller minimum and maximum temperature range than continental stations (see *land-water heating differences*).

Mediterranean shrubland A major biome dominated by shrub formations; occurs in Mediterranean dry summer climates and is characterized by sclerophyllous scrub and short, stunted, tough forests.

Mercator projection A true-shape cylindrical projection on which straight lines are true compass directions (lines of constant bearing)—rhumb lines.

mercury barometer A device that measures air pressure with a column of mercury in a tube that is inserted in a vessel of mercury. The surrounding air exerts pressure on the mercury in the vessel (see *air pressure*).

meridian See *longitude*.

mesopause The upper boundary of the mesosphere.

mesosphere The upper region of the homosphere from 50 to 80 km (30 to 50 mi) above the ground; designated by temperature criteria and very low pressures, ranging from 0.1 to 0.001 mb. The top of the mesosphere, as determined by temperature (−90°C), is the mesopause.

metamorphic rocks One of the three basic rock types; it is existing igneous and sedimentary rock that has undergone profound physical and chemical changes under increased pressure and temperature. Constituent mineral structures may exhibit foliated or nonfoliated textures (compare *igneous rock* and *sedimentary rock*).

meteorology The scientific study of the atmosphere, including a study of the atmosphere's physical characteristics and motions, related chemical, physical, and geological processes, the complex linkages of atmospheric systems, and weather forecasting.

mid-ocean ridges Submarine mountain ranges that extend more than 65,000 km (40,000 mi) worldwide and average more than 1000 km (620 mi) in width; centered along sea-floor spreading centers as the direct result of upwelling areas of heat in the upper mantle.

midlatitude broadleaf and mixed forest A biome in moist continental climates in areas of warm-to-hot summers and cool-to-cold winters; includes several distinct communities. Relatively lush stands of broadleaf forests trend northward into needleleaf evergreen stands.

midlatitude grasslands The major biome most modified by human activity; so named because of the predominance of grass-like plants, although deciduous broadleafs appear along streams and other limited sites; location of the world's breadbaskets of grain and livestock production.

mineral An element or combination of elements that forms an inorganic natural compound; described by a specific formula and crystal structure.

moist adiabatic rate (MAR) The rate at which a parcel of saturated air cools in ascent; a rate of 6C° per 1000 m (3.3F° per 1000 ft). This rate may vary, with moisture content and temperature, from 4C° to 10C° per 1000 m (2F° to 6F° per 1000 ft). Also referred to as wet adiabatic rate (WAR) (see *adiabatic*).

montane forests Needleleaf forests associated with mountain elevations (see *northern needleleaf forest*).

moraine Marginal glacial deposits of unsorted and unstratified material, producing a variety of depositional landforms.

natural levees Long, low ridges that occur on either side of a river in a developed floodplain; sedimentary (coarse gravels and sand) depositional by-products of river-flooding episodes.

niche The basic function, or occupation, of a lifeform within a given community; the way an organism obtains its food, air, and water.

normal lapse rate The average rate of temperature decrease with increasing altitude in the lower atmosphere; an average value of 6.4C° per km, or 1000 m (3.5F° per 1000 ft); a rate that exists principally during daytime conditions.

northern needleleaf forest Forests of pine, spruce, fir, and larch, stretching from the east coast of Canada westward to Alaska and continuing from Siberia westward across the entire extent of Russia to the European Plain; called the taiga (a Russian word) or the boreal forest; principally in the D climates. Includes montane forests.

orographic lifting The uplifting of migrating air masses in response to the physical presence of a mountain, a topographic barrier. The lifted air cools adiabatically as it moves upslope; may form clouds and produce increased precipitation (see *rain shadow*).

outwash plain Glacio-fluvial deposits of stratified drift of meltwater-fed, braided, and overloaded streams with debris beyond a glacier's morainal deposits.

oxbow lake A lake that was formerly part of the channel of a meandering stream; isolated when a stream eroded its outer bank, forming a cutoff through the neck of a looping meander.

paleoclimatology The science that studies climates and the causes of variations in climate of past ages throughout historic and geologic time.

Pangaea The supercontinent formed by the collision of all continental masses approximately 225 million years ago; named by Wegener in 1912 in his continental drift theory.

parallel See *latitude*.

paternoster lake One of a series of small, circular, stair-stepped lakes formed in individual rock basins aligned down the course of a glaciated valley; named because they look like a string of rosary (religious) beads.

paternoster lakes One of a series of small, circular, stair-stepped lakes formed in individual rock basins aligned down the course of a glaciated valley; named because they look like a string of rosary (religious) beads.

ped The smallest natural lump or cluster of particles in soil.

pedon A soil profile extending from the surface to the lowest extent of plant roots or to the depth where regolith or bedrock is encountered; imagined as a hexagonal column; the basic soil sampling unit.

permeability The ability of water to flow through soil or rock; a function of the texture and structure of the medium.

physical geography (Preface) A science that studies the spatial aspects of the physical elements and processes that make up the environment: energy, air, water, weather, climate, landforms, soils, animals, plants, and Earth.

planar A class of map projections based on the geometric shape of a plane. Also known as azimuthal.

planimetric map A basic map showing the horizontal position of boundaries, land-use activities, and political, economic, and social outlines.

plate tectonics The conceptual model that encompasses continental drift, sea-floor spreading, and related aspects of crustal movement; widely accepted as the foundation of crustal tectonic processes.

playa An area of salt crust left behind by evaporation on a desert floor, usually in the middle of a desert or semiarid bolson or valley; intermittently wet and dry.

plumes Columns of heated rock rising through the asthenosphere or mantle, creating hot spots in the migrating overlying lithospheric plates (see *hot spots*).

point GIS vector feature that consists of a location.

polygon GIS vector feature of an area.

polypedon The identifiable soil in an area, with distinctive characteristics differentiating it from surrounding polypedons that form the basic mapping unit; composed of many pedons (see *pedon*).

porosity The total volume of available pore space in soil; a result of the texture and structure of the soil.

potential evapotranspiration The amount of moisture that would evaporate and transpire if adequate moisture were available; the amount lost under optimum moisture conditions—that is, the moisture demand (POTET).

precipitation Rain, snow, sleet, and hail—the moisture supply (PRECIP).

pressure gradient 7 The change in atmospheric pressure horizontally across Earth's surface; measured along a line at right angles to the isobars.

prime meridian An arbitrary meridian designated as 0° longitude; the point from which longitude is measured east or west; agreed on by the nations of the world in an 1884 treaty.

profile view See *topographic profile*.

proxy method A method that finds information about past environments that indicates changes in climate, such as isotope analysis or tree ring dating; also called a *climate proxy*.

pyroclastics An explosively ejected rock fragment launched by a volcanic eruption; sometimes described by the more general term *tephra*.

quadrant compass bearing A compass direction expressed as an arc clockwise or counterclockwise (toward the east or toward the west—whichever is closer) through 90° from the north or the south.

rain shadow The area on the leeward slopes of a mountain range; in the shadow of the mountains, where precipitation receipt is greatly reduced compared to windward slopes (see *orographic lifting*).

raster In GIS, a data structure based on a grid of (usually) square and rectangular "cells."

rating curve A curve that relates stream discharge to gage height in showing the direct relationship between stream discharge and its width, depth (gage height), and velocity.

recurrence intervals The average time period within which a given event will be equaled or exceeded once; also called *return period*.

regime The characteristic seasonal variation in streamflow, reflecting climatic conditions—a stream's annual "personality."

regolith Partially weathered rock overlying bedrock, whether residual or transported.

relative humidity A term that reflects the ratio of water vapor actually in the air (content) compared to the maximum water vapor the air is able

relief Elevation differences in a local landscape as an expression of unevenness, height, and slope variation.

revolution The annual orbital movement of Earth about the Sun; determines year length and the length of seasons.

rhumb line A line of constant compass direction, or constant bearing, which crosses all meridians at the same angle. A portion of a great circle.

rock An assemblage of minerals bound together, or sometimes a mass of a single mineral.

rock cycle A model that represents the interrelationships among the three rock-forming processes: igneous, sedimentary, and metamorphic; shows how each can be transformed into another rock type.

rotation The turning of Earth on its axis; averages 24 hours in duration; determines day–night relationships.

saturated Descriptive of air that is holding all the water vapor that it can hold at a given temperature.

scale The ratio of the distance on a map to the distance in the real world; expressed as a representative fraction, graphic scale, or written scale.

sea cliff Cliffs formed by the undercutting action of the sea.

sea-floor spreading As proposed by Hess and Dietz, the mechanism driving the movement of the continents; associated with upwelling flows of magma along the worldwide system of mid-ocean ridges.

sedimentary rocks One of the three basic rock types; it is formed from the compaction, cementation, and hardening of sediments derived from other rocks (compare *igneous rock* and *metamorphic rock*).

seismic waves The shock waves sent through the planet by an earthquake or underground nuclear test. Transmission varies according to temperature and the density of various layers within the planet.

sensible heat Heat that can be measured with a thermometer; a measure of the concentration of kinetic energy from molecular motion (see *apparent temperature*).

shield volcano A symmetrical mountain landform built from effusive eruptions (low-viscosity magma); gently sloped and gradually rising from the surrounding landscape to a summit crater; typical of the Hawaiian Islands (compare *composite volcano* and *effusive eruption*).

sinkhole A nearly circular depression created by the weathering of karst landscapes; also known as a doline in traditional studies; may collapse through the roof of an underground space (see *karst topography*).

sling psychrometer A device for the measurement of relative humidity using two thermometers—a dry bulb and a wet bulb—mounted side-by-side.

small circle Circles on a globe's surface that do not share Earth's center; for example, all parallels other than the equator.

soil A dynamic natural body made up of fine materials covering Earth's surface in which plants grow, composed of both mineral and organic matter.

soil horizon The various layers exposed in a pedon; roughly parallel to the surface and identified as O, A, E, B, C, and R (bedrock).

soil pH (acidity-alkalinity) The relative abundance of free hydrogen ions (H^+) in a solution. The pH scale is logarithmic: Each whole number represents a 10-fold change. A pH of 7.0 is neutral, and values less than 7.0 are increasingly acidic, while values greater than 7.0 are increasingly basic or alkaline.

soil moisture storage; recharge; utilization; change The retention of moisture within soil; represents a savings account that can accept deposits (*soil moisture recharge*) or experiences withdrawals (*soil moisture utilization*) as conditions change.

soil profile See *pedon*.

soil science An interdisciplinary science that studies soil and involves physics, chemistry, biology, mineralogy, hydrology, taxonomy, climatology, and cartography.

soil structure The size and shape of the aggregates of particles in the soil.

soil-water budget An accounting system for soil moisture that uses inputs of precipitation and outputs of evapotranspiration and gravitational water.

solstice June 20–21 and December 21–22, when the Sun's declination is at the Tropics of Cancer or Capricorn, respectively.

solum A true soil profile in the pedon; ideally, a combination of O, A, E, and B horizons. (See Pedon.)

spatial data In a GIS, the data that are concerned with the location of a feature, as contrasted with *attribute data*.

specific heat The increase of temperature in a material when energy is absorbed. Because water requires far more heat to raise its temperature than does a comparable volume of land, water is said to have a higher specific heat.

specific humidity The mass of water vapor (in grams) per unit mass of air (in kilograms) at any specified temperature. The maximum mass of water vapor that a kilogram of air can hold at any specified temperature is termed its maximum specific humidity.

stability The condition of a parcel of air relative to whether it remains where it is or changes its initial position. The parcel is stable if it resists displacement upward and unstable if it continues to rise; relates to adiabatic and lapse rate processes.

standard atmosphere An agreed-upon model of Earth's atmosphere, with set values for the vertical distribution of temperature, pressure, and density from the surface to approximately 80 km (50 mi).

standard line The line of tangency on a map projection along which the projection surface (cylinder or cone) contacts the globe; along this line the map maintains accuracy, free of distortion.

stereolenses (stereoscope) A device that forces each eye to look straight ahead, thus reproducing binocular vision using a pair of overlapping photographs; permits 3-D viewing of photo stereopairs.

stereoscopic contour map A three-dimensional graphic representation of elevation, viewed through a stereoscope; also known as a stereogram.

storm surge Large quantities of seawater pushed inland by the strong winds associated with a tropical cyclone.

stratopause The upper boundary of the stratosphere.

stratosphere The portion of the homosphere that ranges from 20 to 50 km (12.5 to 30 mi) above Earth's surface, with temperatures ranging from $-57°C$ ($-70°F$) at the tropopause to $0°C$ ($+32°F$) at the stratopause. The functional ozonosphere is within the stratosphere.

subduction zone An area where two crustal plates collide and the denser oceanic crust dives beneath the less dense continental plate, forming deep oceanic trenches and seismically active regions.

subsolar point The only point receiving perpendicular insolation at a given moment—that is, where the Sun is directly overhead.

surplus The amount of moisture that exceeds potential evapotranspiration; moisture oversupply when soil moisture storage is at field capacity.

syncline In a landscape formed by folding, the downfold or trough often creating valleys between the upfolded *anticlines*.

synoptic map A weather map that shows a moment in time over a large area.

tarn A small mountain lake that occupies a basin in a glacial trough or cirque.

temperate rain forest A major biome of lush forests at middle and high latitudes; occurs along narrow margins of the Pacific Northwest in North America, among other locations; a mixture of broadleaf and needleleaf trees and thick undergrowth, including the tallest trees in the world.

temperature A measure of sensible heat energy present in the atmosphere and other media; indicates the average kinetic energy of individual molecules within the atmosphere.

terrestrial ecosystem A self-regulating association characterized by specific plant formations; usually named for the predominant vegetation and known as a biome when large and stable.

thermal equator The isoline on an isothermal map that connects all points of highest mean temperature.

thermopause The upper boundary of the thermosphere.

thermosphere A region of the heterosphere extending from 80 to 480 km (50 to 300 mi), the thermopause, in altitude; contains the functional ionosphere layer.

till plains Large, relatively flat plains composed of unsorted glacial deposits behind a terminal or end moraine. Low-rolling relief and interrupted, unclear drainage patterns are characteristic.

tombolo A landform created when coastal sand deposits connect the shoreline with an offshore island outcrop or sea stack.

topographic map A map that portrays physical relief through the use of elevation contour lines that connect all points at the same elevation above or below a vertical datum, such as mean sea level.

topographic profile A graphic representation of graduated elevations along a line segment drawn on a map—a *profile view*—that is an important method of analysis for a topographic map.

topography The undulations and configurations that give Earth's surface its texture; the heights and depths of local relief, including both natural and human-made features.

township and range A system that the Public Lands Survey System delineated in 1785 that called for the establishment of a grid system based on an initial survey point; most of the United States is subdivided into 36 mi^2 townships, 1 mi^2 sections, and 160 acre homesteads (quarter section).

transform boundary A region where tectonic plates slide past one another, usually at right angles to a seafloor spreading center.

transform faults An elongate zone along which faulting occurs between mid-ocean ridges; produces a relative horizontal motion with no new crust formed or destroyed; strike-slip motion, either left or right lateral.

transpiration The movement of water vapor out through the pores in leaves drawn by their roots from the soil moisture storage.

tropical cyclone A cyclonic circulation originating in the tropics, with winds between 30 and 64 knots (39 and 73 mph); characterized by closed isobars, circular organization, and heavy rains (see *hurricane* and *typhoon*).

tropical savanna A major biome that contains large expanses of grassland interrupted by trees and shrubs; a transitional area between the humid rain forests and tropical seasonal forests and the drier, semiarid tropical steppes and deserts.

tropical seasonal forest and scrub A variable biome on the margins of the rain forests, occupying regions of lesser and more erratic rainfall; the site of transitional communities between the rain forests and tropical grasslands.

tropopause The upper boundary of the troposphere.

troposphere The home of the biosphere; the lowest layer of the homosphere, containing approximately 90% of the total mass of the atmosphere; extends up to the tropopause, marked by a temperature of $-57°C$ ($-70°F$); occurring at an altitude of 18 km (11 mi) at the equator, 13 km (8 mi) in the middle latitudes, and near the poles at lower altitudes.

true north (or south) See *geographic North (or South) Pole*.

true shape A map property that shows the correct configuration of coastlines; a useful trait of conformality for navigational and aeronautical maps, although areal relationships are distorted.

typhoon A tropical cyclone with inward-spiraling winds in excess of 65 knots (74 mph) that occurs in the western Pacific; same as a hurricane except for location.

Universal Transverse Mercator (UTM) grid A special cylindrical projection in which the projection surface is tangent to the globe at a selected pair of opposing meridians or intersects at two selected small circles; scale is constant along those vertical lines. The UTM grid has become the standard for military applications.

vector A GIS data format where point, line, or polygon files are composed of coordinate pairs of x, y locations.

wilting point That point in the soil-moisture balance when only hygroscopic water and some bound capillary water remain. Plants wilt and eventually die after prolonged stress from a lack of available water.

wind-chill factor The enhanced rate at which body heat is lost to the air. As wind speeds increase, heat loss from the skin increases.

zenith The highest point reached by the Sun during its daily path.

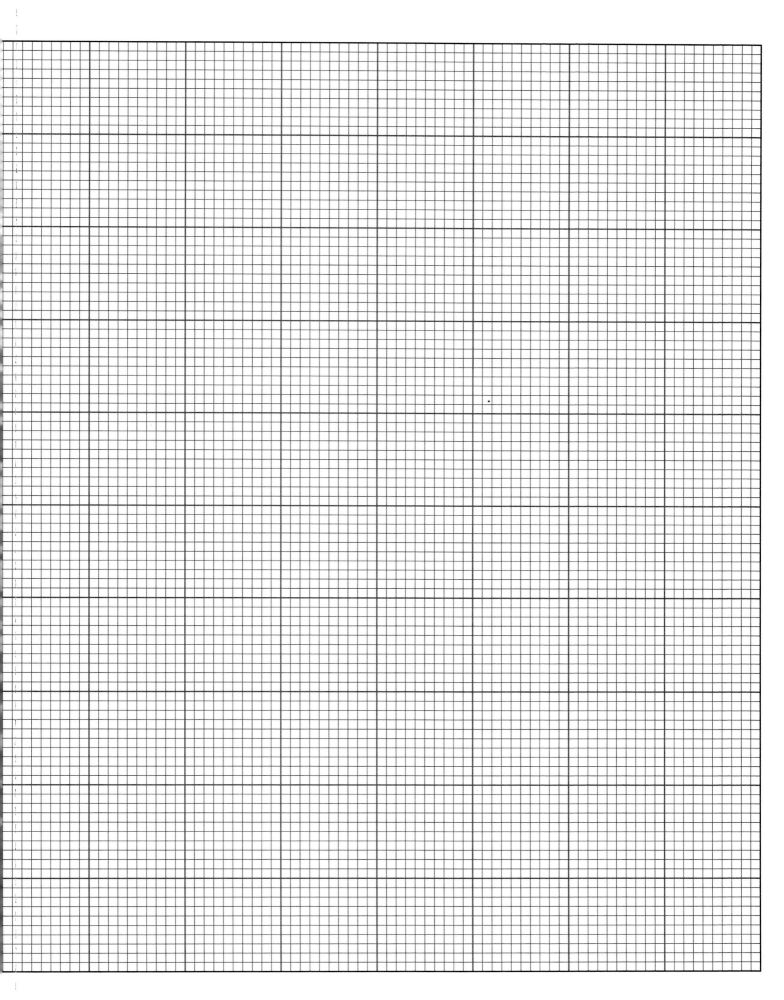

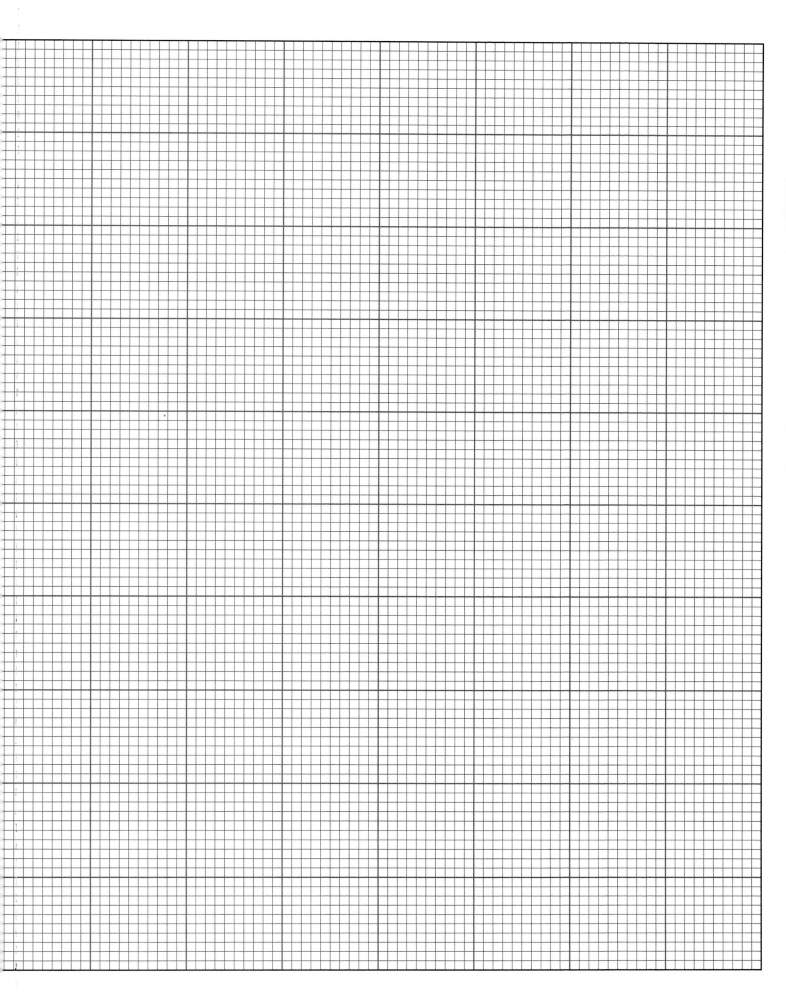

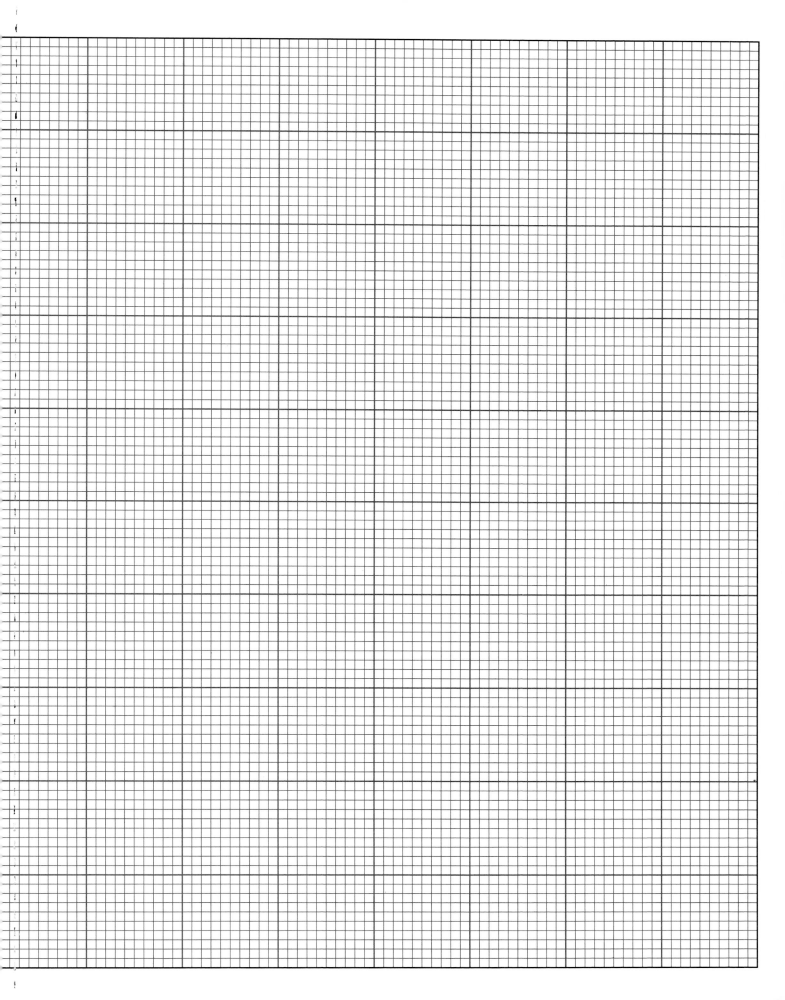

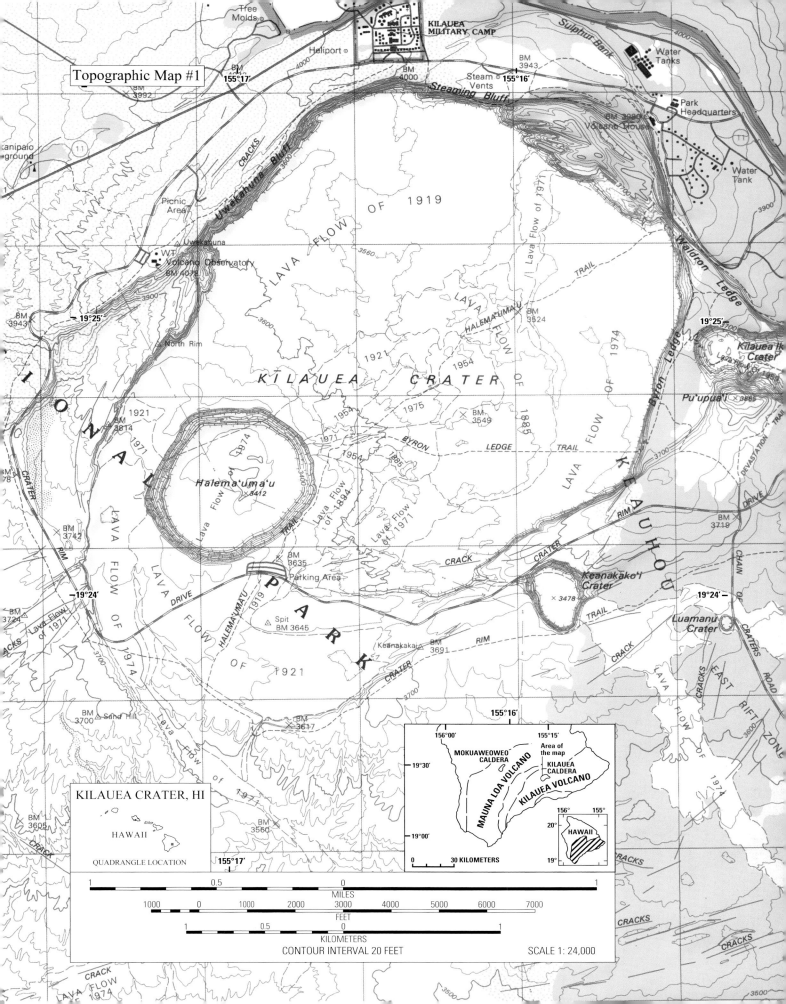

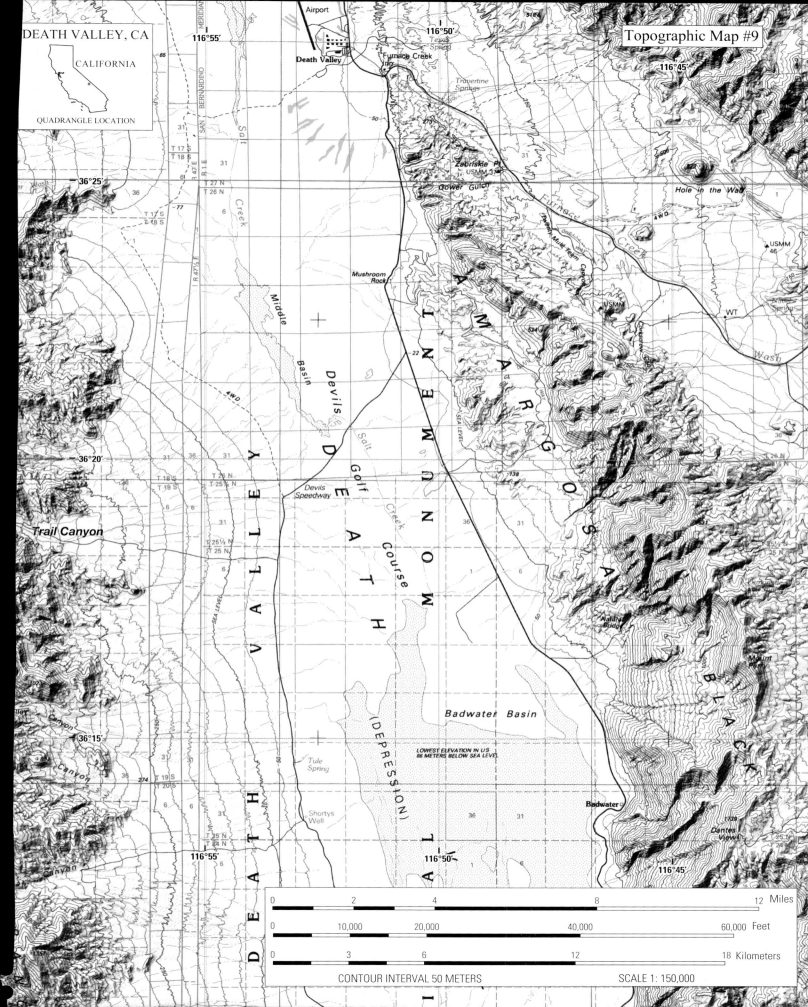

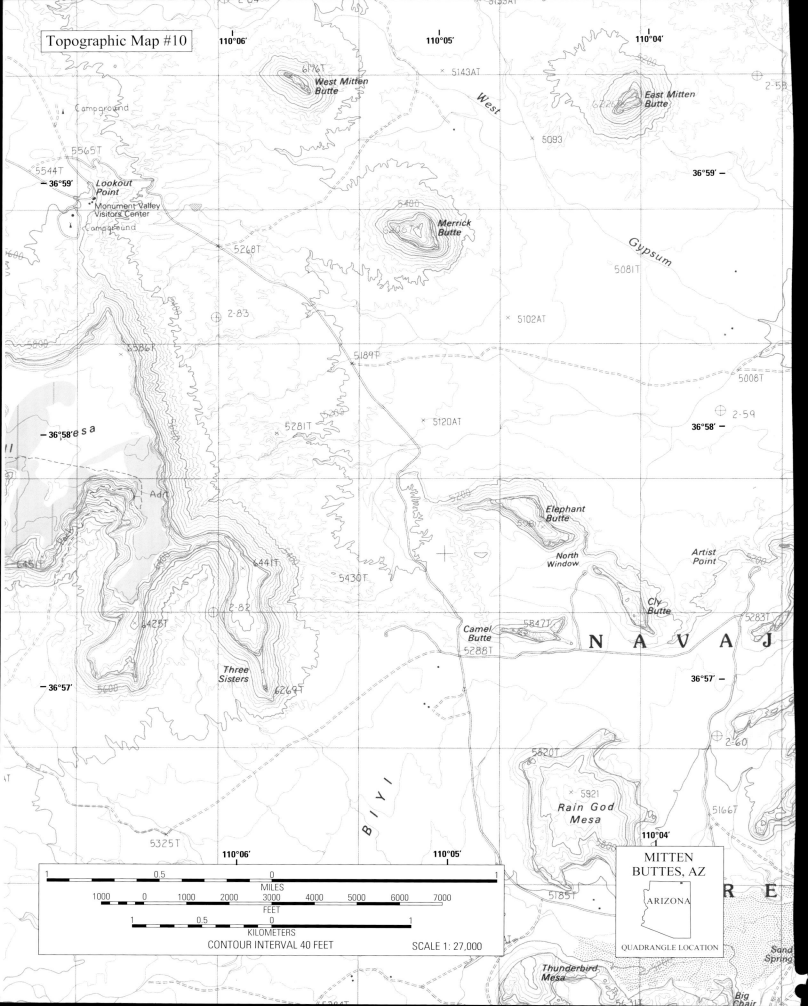

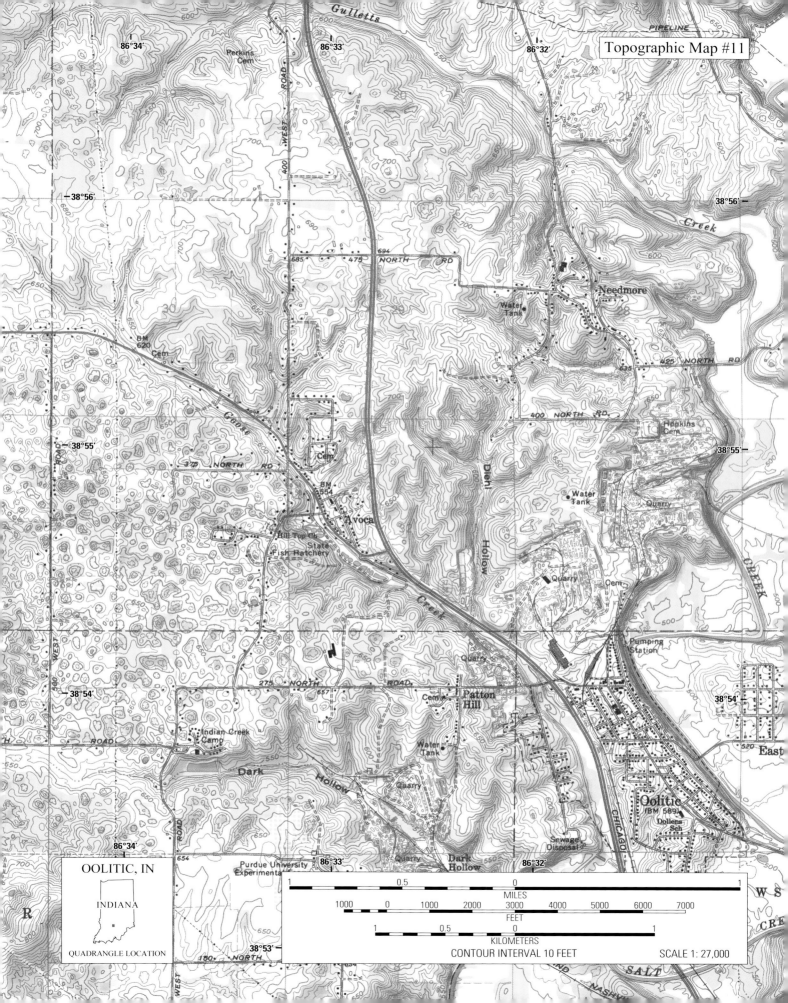